ADVANCES IN
X-RAY ANALYSIS

Volume 23

ADVANCES IN X-RAY ANALYSIS

Volume 23

Edited by

John R. Rhodes

Columbia Scientific Industries
Austin, Texas

and

Charles S. Barrett, Donald E. Leyden, John B. Newkirk,
Paul K. Predecki, and Clayton O. Ruud

University of Denver
Denver, Colorado

Sponsored by
University of Denver Research Institute
and
Department of Chemistry
University of Denver

PLENUM PRESS • NEW YORK AND LONDON

The Library of Congress cataloged the first volume of this title as follows:

Conference on Application of X-ray Analysis.
Proceedings 6th-   1957-   [Denver]

    v. illus. 24-28 cm. annual.
No proceedings published for the first 5 conferences.
Vols. for 1958-   called also: Advances in X-ray analysis, v. 2-
Proceedings for 1957 issued by the conference under an earlier name: Conference
on Industrial Applications of X-ray Analysis. Other slight variations in name of conference.
Vol. for 1957 published by the University of Denver, Denver Research Institute,
Metallurgy Division.
Vols. for 1958-   distributed by Plenum Press, New York.
Conferences sponsored by University of Denver, Denver Research Institute.
    1. X-rays–Industrial applications–Congresses.   I. Denver. University.
Denver Research Institute   II. Title: Advances in X-ray analysis.
TA406.5.C6   58-35928

Library of Congress Catalog Card Number 58-35928

ISBN-13: 978-1-4613-3098-1   e-ISBN-13: 978-1-4613-3096-7

DOI: 10.1007/978-1-4613-3096-7

Proceedings of the Twenty-Eighth Annual Conference on Applications of
X-Ray Analysis held in Denver, July 30–August 3, 1979

Plenum Press, New York
A Division of Plenum Publishing Corporation
227 West 17th Street, New York, N.Y. 10011

Denver, Colorado

FOREWORD

Traditionally the emphasis at each annual Denver X-ray Conference is placed on a particular aspect of X-ray analysis. The past decade has seen a steady expansion of applications of portable X-ray analyzers and probes in the field, in boreholes and in plant process streams. With this in mind, the main theme of the current conference is field applications of X-ray fluorescence with particular reference to analysis of raw materials such as rocks, ores and coal.

The Plenary Session took up this theme with two invited papers reviewing applications of X-ray emission techniques to geochemical, borehole and on-stream analysis, and recent developments in portable instruments for alloy, ore and other analyses. The third paper took us further afield with a review of X-ray spectrochemical analysis on Mars, the Moon and Earth.

It is evident that portable X-ray analyzers are finding more and more applications outside the conventional boundaries of X-ray spectrometry. Users are not analysts and sometimes not even scientists. Until recently this trend has been hindered by the "scientific nature" of the instruments; one needs to understand XRF methods in order to properly operate the instrument. Microprocessor technology has made possible the development of precalibrated, "smart" analyzers with readouts in quantities familiar to the user and interlocks to prevent erroneous operation. Further developments along these lines were reported at this conference.

Contributed papers included 15 on elemental analysis of rocks and ores, 18 on general applications of XRF, 10 on mathematical methods in XRF and a total of 28 covering X-ray diffraction applications.

In 1976 pre-conference workshops were introduced offering in-depth treatment of specific topics. This year the attendance at these workshops reached an all time high. Four one-day workshops were offered on the topics of minicomputers in XRF analysis, techniques in XRF analysis of rocks and ores, computer search methods in XRD and the application of computers in XRPD.

The annual Denver Conference on Advances in X-ray Analysis continues to be a stimulating and exciting affair for all those who attend. The proceedings, published in these volumes, provide an essential source of information for all workers in the field and all people aspiring to use X-ray techniques.

John R. Rhodes

PREFACE

This volume constitutes the proceedings of the 1979 Denver Conference on the Applications of X-Ray Analysis, 28th in the series. The conference was held July 30 - August 3, 1979 at the University of Denver and was sponsored by the Denver Research Institute and the Department of Chemistry, both of the University of Denver. Conference chairmen in residence were D. E. Leyden and C. O. Ruud, with C. S. Barrett and J. B. Newkirk as honorary chairmen. The invited conference chairman, J. R. Rhodes of Columbia Scientific Industries, organized the Plenary Session entitled "Field Applications of XRF Analysis."

In addition to the plenary and regular contributed sessions, a special session on The Use of Computers in Powder Diffraction was organized and chaired by R. L. Snyder of Alfred University and W. Parrish of IBM Research Laboratory.

Several invited speakers were on the program. Their names and the titles of their papers are listed below.

C. G. Clayton and T. W. Packer, "Some Applications of Energy Dispersive X-Ray Fluorescence Analysis in Minerals Exploration, Mining and Process Control."

P. F. Berry, "Development in Design and Application of Field-Portable XRF Instrumentation."

B. C. Clark, "X-Ray Spectrochemical Analysis on Mars, the Moon and Earth."

D. K. Smith, "Some Research Applications of Computer Generated Powder Diffraction Patterns."

G. R. Fischer and W. T. Kane, "Fourier Transforms and Spline Functions in Automated XRPD Data Evaluation."

G. S. Smith, "Advances in the Computer Indexing of Powder Patterns."

R. P. Goehner, "Automation in an X-Ray Diffraction Service Laboratory."

W. Parrish, G. L. Ayers and T. C. Huang, "A Minicomputer and Methodology for X-Ray Analysis."

Tutorial workshops on various topics in fluorescence and diffraction were held during the first two days of the conference. These are listed below together with the names of workshop instructors and organizers.

(1) Search/Match Techniques for Qualitative Powder Diffractometry: (a) Manual Search, (b) Computer Search. Sponsored by JCPDS-International Centre for Diffraction Data, Swarthmore, Pennsylvania

D. W. Beard	R. Jenkins
J. W. Edmonds	T. McGuire
C. M. Foris	J. Messick
R. P. Goehner	D. K. Smith
M. Holomany	R. L. Snyder

(2) Minicomputer Capabilities in EDXRF Analysis

R. P. Gardner, "Introduction"

J. C. Russ, "Minicomputer Characteristics and Capabilities in the Processing of EDXRF Spectra."

R. Jenkins, "Use of Minicomputers for the Control and Optimization of EDXRF Qualitative Analysis Procedures."

H. E. Marr, "Use of Minicomputers in the Interactive and Off-Line Mode for EDXRF Quantitative Analysis."

(3) Elemental Analysis of Rocks and Ores by X-Ray Fluorescence

J. R. Rhodes, "Introduction."

C. O. Ingamells, "Sampling and Standards."

B. P. Fabbi, "Sample Preparation."

R. L. Mykelbust, "Mathematical Methods."

G. R. Lachance, "The Use of Fundamental Coefficients."

J. R. Harris, "Application of XRF in Laboratory Support of Mineral Exploration."

(4)   Application of Computers in X-Ray Powder Diffraction. Spon-
      sored by JCPDS-International Centre for Diffraction Data,
      Swarthmore, Pennsylvania.

      R. L. Snyder, "Introduction to Automated XRD."

      R. A. Sparks, "Improvement of the Performance of the Powder
      Diffractometer with Minicomputer Profile Analysis."

      D. W. Beard, "Applications of the Siemens D500 in Quantitative
      Analysis."

      T. Krapchev, "The Use of Automated Diffraction for Determining
      Particle Size Effects."

      M. Janiak, "Peak Searching Methods Using the Rigaku DMAX
      Microcomputer Based Dystem."

      R. Jenkins, "Applications of the APD 3600 for Qualitative
      Phase Identification."

      R. P. Goehner, "Data Collection, Reduction and Interactive
      Processing of Diffraction Data."

      R. C. Medrud, J. W. Green, R. S. Klaver, "Diffractometer Con-
      trol and Data Reduction Using an HP System 45 Desktop Computer."

      C. M. Foris, "Automated Powder Diffraction Application."

      G. R. Fischer, "Coping with Large Numbers of XRPD Samples."

      M. C. Nichols, "The Pros and Cons of Search Matching with
      Minicomputers."

      G. G. Johnson, "The Johnson-Vand Search Match Program."

      J. W. Edmonds, "The Microcomputer in an Analytical X-Ray Dif-
      fraction Laboratory: Acquisition and Interpretation of High
      Quality Data."

The tutorial workshops have grown considerably in the last few
years and have become an integral part of the conference. We are
particularly indebted to the workshop organizers and instructors
who gave so unselfishly of their time and talent in these workshops, and to the JCPDS-International Centre for Diffraction Data,
for sponsorship.

We are grateful also to the co-chairmen of the conference
sessions and to the conference aids, for their efforts. The
session chairmen were:

R. Baro	G. R. Lachance
C. S. Barrett	W. Lemons
V. E. Buhrke	J. B. Newkirk
D. C. Camp	W. Parrish
J. W. Criss	J. R. Rhodes
B. P. Fabbi	D. K. Smith
R. P. Gardner	D. M. Smith
B. B. Jablonski	R. L. Snyder

Conference aids were:

Herb and Betty Acree	John Cronin
Rose Ann Bellotti	Bob Flowers
Bill Bodnar	Steve Northcott
Linda Butcher	Jim Smith
Barbara Cain	

On behalf of the conference committee, I take this opportunity
to thank Clayton O. Ruud for ten years of outstanding service to
the Denver X-Ray Conference as conference chairman and organizer.
Under this guidance the conference has grown from modest beginnings
to its present status. We wish him well in his new position at
The Pennsylvania State University.

A special word of thanks and appreciation to Mildred Cain,
the conference secretary, for running the whole show.

                                                    Paul K. Predecki

UNPUBLISHED PAPERS

The following papers were presented orally only and are not
published here for a variety of reasons.

"Absorption Corrections in X-Ray Diffraction Dust Analyses: Pro-
cedures Employing Silver Filters," M. T. Abell, B. A. Lange, D. D.
Dollberg and R. W. Hornung, NIOSH, Robert A. Taft Laboratories,
Mail Stop R-7, 4676 Columbia Parkway, Cincinnati, OH 45226

"Heavy Element Analysis in Ore Materials Using X-Ray Fluorescence,"
T. Arai and M. J. Janiak, Rigaku/USA, Inc., 3 Electronics Ave.,
Danvers, MA 01923

"The Determination of ^{88}Y and ^{156}Eu in Radioactive Debris Solutions by X-Ray Fluorescence and Gamma Spectroscopy," Fernando Bazan, Lawrence Livermore Laboratory, P.O. Box 808, L-233, Livermore, CA 94550

"Development in Design and Application of Field-Portable XRF Instrumentation," Peter F. Berry, Texas Nuclear Division of Ramsey Engineering Company, P.O. Box 9267, Austin, TX 78766

"A New Algorithm for Computer Search-Match Identification Techniques," Marvin Burfield, Phoebe L. Hauff and George VanTrump, Jr., U.S. Geological Survey, Mail Stop 917, Box 25046 Federal Center, Denver, CO 80225; and Monte C. Nichols, Sandia Labs (8313), Livermore, CA 94550

"X-Ray Diffraction Analysis of Silicide Coatings (In Situ) on Space Shuttle Attitude Control Rocket Engine Nozzles," Stanley V. Castner and Glenn D. Jones, The Marquardt Company, 16555 Saticoy St., Van Nuys, CA 91409

"X-Ray Spectrochemical Analysis on Mars, the Moon, and Earth," Benton C. Clark, Planetary Sciences Laboratory, Martin Marietta Aerospace, Denver, CO 80201

"Contact Microradiography and Its Applications," Alice M. Davis, Materials Characterization Laboratory, General Electric Company Research and Development Center, P.O. Box 8, Schenectady, NY 12301

"Fourier Transforms and Spline Functions in Automated XRPD Data Evaluation," G. R. Fischer and W. T. Kane, Corning Glass Works, Sullivan Park, FR-1, Corning, NY 14830

"X-Ray Diffraction Analysis of Airborne Asbestos: A Special X-Ray Diffractometer," J. V. Gilfrich, L. S. Birks, and J. W. Sandelin, Naval Research Laboratory, Washington, DC 20390; and J. Wagman, EPA, Research Triangle Park, North Carolina

"Standard Reference Materials for Powder Diffraction," Camden R. Hubbard, Ceramics, Glass and Solid State Science Division, National Bureau of Standards, Washington, DC 20234

"Carbon Analysis Using X-Ray Fluorescence: Principles and Practice," M. J. Janiak, T. Utaka and T. Arai, Rigaku/USA, Inc., 3 Electronics Ave., Danvers, MA 01923

"A Practical Relation between Atomic Numbers and Alpha Coefficients," Gerald R. Lachance, Geological Survey of Canada, Room 757, 601 Booth St., Ottawa, Canada K1A 0E8

"Determination of Inorganics in Carbons and Graphites by XRF,"
Debbie J. Langenfeld and John F. Reilly, NL Industries, 238 N.
2200 W., Salt Lake City, UT 84116

"Thin Glass Films for XRF Calibration Produced by Focused-Ion Beam
Sputtering," Peter A. Pella, National Bureau of Standards,
Chemistry Building, Room A121, Washington, DC 20234

"Determination of Tin in Minerals," W. Ratyński, M. Kisielinski,
J. Kierzek, J. Parus and J. Tys, Institute of Nuclear Research,
Swierk 05-400, Otwock, Poland

"Measurements and Calculations of Welding Residual Stresses,"
C. O. Ruud, Materials Research Laboratory, The Pennsylvania State
University, University Park, PA 16802

"General Rock Analysis by X-Ray Fluorescence," Maynard Slaughter,
Chemistry and Geochemistry Dept., Colorado School of Mines,
Golden, CO 80401

"Some Research Applications of Computer Generated Powder Diffraction Patterns," Deane K. Smith, Dept. of Geosciences, 204 Deike
Bldg., The Pennsylvania State University, University Park, PA
16802

"An In-Depth Study of Energy-Dispersive Emission Spectra," P. Van
Espen, H. Nullens and F. Adams, Dept. of Chemistry, University of
Antwerp, Universiteitsplein 1, B-2610 Wilrijk, Belgium

"X-Ray Diffraction Studies of Natural Zeolite-Molecular Sieves in
Air Separation Processes," H. A. Vincent, Anaconda Copper Co.,
P.O. Box 27007, Tucson, AZ 85726

"New Method of X-Ray Stress Analysis," T. Yashiro and K. Ogiso,
Rigaku Corporation, 2-8 Kanda-Surugadai, Chiyoda-Ku, Tokyo, Japan

CONTENTS

XRF APPLICATIONS IN THE MINERALS INDUSTRY

USE OF COMPUTERS IN POWDER DIFFRACTION

X-RAY DIFFRACTION STRESS (STRAIN) DETERMINATION

SOME APPLICATIONS OF ENERGY DISPERSIVE X-RAY FLUORESCENCE ANALYSIS

IN MINERALS EXPLORATION, MINING AND PROCESS CONTROL

C.G. Clayton and T.W. Packer

Applied Nuclear Geophysics Group
Nuclear Physics Division
AERE-Harwell, U.K.

INTRODUCTION

Energy dispersive X-ray fluorescence (EDXRF) analysis is
finding increasing application in field assay of soils and stream
sediments for geochemical prospecting; in borehole logging and
core analysis for formation evaluation in mine control, and in on-
stream analysis for process control.

The analytical capability of the method is determined by the
choice of source, the energy of the exciting radiation and the
energy resolution of the detector. The nature of the samples,
their atomic number and concentration are also important. However,
the most dominant constraints on limits of detection are often imposed by the environment at the region of measurement: by the
borehole condition, the nature of the rock face or the form of the
particulates and stability of the fluid in a process stream.

Radioisotope sources when used with low resolution detectors
(scintillation or proportional counters) and absorption edge filters
operate only to relatively high limits of detection, but equipment
based on these techniques can often be made compatible with the
environmental conditions and with operational requirements. The
use of monoenergetic exciting radiation, generally derived from a
low-power X-ray tube, and a high resolution Si(Li) detector gives
the ultimate in analytical performance and allows elemental concentrations of a few p.p.m. to be measured, often in the presence of
concentrations (>20%) of other elements of high absorption coefficient. Simultaneous measurement of the concentrations of
several elements is possible.

This paper presents examples of the various types of EDXRF equipment now in use and illustrates the analytical performance which can be achieved.

CHARACTERISTICS OF AVAILABLE RADIATION SOURCES AND DETECTORS

Radiation sources

The highest excitation efficiency results from the use of monoenergetic exciting radiation with an energy just above the absorption edge of the analyte. In addition, by judicious choice of exciting energy, interelement effects can be reduced considerably.

Apparatus for generating quasi-monoenergetic radiation has been developed and shown to be of considerable value in minerals analysis[1]. It consists essentially of an X-ray tube (50 kV, 1mA) and a high voltage unit mounted in an oil-filled, light-alloy enclosure. For the highest X-ray purity the enclosure is coupled to a 'scatter-type' energy selector[1] which can be adapted to generate quasi-monoenergetic radiation from 4.5 keV (Ti K X-rays) to 31 keV (Ba K X-rays) with sufficient intensity for the energy of the exciting radiation to be optimised for many applications.

Radioisotope sources emitting a range of X- and γ-rays are available. They are small and robust and are particularly suitable for use in portable equipment. Although their photon output is relatively low (10^6 - 10^8 s^{-1}) adequate sensitivities can be obtained if the energy of the emitted radiation is just above the absorption edge of the analyte. However, the number of elements for which this occurs is small so that the lowest detectable limits can be achieved in only a few cases.

Radiation detectors

The resolution (FWHM) of a scintillation detector containing a thin NaI(Tl) phosphor is approximately 3 keV for Cu K X-rays (Cu K_α = 8.04 keV) so that it is only possible to isolate X-rays whose energies differ by about 6 to 9 keV. The corresponding resolution of a proportional counter is approximately 1 keV so that this detector can be used to resolve X-rays differing by 2 to 3 keV, approximately. Improved discrimination is achieved by using differential absorption edge filters which enable the K_α X-rays of most elements to be isolated from other characteristic X-rays, except K_β X-rays of adjacent elements. The use of a single channel analyser, adjusted to straddle the 'energy peak' of the analyte, reduces errors from unwanted and scattered characteristic X-rays. It is then possible to correct for matrix absorption by a separate measurement of the "overchannel" reading.

Because proportional counters have the higher resolution they

are generally preferred. However, as their efficiency decreases
rapidly for energies above about 30 keV, their use is usually restricted to analytes with characteristic X-rays below this energy.

   As both proportional and scintillation counters are available
in cylindrical form, they are ideal for use in borehole probes.
They are also used for the analysis of elements from Z = 22 (Ti)
to Z = 82 (Pb) in solid and powdered borehole cores. Limits of
detection in these applications vary from 0.01% to 0.3% in measurement times of 20s, depending on the analyte and on the matrix
elements.

   High resolution semi-conductor detectors are finding increasing use in spite of a requirement for cryogenic cooling. As these
detectors have a resolution of about 150 eV (for Mn K X-rays), it
is possible to isolate characteristic X-rays of most analytes although there may be some overlap from adjacent, lower-Z elements.
Hence, considerably lower limits of detection can be obtained. By
using an X-ray tube source, simultaneous analysis of elements can
be made with limits of detection of a few p.p.m.

ADVANTAGES OF OPTIMISING THE EXCITATION ENERGY

   Experience gained from using an X-ray tube excitation
system[1,2] has shown that secondary targets of Rh, Ge and Fe are
effective for exciting a range of elements of geochemical interest.
For example, Fe K X-rays excite Ti, V and Cr efficiently; Ge K
X-rays are optimum for Ni, Cu and Zn, and Rh K X-rays effectively
excite Mo K X-rays and also excite the L X-rays of heavy elements
such as U, Th and Pb.

   The importance of using exciting radiation of optimum energy
is illustrated by considering the determination of Cr in samples
of SiO_2 containing 20% Fe using Rh K, Ge K and Fe K excitation in
turn. Excitation by Fe K X-rays produces approximately 75 times
the intensity of Cr K X-rays compared with Rh K X-rays, and 6
times more than Ge K X-rays with the X-ray tube operated at 45 kV,
0.4 µA.

   In general, the magnitude of matrix absorption varies considerably depending on whether absorption coefficients for the exciting and characteristic radiation of interfering elements are
significantly different from absorption coefficients of the principal rock-forming elements. Significant differences in the
magnitude of matrix absorption also arise according to whether the
absorption edges of interfering elements are above or below the
energies of the characteristic X-rays of the analytes.

   The intensity of characteristic X-rays excited in geochemical samples of saturation thickness (generally the case) is given by

$$I_{f1} = K_1 \, I_o \, \omega_K \, \tau_K \, r / \Sigma_i (\mu_e + \mu_{f1}) \, r_i$$

where K_1 is the overall detection efficiency for the characteristic
X-radiation, $\omega_K \, \tau_K$ is the excitation efficiency, μ_e and μ_{f1} are the
mass absorption coefficients of element i for the incident and
fluorescence radiation respectively, assuming normal incidence;
r_i is the concentration of element i, and r is the concentration
of the wanted element.

The corresponding intensity of scattered radiation is given by

$$I_{scat} = \left[I_{coh} + I_{Comp} \right] = K_2 \, I_o \left[\frac{\sum\limits_i (\sigma_{coh})_i \, r_i}{\sum\limits_i (2\mu_e)_i \, r_i} + \frac{\sum\limits_i (\sigma_{Comp})_i \, r_i}{\sum\limits_i (\mu_e + \mu_{Comp})_i \, r_i} \right]$$

where I_o is the incident radiation intensity, K_2 is the overall
detection efficiency of the system (including geometrical efficiency), σ_{Comp} and σ_{coh} are the Compton and coherent scatter
coefficients, and μ_{Comp} is the mass attenuation coefficient for
scattered radiation for element i.

As the Compton scatter coefficient is nearly independent of
atomic number and $K_1 = K_2$ (usually), it follows that:

$$\frac{I_{f1}}{I_{Comp}} \alpha \left[\frac{\sum\limits_i (\mu_e + \mu_{Comp})_i \, r_i}{\sum\limits_i (\mu_e + \mu_{f1})_i \, r_i} \right] r$$

Provided that the energy of the exciting radiation is nearly the
same as the energy of the characteristic X-rays, $(\mu_{f1})_i \simeq (\mu_{Comp})_i$
except for the element being excited, and the ratio of I_{f1}/I_{Comp}
is then proportional to r and independent of the matrix. As it is
usual for the energy of the absorption edge of the element being
excited to lie between the energy of the Compton radiation and the
excited characteristic X-ray, this ratio gives over-compensation
at high concentrations of wanted element. However, in geochemical

samples, most elements of interest occur at low concentrations
(<0.1%) so that this error in the measurement is not important.

The ability to compensate for matrix absorption effects in an
element with a lower Z than the analyte by separate measurement of
the intensity of the Compton scattered radiation is an important
advantage of combining high resolution detectors with monoenergetic
exciting radiation. Best results are obtained by using backscatter
geometry since the Compton 'peak' is then furthest removed from the
coherent scattered X-rays so allowing them to be resolved at
lower excitation energies. If the major interfering elements have
absorption edges above that of the analyte, matrix absorption can
be minimised by selecting the exciting radiation energy to be
immediately above the absorption edge of the analyte but below
that of the interfering elements.

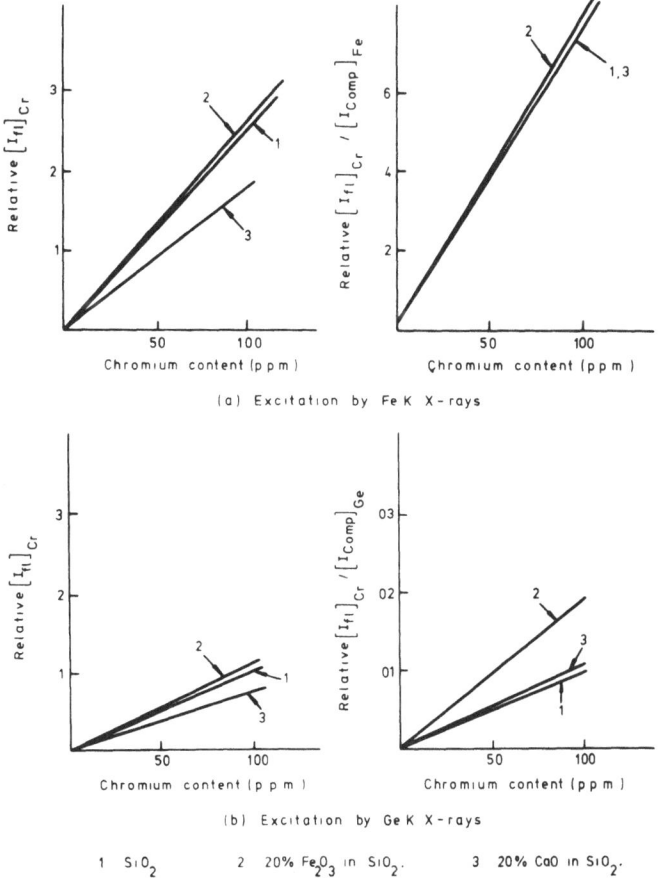

(a) Excitation by Fe K X-rays

(b) Excitation by Ge K X-rays

1   SiO$_2$ 2   20% Fe$_2$O$_3$ in SiO$_2$.   3   20% CaO in SiO$_2$.

Fig. 1   Comparison of effectiveness of Fe K and Ge K X-rays
for analysing Cr in SiO$_2$ containing Fe$_2$O$_3$ and CaO.

The preferred approach can be demonstrated by considering the case when Cr K X-rays are excited in samples containing high and variable concentrations of Fe, whose K absorption edge (7.11 keV) is just above the energy of the Cr K_α X-rays (5.40, 5.41 keV) and Ca, whose K absorption edge (4.07 keV) is below.

Figure 1 shows the calculated intensities of Cr K X-rays excited in samples containing different concentrations of Fe_2O_3 and CaO by Ge K X-rays with energies above the Fe K absorption edge and Fe K X-rays with energies below. It is seen that matrix effects caused by the presence of Fe_2O_3 are minimal, but as expected 20% CaO produces considerable matrix absorption. Figure 1 also shows the ratio of Cr K X-rays to the intensity of Compton scattered Ge K_α and Fe K_α X-rays. Whereas this ratio is nearly independent of CaO content when exciting with either energy of X-rays, it is only independent of Fe_2O_3 content when using Fe K X-ray excitation.

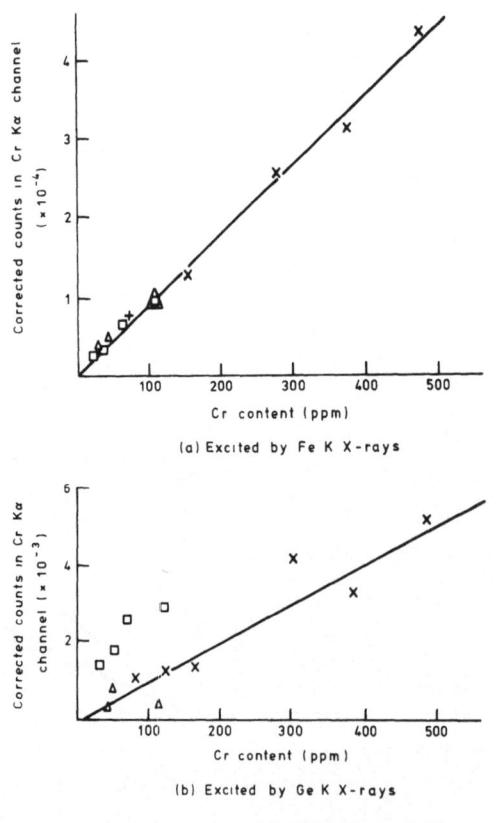

(a) Excited by Fe K X-rays

(b) Excited by Ge K X-rays

Δ 3-5% Fe_2O_3 X 7-10% Fe_2O_3 □ 11-15% Fe_2O_3

Fig. 2 Comparison of calibration curves for Cr in SiO_2 containing Fe_2O_3 and CaO with and without effective matrix compensation.

Corresponding experimental results are illustrated in Fig. 2 which shows the calibration curves obtained when Cr K_α X-rays are excited in samples containing up to 15% Fe_2O_3 and 12% CaO by Ge K X-rays and by Fe K X-rays. It is seen that by using Fe K X-rays not only can the measurement sensitivity be increased, but compensation for matrix absorption due to variations in the concentrations of both Fe and Ca (based on detecting a constant number of Compton scattered X-rays) can be achieved.

SIMULTANEOUS MULTI-ELEMENT ANALYSIS OF SOILS AND STREAM SEDIMENTS

A complete energy dispersive X-ray fluorescence spectrometer[2] suitable for carrying out simultaneous multi-element analysis of soils and stream sediments in the laboratory and in the field is shown in Fig. 3. It includes a Si(Li) detector, X-ray generator and excitation energy selector.

Fig. 3 Energy dispersive X-ray spectrometer showing the monoenergetic X-ray generator and sample presenter. The couplings on the side enable the sample presenter to be evacuated when analysing elements of low atomic number.

For simultaneous multi-element analysis it is not possible to use the optimum excitation energy for all elements. Results obtained using near-optimum excitation energies for only a few elements are presented in Table 1 and these are compared with results obtained when exciting up to sixteen elements simultaneously. Correction for matrix absorption effects was achieved in each analysis by adjusting the measurement time automatically in order to derive a constant number of Compton scattered X-rays.

Table 1   Variations in measurement errors in simultaneous multi-element analysis of soil and stream-sediment samples according to energy of exciting radiation.

Analyte	Errors[†] due to counting statistics (95% c.l.) in measurement times varying between 100 and 300 s p.p.m.		
	Excitation X-rays		
	Fe K	Ge K	Rh K
Titanium	11 – 15	~ 50	120
Vanadium	8 – 10		50
Chromium	4 – 6	~ 20	35
Manganese		~ 50	90
Iron		200	300φ
Nickel		2 – 7	20
Copper		3 – 7	20
Zinc		3 – 7	20
Tungsten (L)			60
Arsenic			20
Lead (L)			20
Rubidium			7
Uranium (L)			15
Strontium			8
Yttrium			8
Zirconium			8
Niobium			8
Molybdenum			7

[†] all elemental concentrations approximately 200 p.p.m.

φ at 8% Fe_2O_3

It is clear that as the excitation energy is brought closer to the absorption edge of each analyte a lower limit of detection is achieved. In some applications multiple simultaneous excitation

energies (obtained by means of a compound energy selector) can be
used and this technique can result in limits of detection within a
factor two of the optimum for a single element.

CHARACTERISTICS AND PERFORMANCE OF EQUIPMENT USED IN MINE CONTROL

In general, equipment used for grade control is characterised
by a need to measure only a few elements at concentrations >0.03%.
There are notable exceptions; in particular Au (~0.001%) and U
(~0.01%). For most applications portable instruments using radioisotope sources with scintillation or proportional counters are
generally acceptable although for analysis of Au and U high resolution detectors are required. For some applications excitation
by monoenergetic X-rays from a low power X-ray generator is
preferred. However, radioisotope sources have the advantage of
allowing K X-rays of high-Z elements to be excited. Applications

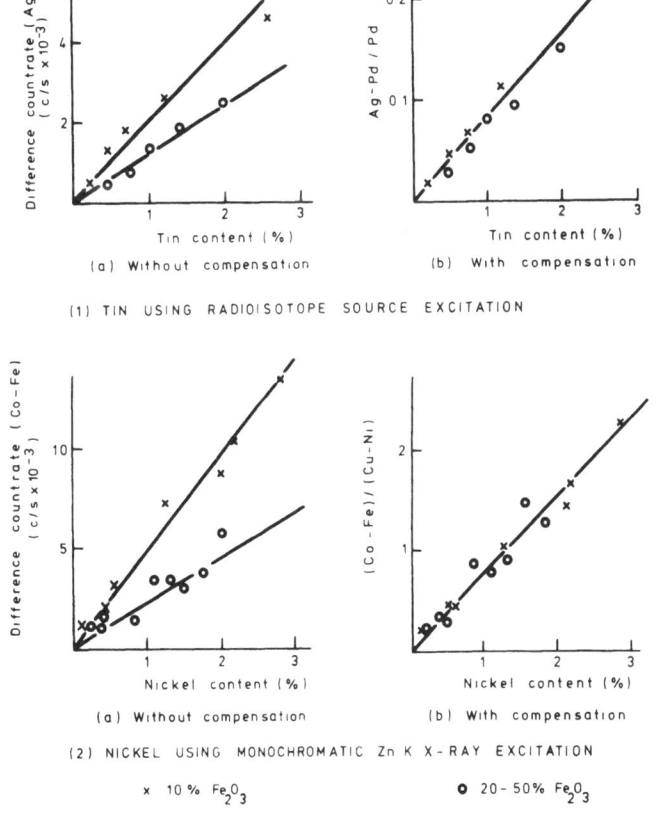

Fig. 4 Compensation techniques using radioisotope and mono-
        energetic X-ray sources with low resolution detectors.

occur in borehole logging and for analysis of borehole cores, rock
faces and powdered rock samples. Equipment presently available
includes facilities for selecting characteristic X-rays of up to
four elements by incorporating suitable pairs of differential
filters.

    Although it is not normally possible to isolate Compton and
coherent scattered radiation using this type of equipment, a separate measurement of the intensity of the total scattered radiation,
given by the overchannel count when using a single channel analyser,
can often be used to correct for matrix absorption effects.

    Results of measurements with a portable analyser on powdered
core samples containing Sn or Ni with varying concentrations of Fe
are shown in Fig. 4. Measurements of Sn content is made by using
an ^{241}Am source with correction for matrix absorption by Fe based
on the reading with the Pd filter. Measurement of Ni content is
by means of monoenergetic X-ray excitation from an X-ray generator

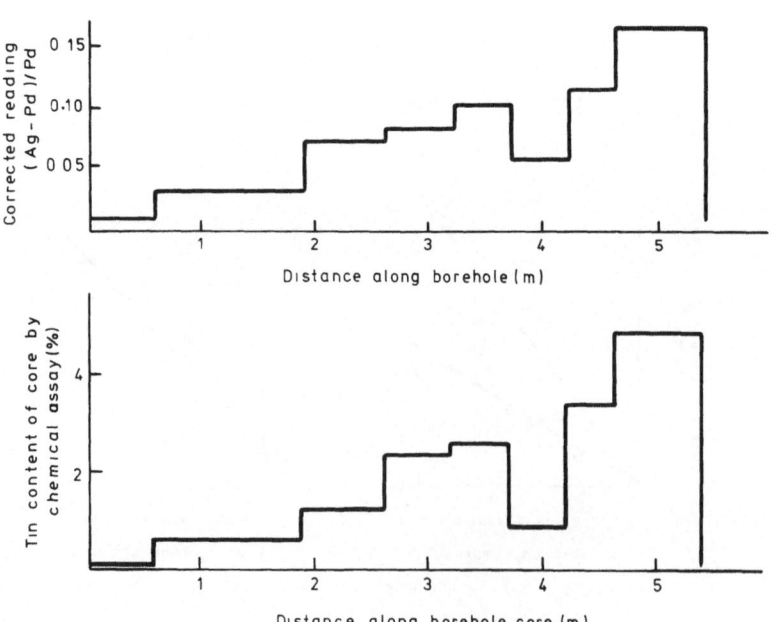

Fig. 5   Comparison between tin content of rock by borehole
         logging and chemical analysis of core.

and by separate measurement of the scattered Zn K X-rays, the latter being isolated by balanced filters of Cu and Ni. Good compensation is demonstrated by both techniques. The same methods
are in use for a range of analyses of solid cores and powders.

In general the restricted diameters of boreholes and the
depths of operation prohibit the use of X-ray generators and semi-
conductor detectors although some success in the development of
probes incorporating high resolution Ge detectors have been
reported[3].

The relatively low excitation and fluorescence radiation energies in EDXRF, especially for elements of low and medium Z (Z<40
approx.) result in low penetration (generally <1cm) in rock and
this restricts applications to virtually dry and shallow boreholes.
Most applications therefore occur in open pit mines. For measurement of Sn and higher-Z elements operation in partially or wholly
water-filled boreholes is possible.

In spite of these limitations borehole loggers incorporating
balanced filters are being used throughout the world for determining concentrations of Sn, Cu, Ni, Zn and Pb. A comparison between a borehole log for Sn and the corresponding chemical analysis
of the core is shown in Fig. 5.

The environmental problems encountered in borehole logging
are removed when measurements are made on borehole cores and
equipment developed for the analysis of tin in cores in a field
laboratory is shown in Fig. 6. Excitation of Sn K X-rays is by
eight [241]Am sources and the detector is a special proportional
counter. The tin concentration is averaged over each 20 cm length
of a 1 metre core. Compensation for matrix effects is included
automatically and the equipment prints the tin concentration (in %
Sn) and other derived data directly on a paper tape.

ON-STREAM ANALYSIS

The use of EDXRF techniques in on-stream analysis is now established and an increasing number of successful installations are
being reported. A comprehensive review of this area of application
has been given by Watt[4]. In general the arguments in favour of
EDXRF compared with WDXRF systems are strongest when the number of
streams to be analysed is restricted (<10) and a constraint on
measurement time (<100 s) is not imposed.

The installation strategy for EDXRF systems varies considerably and depends strongly on the position in the circuit where
measurements are required (eg. in the feedstock, middlings, concentrates or tailings), on the slurry density, on the mineral to
be analysed and on the accuracy required. For currently installed

systems accuracies vary from about 0.01% to 2.0% element content
depending on the application. To meet the various operational
requirements both in-stream probes [NaI(Tl)] and by-line systems
[Si(Li), NaI(Tl) and proportional counters] are in use or under
development.

   A summary of reported applications and performances of on-
stream EDXRF systems is given in Table 2.

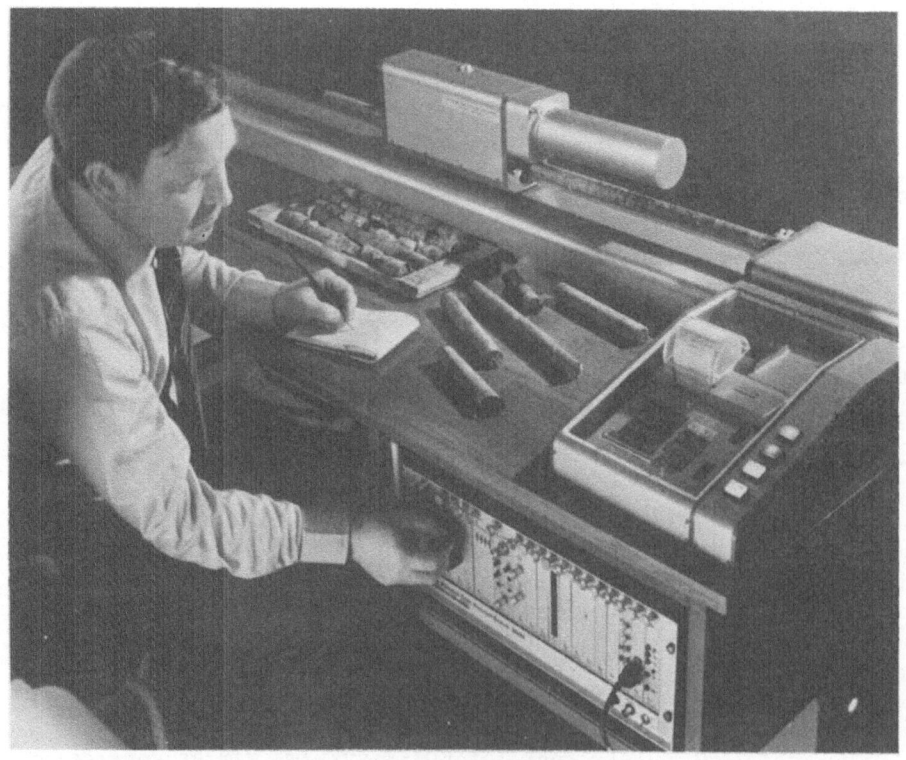

Fig. 6   Borehole core analyser designed to measure
         the concentration of tin.

Table 2　Operational performance of some EDXRF on-stream
(in stream and by-line) systems[4]

Element	Mean Concentration in Slurry wt %	Relative Standard Error of Analysis %	Element	Mean Concentration in Slurry wt %	Relative Standard Error of Analysis %
Pb	8.3	5	Cu	1.0	8
Pb	0.75	9	Cu	0.1	14
Pb	13	6	Cu	1.6	6
Pb	0.85	20	Cu	27	3
			Cu	0.1	14
Zn	1.8	15	Cu	3.8	7
Zn	52	1	Cu	36	3
Zn	0.82	21	Cu	0.23	11
Zn	3.6	13			
Zn	2.1	9	Sn	1.84	10
Zn	3.1	10	Sn	0.86	7
Zn	10.5	6	Sn	2.9	9
Zn	52	1	Sn	0.34	6
Zn	0.65	26	Sn	0.12	17
Zn	1.8	5	Sn	0.16	13
Zn	49	-	Sn	20	10
Zn	0.5	-	Sn	0.2	5
			Sn	0.4	13
			U	0.25	2.5

REFERENCES

1. C.G. Clayton, T.W. Packer and J.C. Fisher, Primary energy selection in non dispersive X-ray fluorescence spectrometry for alloy analysis and coating thickness measurement. Nuclear techniques in the basic metal industries, I.A.E.A., Vienna (1973).

2. C.G. Clayton and T.W. Packer, Analysis of stream sediment and soil samples using a high resolution energy dispersive X-ray fluorescence spectrometer. Nuclear techniques and mineral resources, I.A.E.A., Vienna (1977).

3. G.R. Boynton, Canister cryogenic system for cooling germanium semi-conductor detectors in borehole and marine probes. Nucl. Instrum. Methods (23: 599 (1975)).

4. J.S. Watt, Nuclear techniques for on-line measurement in the control of mineral processing. Nuclear techniques and mineral resources, I.A.E.A., Vienna (1977).

ANALYSIS OF MAGNESIUM BRINES BY ENERGY DISPERSIVE XRF

John F. Reilly and Debbie J. Langenfeld

NL Industries, Inc.

Salt Lake City, Utah 84116

INTRODUCTION

The analysis of magnesium brines can be performed by atomic absorption and classical wet chemical techniques. These methods are subject to dilution errors and are time consuming. This paper describes a method using energy dispersive x-ray fluorescence to determine magnesium, sulfur, chlorine, potassium and calcium in concentrated magnesium brines.

PROCEDURE

The x-ray spectrum produced by a typical brine is shown in Fig. 1. Interelement corrections using the Lucas-Tooth and Pyne[1] (or Delta I) model were applied. Chlorine is, by far, the element of highest concentration and influences all other elements. The concentration of chlorine is proportional to the total dissolved solids and therefore proportional to the density. A chlorine slope correction is made for all other elements and a chlorine background correction is necessary for both magnesium and sulfur. The potassium Kβ appears as background to the Ca Kα, so this interference must also be taken into account.

The brines analyzed by this method fall into two categories; high total dissolved solids and very high total dissolved solids. Those of the first category remain in the liquid state at laboratory temperatures and can be analyzed directly. Those of the second category develop a solid phase and must be diluted before analysis. The analysis is done using a rhodium x-ray tube at 15 KV and 900 μa.

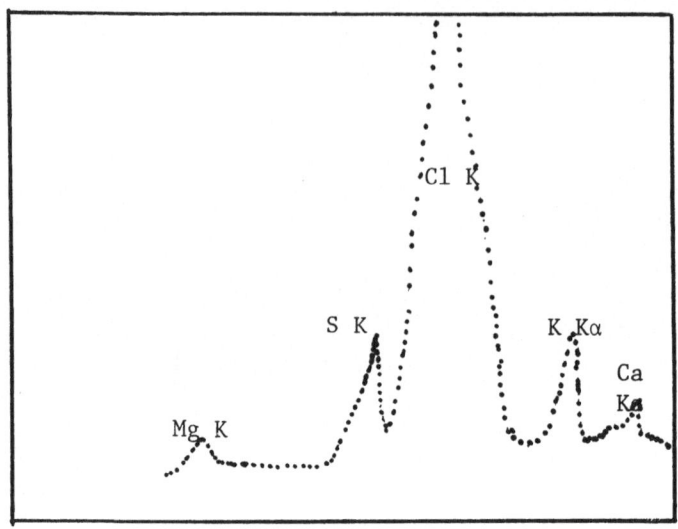

Fig. 1. Typical Brine Spectrum

The counting time is 100 seconds for the undiluted brines and 200 seconds for the diluted samples.

     The majority of analyses reported by x-ray fluorescence in-volve heavier elements in solid form. The elements discussed here require either a vacuum or helium path. Liquid samples require a helium path. The mylar film used to prepare the sample cups is 0.00015" (< 4 microns) to optimize the magnesium count rate. Even thinner films have been used, but the frequency of leaks was too great to tolerate. A leak in the film can be catastrophic since the beryllium windows are extremely sensitive to chlorides. The analysis of leaky samples can be avoided by placing the filled cup on an absorbant surface for a few minutes before inserting into the instrument. The presence of any leak will be apparent before analy-sis. Precautions must also be taken not to allow the sample chamber to be evacuated while, unknown to the operator, a brine sample is inside.

RESULTS AND DISCUSSION

     A comparison of the results reported by the XRF method and classical methods is shown in Table 1. The range of concentrations is small, which makes this application particularly well suited to the Delta I model. The concentration of chlorine is not precisely known, and is included only to satisfy the requirements of the model.

Table 1.

UNDILUTED BRINES

	Magnesium		Sulfur ($SO_4^=$)		Chlorine		Potassium		Calcium	
	Wet	XRF	Wet	XRF	Wet	XRF	Wet	XRF	Wet	XRF
	6.9	6.85	5.7	5.6	18.2	18.5	1.71	1.63	0.003	0.005
	6.4	6.41	7.5	7.4	20.4	20.7	2.55	2.63	0.003	0.003
	5.6	5.50	5.7	5.7	19.8	19.7	2.39	2.32	0.11	0.11
	8.2	8.23	3.5	3.4	21.9	21.6	1.43	1.49	0.004	0.003
	7.4	7.0	2.4	2.3	17.1	17.4	0.09	0.10	0.110	0.108

Table 2.

BRINES DILUTED 1:1

	Magnesium		Sulfur ($SO_4^=$)		Chlorine		Potassium		Calcium	
	Wet	XRF	Wet	XRF	Wet	XRF	Wet	XRF	Wet	XRF
	7.7	7.6	2.3	2.3	21.1	20.9	0.29	0.28	0.04	0.04
	7.0	7.1	1.1	1.1	21.7	21.6	0.25	0.25	0.12	0.14
	10.4	10.4	0.1	0.1	31.4	31.6	0.21	0.21	0.28	0.24
	8.8	9.0	0.09	0.1	27.1	26.9	0.32	0.32	0.41	0.40
	8.9	8.8	0.07	<.1	25.7	25.4	0.21	0.21	0.22	0.22
	7.2	7.3	0.08	<.1	21.6	22.0	0.26	0.26	0.26	0.28

Many brine samples arrive at the lab in two phases. These contain very high total solids, and precipitation occurs at room temperature. In order to provide the spectrometer with a homogeneous and representative sample, they are diluted 1:1 by volume with water before analysis. Table 2 shows the results of comparative analyses between XRF and classical techniques. The concentrations reported are on the basis of undiluted brines. This is necessary in order to report the sulfate reliably down to the level of 0.1% in spite of the influence of the chlorine K line. The standards are prepared in the same manner.

There was some concern that the volume of brine placed in the sample cup might affect the analytical results. Table 3 put this fear to rest, as well as providing a good idea of the precision of the technique. Various volumes of the same sample were placed in different sample cups and analyzed. No reasonable volume change had any significant effect on the reported concentrations.

CONCLUSION

The agreement between the classical methods and the XRF technique is satisfactory and provides a significant savings of time. The variation in replicate analyses is equivalent to that of other techniques employed.

Table 3.   Analytical Variation

Sample Vol. (ml)	Thickness (mm)	Mg	$SO_4^=$	Ca
1	1.3	7.4	2.6	0.02
2	2.5	7.3	2.8	0.02
3	3.8	7.3	2.5	0.02
4	5.1	7.3	2.8	0.02
5	6.3	7.2	2.6	0.02

REFERENCE

1. J. Lucas-Tooth and C. Pyne, "The Accurate Determination of Major Constituents by X-Ray Fluorescence Analysis in the Presence of Large Interelement Effects," *in* Advances in X-Ray Analysis, 7:523-41, Plenum Press, 1964.

PORTABLE X-RAY FLUORESCENCE ANALYZERS AND THEIR USE IN AN UNDERGROUND EXPLORATION PROGRAM FOR TIN

Mark W. Springett

Gold Fields Mining Corporation

Lakewood, Colorado 80228

In 1968 Consolidated Gold Fields commenced exploration below adit level at the Wheal Jane tin prospect in south-west England. During the surface exploration program the relatively recently introduced Portable X-ray Fluorescence Analyzer ("P.I.F.") had been used for semi-quantitative scanning of drill core and for providing preliminary assays for tin on prepared core samples. The P.I.F. was introduced to the mining industry in 1965 (Bowie et. al. (1) and had originally been equipped specifically for tin analysis due to the current interest in tin exploration, the difficulties of chemically assaying for tin and the suitability of tin to X-ray analytical techniques. The instrumentation and principals of operation have been described by Bowie (1,2). At Wheal Jane the need to obtain rapid analyses to help control underground development and difficulties in visually recognizing the tin mineralization made the prospect an excellent testing ground for the practical aspects of the instrument.

The P.I.F. analyzer makes use of a radioisotope as the source of exciting radiation with balanced filter pairs for energy selection. The filters have similar transmissivities excepting for a passband bracketing the fluorescent wavelength characteristic of the element sought. A scintillation crystal with photomultiplier and counter detect the fluorescent emissions over set counting periods. Display was in ratemeter form in the original model but digital displays have been used in subsequent models. The equipment is typically about 12 to 20 lbs in weight and can be either battery or mains operated.

GEOLOGY AND MINERALIZATION

The tin at Wheal Jane occurs in a complex cassiterite-sulfide ore located in shear zones along the footwall of the lowest of a series of shallow-dipping porphyry dikes. Minor tin mineralization also occurs on other porphyry dike contacts and within sets of steeply-dipping fractures. The major zone of mineralization essentially consists of brecciated slate and vein quartz mineralized mainly by cassiterite and tourmaline and in some places by massive pyrite and sphalerite. The cassiterite is typically fine-grained and difficult to recognize visually. The average plan width of the lode is 12 feet and tin content averages 1.25% Sn (Rayment et. al. 3).

SAMPLING METHODS

Underground sampling of the main or 'B' lode was carried out by several methods. An exploration drift was driven on the mineralization from which crosscuts, raises or drill holes were put out every 50 feet to expose the full width of the mineralization for sampling purposes (Figure 1). Samples were either taken by channel sampling the crosscut or raise walls or by sampling the split half drill core. In addition to this systematic sampling of the ore body, grab samples were taken from each car of development muck to provide a composite sample of each round blasted. Development continued two shifts per day, six days a week and speed in obtaining assay results was essential for the close control of drifting.

SAMPLE PREPARATION

All drill core was scanned with a P.I.F. equipped for tin. Readings were taken roughly every foot on apparently unmineralized core and approximately every two inches on mineralized sections. Difference count rates (DCR) were marked on white cards and placed in the core tray to assist the geologist logging the core and marking up sections for splitting and sampling. The difference count readings were regarded as semi-quantitative and were entered on the geological log as a permanent record. Core sections selected as samples were halved and one half was crushed, split and pulverised in a tema mill to -100 mesh. Three portions of approximately 30 grams were placed in sample dishes for P.I.F. analyses and a fourth portion was held for dispatch for chemical assay. After the three separate replicates had been analyzed on the P.I.F. instruments the material was stored.

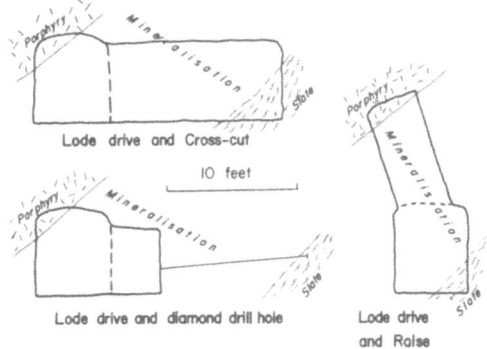

Figure 1. Alternative Methods of Sampling 'B'-lode Mineralization

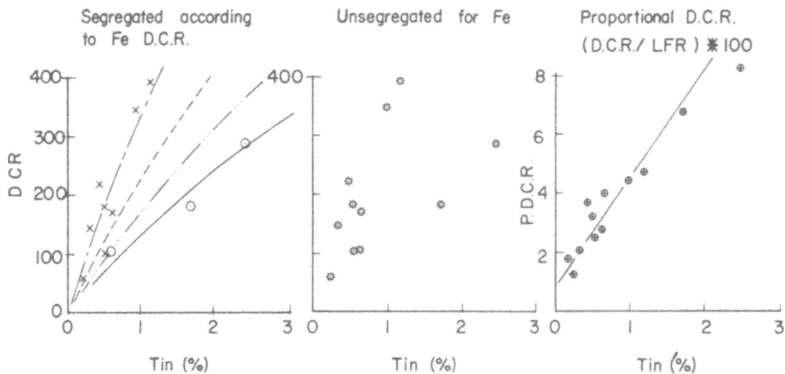

Figure 2. Calibration curves for tin showing effect of iron correction by segregation and by ratio of fluorescent to back-scattered radiation.

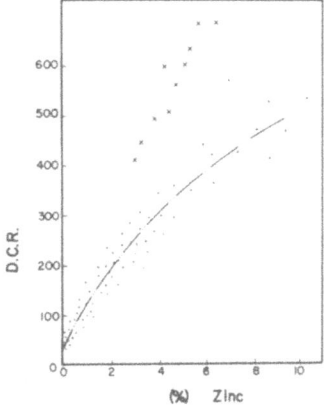

Figure 3. Zinc calibration curve showing suspect assays.

Channel samples were prepared and treated similarly to the core samples except that a drying stage was required prior to pulverizing. The bulky composites required an additional stage of crushing to -3/8 inch and then quartering to a quantity that could follow the standard core or channel sample routine. The same P.I.F. analysis routine was followed on all three types of samples. The bulk samples received priority as the rounds they represented were held in temporary 50 ton bins at the shaft head for assignation either to waste, low grade or high grade stockpile on the basis of the P.I.F. result.

Samples were analyzed, in triplicate for tin, iron and zinc. Tin was the element of primary importance, zinc was present in significant quantities and was considered to possibly be of economic importance to the project. Iron was analyzed in order to make adjustment to the tin estimates due to the effects of absorption caused by high iron contents. A suite of calibration curves was used initially for estimating tin content based on different iron DCR ranges. Zinc lower and upper filter readings were recorded but for a period of over a year no attempt was made to prepare a zinc calibration curve. A turn round of less than 36 hours was normal for all samples. A staff of two technicians was employed fulltime on sample preparation and analysis.

CALIBRATION

Calibration curves can be prepared in several ways. One commonly favored practice is to use a few carefully prepared standard samples which have been very precisely and accurately analyzed and that cover a suitable range of values in convenient increments. This technique was used initially but a second method was found more useful in practice, although the standards were retained for check purposes.

All samples were sent for chemical analysis during the first six months of operation providing in excess of 2500 paired results for which upper and lower filter readings for tin, zinc and iron and chemical assays for tin and zinc were known. A scattergram of tin DCR's versus assay result provided a calibration curve that gave a more realistic indication of the actual comparison between predicted tin content and chemical assay, taking into account the combined errors due to firstly the X-ray fluorescence technique, secondly the sub-sampling errors in splitting out the portions for chemical as opposed to X-ray analysis and thirdly the errors inherent in the chemical assay. The calibration curve prepared in this way was improved by segregating the results according to iron DCR. Much later in the exploration program the suite of curves for correcting for the effects of iron was superseded by a single calibration curve based on the ratio of the

difference count rate to the lower filter reading which was found
to provide a simple and effective correction for matrix absorption
effects that made it no longer necessary to analyze for iron (Figure 2). The use of the ratio of fluorescent to backscattered
radiation has been described by several authors and becomes of
particular use with sources of short half life and for situations
where samples with a wide variation in matrix are being analyzed.
The proportional difference count rate (PDCR) can also be useful
for in situ borehole analyses where it is impractical to analyze
for additional elements such as iron and where irregular bore hole
wall geometry increase the matrix effects.

When a scattergram of zinc assays versus difference count
rates was prepared there were a significant number of unexplained
outliers from the cloud of points (Figure 3). The outliers would
have been predicted to have a higher zinc content than reported by
the assayer. When checked the X-ray results were reproducible but
the atomic absorption results reported some but not all as previously. On further investigation the laboratory was found to be
failing to report dilution factors for high zinc samples that went
off-scale. This error would have introduced a bias of economic
significance to ore reserve calculations for zinc and it had been
recognized by use of the P.I.F.

BORE HOLE PROBE

In 1969 Consolidated Gold Fields started testing a borehole
probe for tin. The probe had been designed by the Nuclear Techniques in Mining and Quarrying division of the U.K. Atomic Energy
Authority and was being tested with the assistance of the Institute of Geological Sciences (I.G.S.). The probe is designed for
use in BX boreholes and is connected to the standard surface
instrumentation by means of a flexible rubber hose containing a
compressed CO_2 gas line for pneumatically switching filters and a
multicore electric cable for transmitting the information from the
photomultiplier to the surface instrumentation. Probe configuration is shown in Figure 4. Plots of two holes are shown with a
five term moving average of the DCR lower filter reading (LFR) and
the PDCR results from the probe (Figure 5 and 6). Assayed tin
content of the core is shown in each case. The general correlation of DCR peaks with the sections of highest tin content is
immediately apparent. The logs illustrated here show some marked
discrepancies as well as a generally good correlation. In some
instances these are attributable to sampling problems in the core,
for example in DD 305 a complete assayed section was not available
due to poor core recover. Tin in significant quantities is clearly
present at 31 feet down the hole. It should be noted that the
correlations between assayed tin content and probe results are all

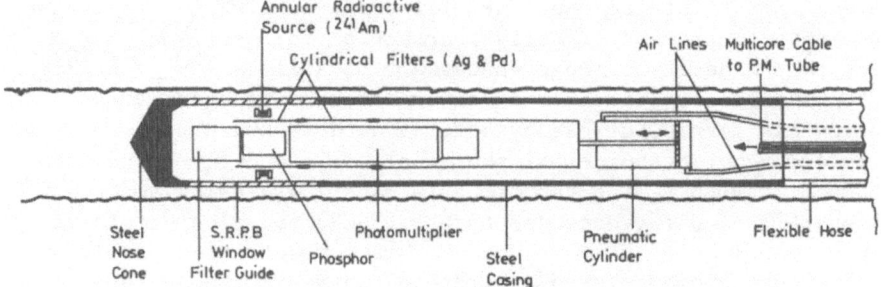

Figure 4.   Configuration of components of borehole probe.

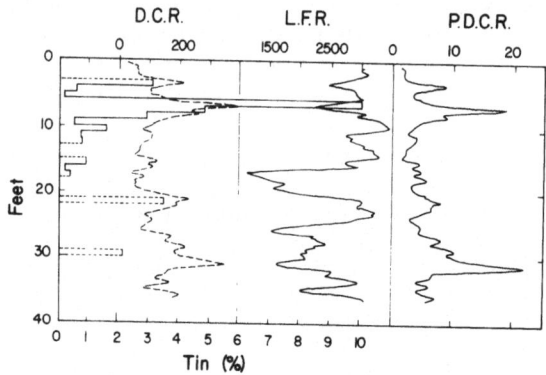

Figure 5

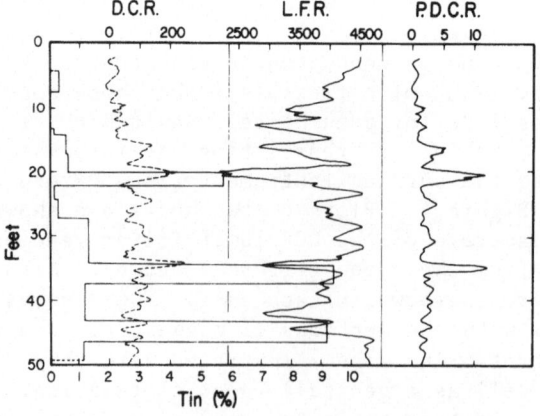

Figure 6

Figures 5 and 6.   Core assays for tin and five term moving average
of lower filter reading, difference count rate and proportional
difference count rate for hole 305 (Figure 5) and 48 (Figure 6).

comparing the X-ray analysis of the hole walls against the core
taken from the hole. During the two year final phase of underground exploration approximately 15,000 samples were analyzed by
P.I.F. of which roughly one third were selected for chemical
assay. The ability to select material for assay represents a
saving of approximately $50,000 in annual direct assay costs
alone. The rapid turnround with the P.I.F. made the creation of
the stockpiles a practical proposition and at today's metal prices
the contained value of tin in the stockpile would be in excess of
$500,000. The additional control that rapid tin assays provide is
an intangible advantage of considerable significance in a major
underground exploration program.

ACKNOWLEDGEMENTS

The author would like to thank Consolidated Gold Fields for
permission to publish this paper, Barbara Cox for her assistance
in typing and Stephen Howse for help with the graphics.

REFERENCES

1. S. H. U. Bowie, A. G. Darnely and J. R. Rhodes, "Portable
    Radioisotope X-ray Fluorescence Analyser," Trans. Instn.
    Min. Metall., 74, 361-79 (April 1965).

2. S. H. U. Bowie, "Portable X-ray Fluorescence Analysers in
    the Mining Industry," Min. Mag. Lond. 118, 230-9 (April
    1968).

3. B. D. Rayment, G. R. Davis and J. D. Willson, "Controls
    to Mineralization at Wheal Jane, Cornwall," Trans. Instn.
    Min. Metall. (Sect. B: Appl. earth sci.) 80, B224-237
    (August 1971).

APPLICATIONS OF A NEW MULTIELEMENT PORTABLE

X-RAY SPECTROMETER TO MATERIALS ANALYSIS

C. von Alfthan and P. Rautala

Outokumpu Oy
Espoo, Finland

J. R. Rhodes

Columbia Scientific Industries
Austin, Texas 78766

ABSTRACT

A new hand-portable, microprocessor-based, multielement X-ray fluorescence analyzer is described. The instrument is light in weight (19 lb), completely self-contained and powered by internal rechargeable batteries. The detector is a special proportional counter whose room temperature energy resolution is sufficient to enable adjacent element X-rays to be deconvoluted with the help of the microprocessor. The instrument yields concentration readout of elements, corrected for matrix effects, in groups of four at a time. A number of field and laboratory applications to ore, solution and alloy analysis are described.

INTRODUCTION

Since their first availability over a decade ago portable X-ray fluorescence analyzers[1,3] have found application in the field, in industrial plants and in the laboratory for determination of minor and major elements in a wide range of materials. These first-generation instruments employ a small, sealed radioisotope source for excitation of the sample, and a scintillation or proportional counter with a single channel analyzer and reversible scaler for detecting and counting the emitted X-rays. Discrimination between X-ray lines too close to be resolved by the detector is done by balanced, absorption-edge filters. Although these instruments will continue to be employed for simpler analyses requiring essentially

27

single-element, sequential measurements, they have two basic drawbacks that have prevented them from coming into more widespread
use. The subtraction of two counts inherent in the balanced filter
method causes an escalation of statistical errors at low element
concentrations, which results in a significantly worse sensitivity
than would be obtained if balanced filters could be avoided. The
second difficulty has arisen through the very versatility of the
instrumentation. In contrast to laboratory X-ray fluorescence
systems, portable analyzers have many potential uses where the
operator is not a scientist or technician. In our experience,
successful operation of first-generation portable analyzers requires
significant technical training and a working knowledge of X-ray
fluorescence techniques.

The new instrument overcomes these drawbacks without compro-
mising portability and "field worthiness." At the same time it
extends the capability to multielement analysis and to direct readout of element concentration. Two technological advances have made
this possible, (1) significant improvements in the X-ray energy
resolution of proportional counters and (2) the incorporation of
microprocessors. Lehto[4] and Hietala et al.[5] have shown that if the
X-ray energy resolution of sealed proportional counters can be
improved only moderately, elements adjacent in the Periodic Table
can be determined with satisfactory accuracy using simple window
integration methods. Such improvements in energy resolution of
sealed proportional counters have been obtained by operating at
very low gas gains (10 to 100) and by utilizing new rare gas mixtures
as fillings.[4,6]

DESCRIPTION OF INSTRUMENT

The instrument consists of an analyzer unit and one of three
probes, a surface probe, a sample probe and a light element probe.
The analyzer unit and one probe are usually housed in a leather
carrying case along with accessories such as sample cups and calibration update standards. Figure 1 shows a photograph of the
analyzer unit with the three probes.

A simplified electronic block diagram is shown in Figure 2.
The preamplifier and high voltage supply are housed in the probe.

The pulse processor includes a linear pulse amplifier with
RC pulse shaping, gated baseline restorer and dead-time corrector.
The analog-to-digital convertor is an 8-bit device that produces
a 256 channel spectrum for storage in the microprocessor unit (MPU).
The latter includes 6K of PROM and 2 K of RAM. The total power
consumption is about 3 W giving the integral rechargeable batteries
a life of 8 to 12 hours between charges.

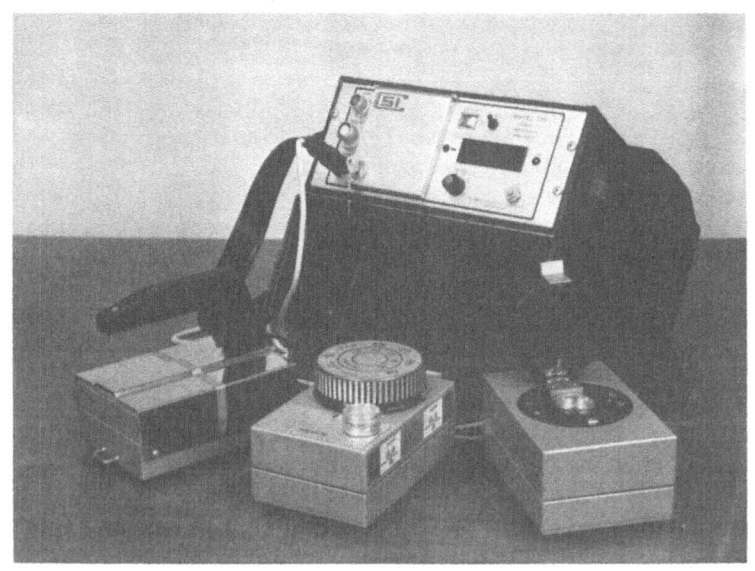

Figure 1.   Photograph of Instrument

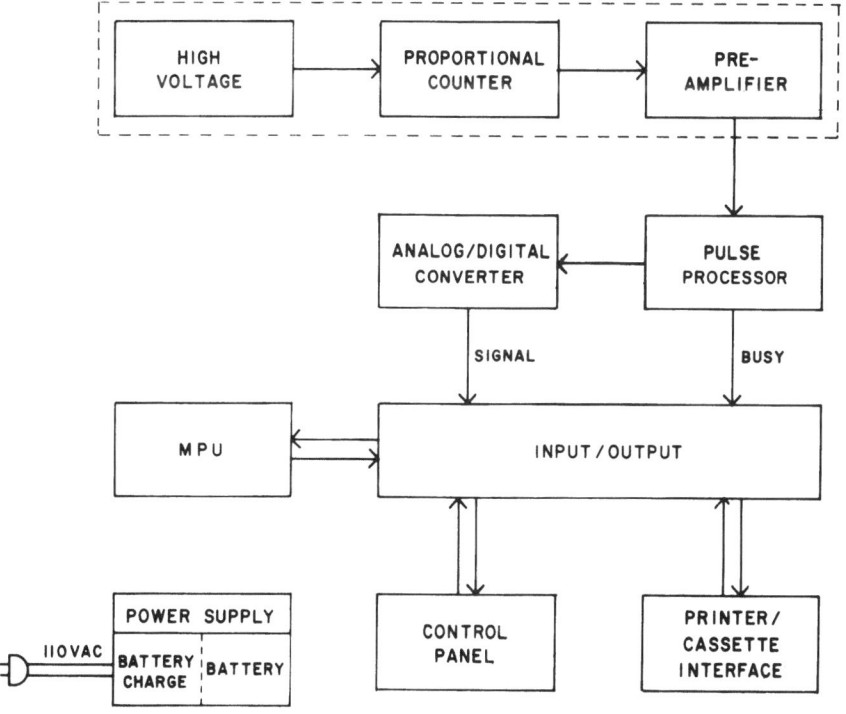

Figure 2.   Simplified Electronic Block Diagram

In both the sample probe and surface probe "central source geometry" is used, with the 8 mm diameter x 5 mm thick source capsule on the common axis of the detector window and sample aperture. The aperture of the surface probe is flush with the flat face of the probe and is 2.8 cm in diameter. When the probe is placed on the material to be analyzed a pin is depressed and the source exposed. A LED in the probe handle glows red when the shutter is closed, goes out when the shutter is properly open and glows green during data accumulation. Uneven surfaces can be measured by this probe as long as they are not too re-entrant or so sharply curved that the shutter pin cannot be operated. Compensation for geometrical errors is incorporated into two of the calibration models (Models 7 and 8, see below). The correction comprises ratioing each characteristic X-ray intensity to the backscattered intensity measured on the same sample. The light element probe employs a more compact geometry than the other two. This permits X-rays down to Al K to be measured without resorting to a helium or vacuum path. The source interlock on both the light element and the sample probe is a shutter which must be closed when the lid is opened for sample handling. Both probes take samples of maximum diameter 3.18 cm and thickness 1.6 cm, in the form of powders or liquids contained in sample cups or cast, pressed or machined solids.

Table 1 lists the sources used, their relevant properties and the ranges of elements normally measurable with each source.

Table 1. Relevant Properties of Sources Used

Isotope	Half-Life	Emission*	Preferred Elements Excited	
			K X-Rays	L X-Rays
Fe-55	2.6 yrs	Mn K X-Rays 6 keV	Al-V	–
Cm-244	17.6 yrs	Pu L X-Rays 14-21 keV	Ti-Zn	Ta-Pb
Cd-109	1.3 yrs	Ag K X-Rays 22 keV	Ti-Mo	Ta-U
Am-241	433 yrs	Gamma-Rays 59.6 keV	Zr-Ba	Ta-U

*Source activities are in the range 1-10 mCi with photon outputs in the range 10^7-10^8/sec (over 2π ster.).

Three controls are available to the operator of a calibrated instrument, ON/STAND BY/OFF, MODEL SELECTOR and START (See Figure 1). The red/green LED to the right of the digital display panel indicates the measurement status; flashing green is standby and steady green is accumulating data, at the end of which period it goes out and the results are displayed. The left hand digit of the display indicates the element number (1 to 4, designated during calibration) and the other four digits the element concentration (range -99.99 to +99.99). The display is also used for diagnostics and memory readout. The rotary switch marked "MODEL" is used by the operator to select the calibration model to be used. Eight models are available and in each one all the data required for a given four-element analysis is stored. The models can be used independently of one another and/or with different probes successively. For example, a given sample can be analyzed for many elements (four at a time) after a single data acquisition, by successively selecting the appropriate models.

Four rotary switches and five thumbwheel switches are used for instrument calibration. These controls are behind a lockable door on the front panel. The first stage of calibration is to set up the element windows about the appropriate peaks in the X-ray spectrum and to measure and store the spectral overlap factors. This is done using four single element standards and one scattering standard. Each standard can be a compound of the element mixed with epoxy resin or other suitable potting compound and cast into a small disc. With the proper switch positions, the scatterer and the four single-element standards are counted in order. The instrument automatically sets a window approximately one fwhm wide about each peak. These window selections can be manually overridden by the operator if he chooses. The entire X-ray spectrum can also be read out channel by channel, or in successive channel groups. The spectral overlap factors are now measured, calculated and stored by recounting the five standards in the same order as before but with different switch settings. At this point the instrument readout after counting a given sample is net counts in each of the four element windows, corrected for background and spectral overlap and normalized to a 15 second count time. The second stage of calibration is to provide normalization of all net counts to a fixed, stable reference standard. This is done both at calibration time and prior to every subsequent period of measurement activity, in order to eliminate changes in calibration due to source decay and errors due to small instrument drifts. A convenient reference standard is a mixture of compounds of the four elements cast in epoxy or other suitable potting material. Again, the instrument performs the normalization automatically when the standard is counted using the proper rotary switch settings.

The third stage of calibration is to generate and store the appropriate coefficients for conversion of the normalized count rate I to element concentration C. The mathematical model used is

$$C_i = \alpha_{io} + \sum_j \alpha_{ij} I_j \qquad (1)$$

where C_i is the element concentration (i = 1 to 4), I_j is the
normalized count rate in each of the four element windows (j = 1
to 4), α_{io} are the zero intercepts and α_{ij} are the slope coefficients. The procedure for measuring and entering the α coefficients
is as follows. First the normalized count rates I are measured and
noted for a suite of calibration standards. Then the α coefficients
are calculated off-line using any convenient calculator and mathematical procedure. Thirdly, the coefficients are entered via the
thumbwheel switches. This mathematical model can be used in various
ways depending on the analysis accuracy required, the available
standards and the interelement effects expected. The simplest
approach employs one standard (e.g., the reference standard) to
obtain a single point calibration. Other approaches commonly used
are linear regression and multiple regression, each of which require
a number of calibration standards covering a range of concentration
for each element.

APPLICATIONS

     Potential applications of this instrument cover the whole range
of analyses possible to do by X-ray fluorescence, both in the field
and in the laboratory. Only a few have been investigated at this
time and the results described here are typical.

Light Element Determinations

     The silicon content of ferrochrome was determined in the range
1 to 14% Si with a standard deviation of ± 0.9% Si using the light
element probe, a 2 mCi Fe-55 source and a 4 minute measurement time.
Spectral windows for three "dummy" elements, Al, Ca and Ti, were
entered and their α coefficients set at zero to complete the four-
element calibration matrix. Fe K X-rays are not excited by Fe-55.
Cr K X-rays are excited by the MnK_β X-rays from Fe-55 and were
included in the backscatter peak. No interelement effects were
observed, the main error being due to counting statistics. Other
light element analyses studied include P in phosphates (range 1 to
35% P_2O_5, standard deviation ± 1.4% P_2O_5); S in liquids (range 0 to
2% S, standard deviation 0.07% S) and Cl in solution (range 0 to 1%
Cl, standard deviation ± 0.004% Cl).

Rock and Ore Analysis

     Nine elements, Al, Si, S, K, Ca, Fe, Cu, Zn and Pb, were
determined in powdered samples of drill hole cuttings using two

probes and three calibration models. The light element probe with
a 2 mCi Fe-55 source was used for two groups of four elements;
(1) K, S, Si and Al and (2) Si, S, K and Ca. The sample probe with
a 10 mCi Cm-244 source was employed for the third group, Pb, Zn, Cu
and Fe. In all three cases the measurement time was four minutes
and multiple regression analysis was used to find the α coefficients.
The element concentration ranges and standard deviations obtained
were: Al_2O_3, 9 to 27 ± 3.0%; SiO_2, 45 to 78 ± 2.0%; S, 0 to 2.8 ±
0.11%; K_2O, 2 to 8 ± 0.5%; CaO, 0.1 to 6.9 ± 0.3%; Fe, 1.3 to 9.4 ±
0.3%; Cu, 0 to 0.35 ± 0.015%; Zn, 0 to 0.51 ± 0.009%; and Pb, 0 to
0.29 ± 0.025%. Figures 3 and 4 show the calibration curves obtained
for sulphur and zinc.

   Examples of ore analysis studied are Ni and Cu in nickel ores
and Fe, Cu, Zn and Pb in zinc concentrates. The nickel ores were
analyzed using a 10 mCi Cm-244 source, a 4 minute measurement time
and element windows for Fe, Ni, Cu and Pb. Typical values of
standard deviations obtained were ± 0.02% Cu and ± 0.03% Ni in
tailings, and ± 0.16% Cu and ± 0.21% Ni in heads, tailings and
concentrates combined (range 0 to 7% Cu and 0 to 11% Ni). Again
powder samples were measured and multiple regression used for es-
tablishing the calibration coefficients. Note that all the quoted
standard deviations are calculated from the calibrations of instrument concentration readout against wet chemical analysis and so
include all sources of error.

Alloy Analysis

   The sample probe with a 30 mCi Cm-244 source was used to determine copper, zinc and lead content of brass samples in the form of
machined discs. The fourth element was iron which was present in
unknown amounts. Again, the measurement time was 4 minutes. Multiple regression analysis of the X-ray intensity data from thirteen
standards showed the interelement effects to be quite large. However, they were satisfactorily overcome as is seen by the values of
standard deviation obtained: copper, ± 0.15% Cu (range 57 to 70%);
zinc, ± 0.13% Zn (range 29 to 40%); and lead, ± 0.06% Pb (range 0 to
3%). Such accuracies are useful not only for alloy identification
but also for composition control. Figure 5 shows the calibration
curve obtained for copper.

   The surface probe with a 3 mCi Cd-109 source was calibrated for
determination of Mo, Ni, Fe and Cr in stainless steels, using
analyzed steel plates as standards. In a measurement time of 4
minutes the following values of standard error were obtained: molybdenum, ± 0.03% Mo (range 0 to 3%); chromium, ± 0.2% Cr (range 16 to
18%); nickel, ± 0.16% Ni (range 8 to 14%); and iron, ± 0.3% Fe
(range 63 to 72%). Figure 6 shows the calibration curve obtained
for molybdenum. Multiple regression analysis was necessary to
minimize matrix effects between chromium, iron and nickel but linear

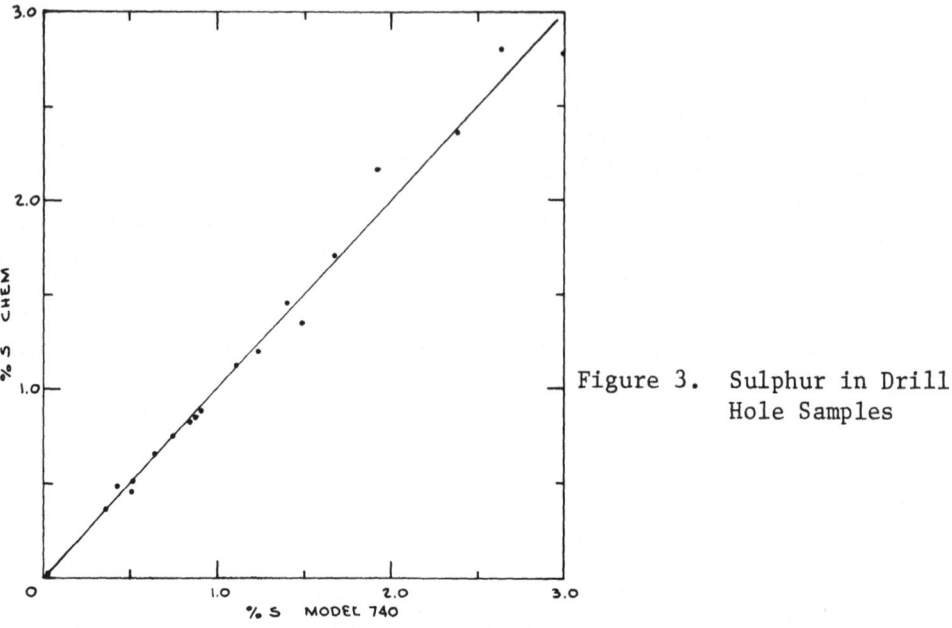

Figure 3.   Sulphur in Drill
            Hole Samples

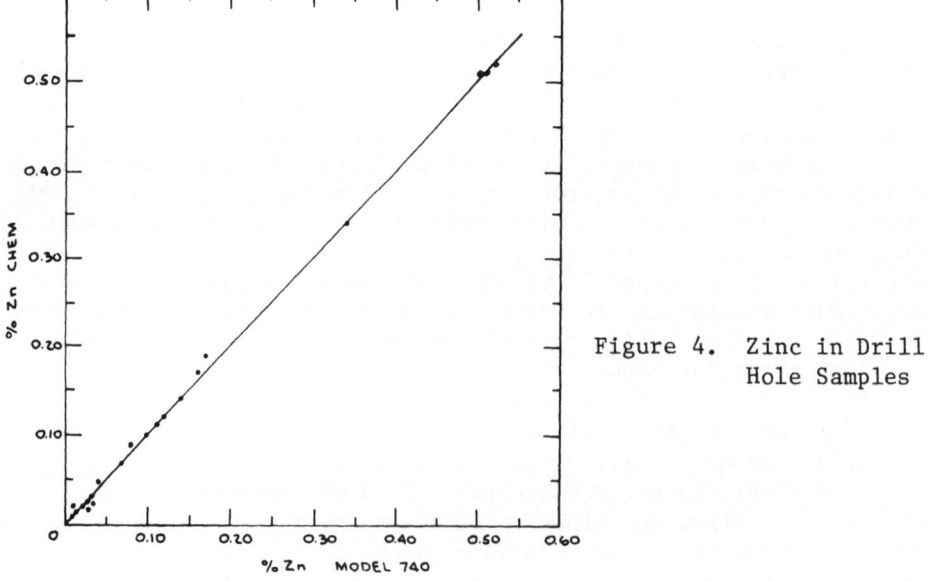

Figure 4.   Zinc in Drill
            Hole Samples

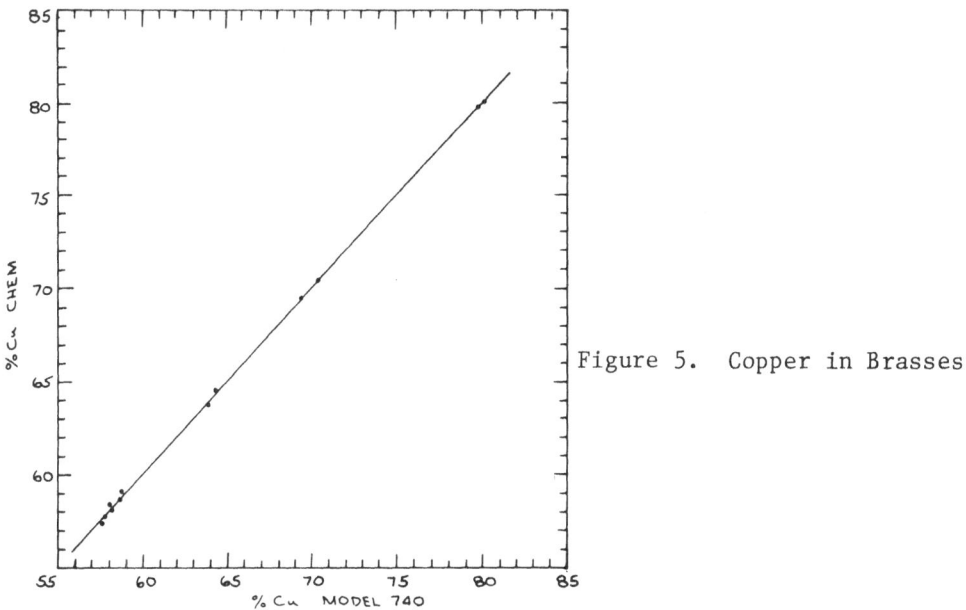

Figure 5.   Copper in Brasses

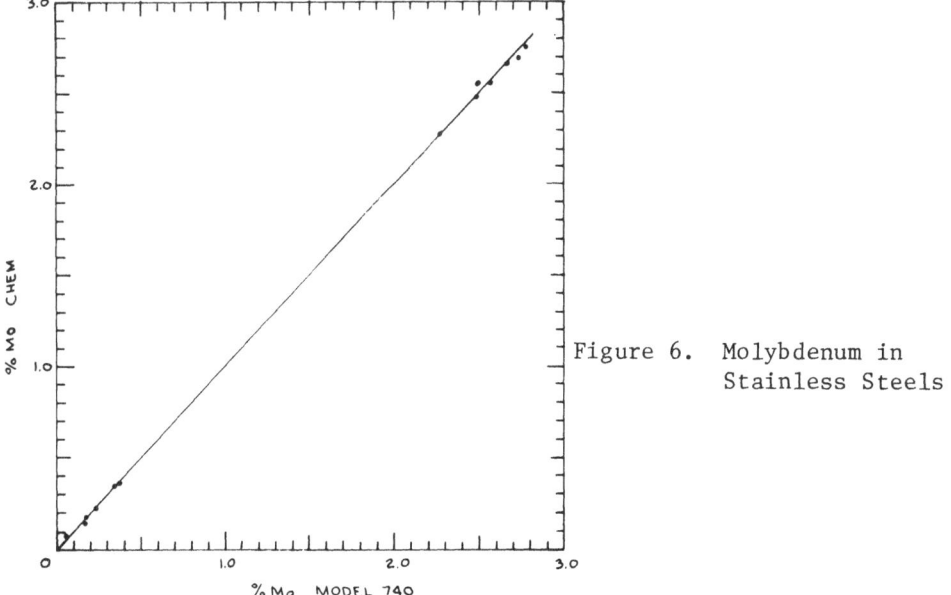

Figure 6.   Molybdenum in
            Stainless Steels

regression gave best results for molybdenum.

Other alloy analyses studied include inconels for Cr, Fe, Ni and Mo using Cd-109; cupronickels for Mn, Fe, Ni and Cu using Cm-244; low alloy steels for V, Cr, Co, W and Mo using Cd-109; brasses for Cu, Zn, Pb and Sn using Am-241; and titanium alloys for Ti, V, Zr, Mo and Sn using Cd-109 and Am-241.

CONCLUSIONS

A basic advance in proportional counters combined with the latest microprocessor technology has made possible the first practical multielement, hand-portable X-ray fluorescence analyzer with concentration readout. Experimental results show that excellent values of sensitivity and accuracy are obtainable for a wide range of ore, alloy and solution analyses in field, plant and laboratory conditions.

REFERENCES

1. S. H. U. Bowie, A. G. Darnley and J. R. Rhodes, Trans. Instn. Mining Met. 74, 361 (1965).
2. J. R. Rhodes, Analyst 91, 683 (1966).
3. J. R. Rhodes and T. Furuta in "Advances in X-Ray Analysis", Vol. II, Eds. John B. Newkirk, Gavin R. Mallett and Heinz G. Pfeiffer, 249, Plenum Press, New York (1968).
4. J. Lehto, I.E.E.E. Trans. Nucl. Sci., NS-25, 777 (1978).
5. M. Hietala and J. Viitanen in "Advances in X-Ray Analysis", Vol. 21, Eds. C. S. Barrett, D. E. Leyden, J. B. Newkirk and C. O. Ruud, 193, Plenum Press, New York (1978).
6. H. Sipila and E. Kiuru, loc cit, 187.

IN SITU ROCK ANALYSIS

J. Lantos and D. Litchinsky

Inax Instruments Limited

Ottawa, Ontario, Canada

ABSTRACT

Recent advances in solid state electronic circuitry and cryostat design have allowed the development of a low power consumption, fully portable XRF analyzer. The potential of such an instrument in geochemical surveying was investigated using the Inax 540 portable XRF pulse height analyzer/computer combined with the Inax 600 portable Si(Li) detector/dewar/LCD display. These two units comprise a total XRF analyzer and will determine and display the analyses of up to 36 elements simultaneously. Apart from providing a freshly exposed surface, no sample preparation is required.

Data illustrating the problems in the calibration of the unit for whole rock analysis are presented. Specifically data are presented on Uranium bearing rocks where the Uranium occurs along fracture lines in the medium. Results are compared with known assay and sampling problems are discussed.

INTRODUCTION

X-ray fluorescence spectrometry is an established technique for non-destructive geochemical analysis. The advantages of this technique over others include minimal sample preparation and rapid low cost sample throughput. Until recently XRF geochemical analyses have been limited to central laboratories due to the power requirements and the bulk of the instrumentation. Essentially one would prefer that the assay lab be in the field where the time lag between sampling and analysis would be greatly reduced, allowing

37

a survey team to make more effective decisions than could be made when data arrive days or weeks after the fact.

The development of low power analogue circuitry, coupled with developments in microprocessor technology and advances in cryostat design, has allowed the development of fully portable XRF instrumentation for in situ analysis of geological samples (1). This paper describes one such instrument produced by Inax Instruments Limited, and its application to the analysis of Uranium in powders and cleaved solid specimens.

The portable XRF spectrometer system developed in our laboratories is a microprocessor based system specifically designed for low power drain and reliability. It is a Si(Li) detector energy dispersive system which provides the convenience of simultaneous multiple element analyses. The various elements of the system are shown in Figure 1. The unit consists of two packages: a liquid nitrogen dewar/cooled FET lithium drifted silicon detector suitable for energies <60keV/preamp/LCD display weighing 3.6 kg and an amplifier/PHA/calculation unit weighing 5 kg. The two units are linked through a 1/4" diameter shielded cable. Features of the system are:

a)   Full 1024 channel spectrum storage with access
     through computer instruction set.

b)   Pulse processing speed to 8000 counts per
     second with automatic dead time correction
     through an internal timing register.

c)   Operation controlled by a calculator-like
     instruction set. I/O routines available for
     LCD and/or EIA RS232C interface.

d)   Fixed or changeable multiple program options
     for totally automatic operation or user
     program development. Programs are selected
     via a front panel switch.

e)   Volatile memory standby power feature so
     that data or programs will be maintained
     in memory when the unit is switched off.

f)   Data acquisition initiated by exceeding a
     count rate threshold, thus eliminating most
     external controls.

g)   Program memory expansion to 8000 steps of
     high level calculator-like instructions.

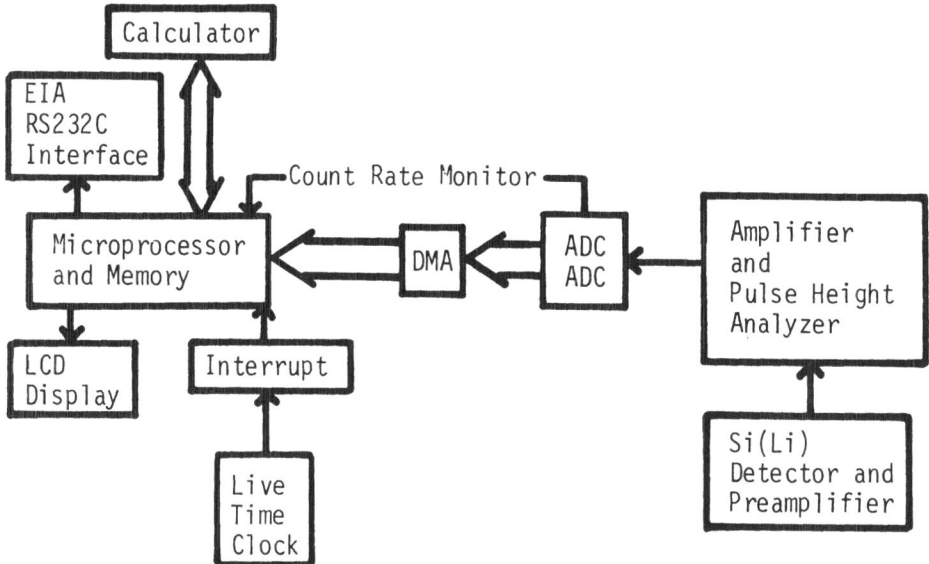

Figure 1.   Inax portable XRF system

h) Battery life 6 hours continuous operation.
    Batteries are rechargeable.

i) Liquid N_2 holding time 20 hours.

EXPERIMENT

For these experiments the unit was configured with an I^{125}
radioisotopic X-ray source having a characteristic X-ray of
27eV (Te K_α). This isotope was chosen for its "clean" spectrum
below the characteristic and an incident energy allowing analysis
of a wide range of elements in the range, Titanium through
Silver. The higher Z materials are generally of more interest to a
geological survey team.

Equations were derived for Uranium in varying matrices using
powder standards in the concentration range .05% to 7% U. Corrections
were included for Thorium and Lead which are likely to occur
frequently in Uranium bearing samples. Rubidium was not an interfering element for these samples, and thus no corrections were
included. However, for substantial amounts of Rb a simple test was
included based on the intensity of the $U_{L\gamma}$ line to see if a major

contribution from other interferences were operating.

The form of the equation used was:

$$\%U = A \frac{U}{BSc} - B(\frac{Th}{BSc} - C \frac{U}{BSc}) - D$$

Where A, B, C and D are constants. BSc refers to the back scatter peak and is used as a first order matrix correction as well as an isotopic X-ray source decay correction.

Solid samples were measured at as many different points as possible on the cleaved surfaces, each reading taking 65 seconds and the results averaged. Powders were analyzed for 65 seconds also.

Table 1a shows the reproducibility of the instrument for a given sample; these are readings taken over a period of one day. Table 1b shows a similar sample of Uranium loaded with Thorium showing the stability of the Uranium reading when measuring mixed deposits of U/Th. The measured detection limit for U is 19 ppm.

RESULTS AND DISCUSSION

Uranium occurs, broadly speaking, in three main types of deposits (2,3,4).

1) Syngenetic deposits, i.e. quartz pebble
    conglomerates

2) Epigenetic deposits, i.e. Colorado type
    sandstones unconformity type deposits

3) Igneous deposits, i.e. pitchblende vein
    type

The type of deposit to be measured is significant in that the pattern of Uranium concentration within the sample will differ from one type to another and also interfering elements will have significantly different concentrations from deposit to deposit.

For example: Igneous and Epigenetic deposits, for which the primary ore bearing mineral will be pitchblende, may tend to contain less Thorium than Syngenetic deposits containing the mineral Monazite. Certain ores will also contain appreciable amounts of As, Pb, Zn, Ni and V, etc.

In terms of Uranium distribution within Igneous and Syngenetic deposits, concentration will occur within veins and along fractures, whereas in quartz pebble conglomerates the concen-

REPRODUCIBILITY

Powders

Table 1a	Table 1b
6% U by weight in SiO_2	1.88% U weight 3.5% Th by weight in SiO_2

Run #	Observed % U	Run #	Observed % U
1	6.00	1	1.90
2	6.14	2	1.78
3	6.27	3	1.79
4	6.01	4	1.87
5	5.97	5	1.83
6	6.11	6	1.84
7	6.14	7	1.73
8	6.04	8	1.83
9	6.03	9	1.84
10	6.03	10	1.83

Average: 6.08   Average: 1.82%

SD $\pm$ .09   SD $\pm$ .05

Relative error   Relative error
per 65 second run: .8%   per 65 second run: 1.2%

Table 2

Sample: Sandstone, Drill Core Material

Integration Time: 65 second

Source: I^{125}

Run#	Observed U	Given U	Run #	Observed U	Given U
1	3.08	1.20	1	.70	.30
2	2.52		2	.79	
3	1.46		3	.88	
4	.08		4	.14	
5	2.64		5	.04	
6	.63		6	.04	
7	1.55		7	1.60	
8	.59		8	.87	
9	.64		9	.16	
10	1.43		10	.14	

Average: 1.46   Average: .54

tration is random within the matrix. Thus, it is clear that the choice of sample area must be well considered, and care taken to choose a representative cross-section, rather than one of obvious concentration.

The samples chosen for this study were Uranium-bearing sandstones from the Athabasca formation of Northern Saskatchewan, and are believed to be representative of the unconformity type Epigenetic deposits.

In many respects these samples posed the least amount of difficulty in that their chemistry is relatively simple - Th concentrations are generally low, and due to their relatively young age, they are Pb poor. Furthermore, they are fairly fine grained in nature, which leads to greater homogeneity of Uranium within the sample. Nevertheless, it will tend to concentrate within high porosity channels, such as fractures, bedding planes, etc.

This tendency to concentrate along fracture lines and bedding planes makes in situ analysis very difficult in these cases. It is particularly difficult for the sandstones, as the usual method of sampling for the geologist is to provide a fresh surface for analysis. There is a high probability that breaking of the sample will occur along fracture lines so that measurements of the cleaved surface will most likely yield surface values higher than the bulk concentration. Table 2 shows two such samples with measurements taken over the cleaved surface. Note the large fluctuations in readings with averages higher than the crushed rock analysis as obtained by an independent laboratory. With few exceptions averages over a cleaved surface will read higher than a bulk sample measurement.

For field applications two approaches are used to provide results that are accurate. The first solution is to require that all samples are crushed and mixed prior to measurement. This approach is a time penalty on the ability of the instrument to make a series of measurements. The second solution is to program the instrument to force the operator to make a minimum of say 10 readings of the surface to be measured before a display of the Uranium concentration is obtained. For the sake of accuracy, the instrument can also make sure that enough data are collected before a reading is presented so that the analysis is statistically significant.

CONCLUSION

Relatively rapid and precise analyses of low concentrations of Uranium in various matrices are possible. The instrument is a low cost effective device for Uranium exploration as it allows a survey

team to obtain a large number of assays quickly and thus make
accurate decisions about the geology of the sample which would previously have taken days or weeks of waiting for results from a
central analytical laboratory.

BIBLIOGRAPHY

1. S.H.U. Bowie, A.G. Darnley and J.R. Rhodes,
    "Portable Radioisotope X-ray Fluorescence
    Analyzer". Trans. Inst. Mining Met. 74;
    316, 1965

2. R.H. DeVoto, "Uranium Geology and Explora-
    tion". Colorado School of Mines, Golden
    Colorado, March 1978

3. M.M. Kimberley, "Short Course in Uranium
    Deposits; Their Mineralogy and Origin",
    Mineralogical Assoc. of Canada, October 1978.

4. R.S. Munday, "Uranium Mineralization in N.
    Saskatchewan", C.I.M. Bulletin, April 1979

RAPID DETERMINATION OF ASH IN COAL BY COMPTON SCATTERING, Ca,

AND Fe X-RAY FLUORESCENCE

Jacques Renault

New Mexico Bureau of Mines & Mineral Resources

Campus Station, Socorro, NM, 87801

ABSTRACT

The intensity of Compton scattering, Fe, and Ca characteristic radiation can be used to estimate the amount of ash in coal by X-ray fluorescence spectroscopy. Mo, W, and Cr radiation were used to study a suite of New Mexico coals, and the best results were obtained with Mo and W X-ray tubes. If the actual concentrations of Fe_2O_3 and CaO and the mass absorption coefficient, μ^*, at the Compton wavelength of scattered Mo K radiation can be determined, the regression equation:

$$\%Ash - 24.2\mu^* - 6.28(\%Fe_2O_3) - 1.96(\%CaO) - 3.4$$

estimates the ash content with an average error of 0.5% ash at 0.71Å.

INTRODUCTION

The percent ash in coal is usually determined by the standard ASTM (1) method. This is a gravimetric procedure which involves programmed combustion of the coal and weighing the residual ash. It is time consuming, and inter-laboratory variations of 10 percent are acceptable for coals with more than 12% ash containing pyrites and carbonates. X-ray fluorescence analysis for ash can reduce analysis time to a few minutes and provide acceptable reproducibility.

The notion that the mass absorption coefficient of a coal is an estimator of its ash content has been explored by a number of workers particularly with regard to the application of radioisotopes to on-line analysis. Fookes, Gravitis, and Watt (2) and Boyce, Clayton, and Page (3) cite the early work and show how analytical conditions

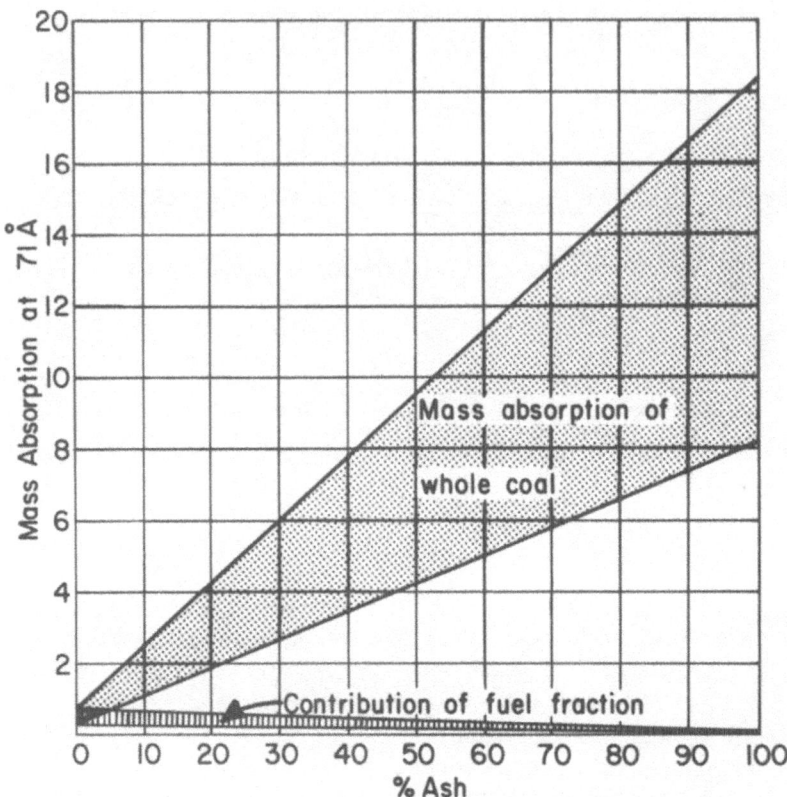

Figure 1. Variation in the mass absorption coefficient of New
Mexico coals as a function of ash content.

Table 1. Concentration of major
ash components as estimators of
mass absorption coefficient

Component	Correlation Coefficient
Ca	.51
Si	.72
Al	.73
Fe	.70
Fe & Ca	.99

can be optimized for radioisotope sources.

The purpose of the present study is to show how Compton scattering combined with both Fe and Ca intensity measurements obtained from a wavelength dispersive X-ray spectrometer can yield useful ash determinations.

RATIONALE

In very simple terms, a coal can be thought of as a two-component mixture consisting of a fuel fraction and a non-combustible ash fraction. The fuel fraction is composed of carbon and volatiles, and the ash fraction of mineral matter. At a wavelength of 0.7 angstroms, the mean and standard deviation of the mass absorption coefficients are 0.6 and 0.2 for the whole fuel fraction of a variety of New Mexico sub-bituminous coals. For the ash fraction they are 13. and 5. respectively.

The mass absorption coefficient of the fuel fraction in the various coals is quite uniform; however, the ash is not, and this prevents the mass absorption coefficient from being used alone as a good estimator of ash concentration.

From Fig. 1, it is easy to see that the mass absorption of coal is strongly dependent on the ash concentration. Much of the variability is due to differences in the environment of deposition and consequently the Ca and Fe concentrations. In addition, the secondary introduction of Ca and Fe into the coal is particularly important.

Table 1 shows the relative influence that major ash components have on the mass absorption coefficient of the ash fraction. This influence is indicated by the correlation coefficients of the components when they are used as estimators of the mass absorption coefficients. Al and Si concentrations are characteristic of the environment of deposition and tend to be inversely correlated with each other, so multiple regression on them does not materially improve their ability to estimate the mass absorption coefficient. Ca is often secondarily introduced, so it is not strongly correlated with the other components.

Because Fe and Ca are not correlated with each other, and have the highest Z of the major ash components, they are good estimators of the mass absorption coefficients of ash. Furthermore, Ca, often being a secondarily introduced component, is largely responsible for heterogeneity of ash compositions within a single depositional environment.

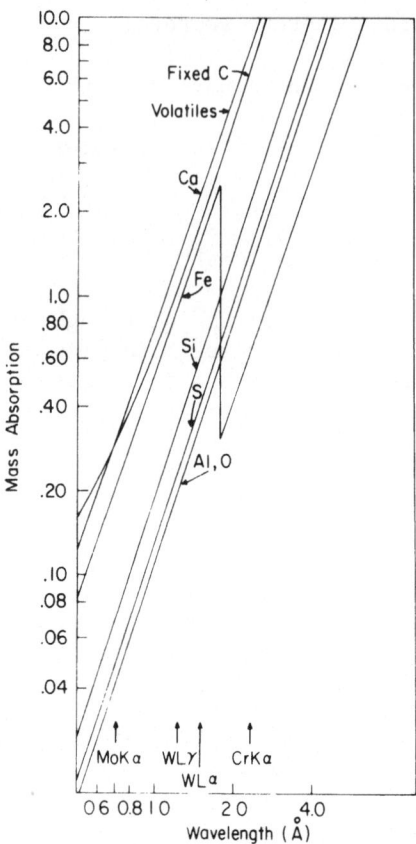

Figure 2. Variation in fractional mass absorption as a function of
    wavelength for the principal components of a sub-bituminous New
    Mexico coal with 10% ash.

    If mass absorption coefficients are to be used as estimators
of ash concentration, the effectiveness of the Compton scattering
from a variety of X-ray tubes needs to be evaluated. Fig. 2. shows
the variation in fractional mass absorption as a function of wavelength for a coal with 10% ash. The wavelengths of the Compton
peaks we used for this study are indicated. For this particular
coal, the absorption by the volatiles is equal to that by the Ca.
Note that the Fe K absorption edge lies below the wavelength of Cr.
This reduces the influence of Fe on the determination of mass
absorption coefficient at the Cr wavelength.

    Fig. 3 shows calculated ash versus reported ash on 13 New Mex-
ico coals from a variety of environments. Not only do the geological environments vary, but so does the way in which the coals were
collected. Some were obtained from outcrop, others from active and
abandoned mine workings and stock piles. For this suite of coals,

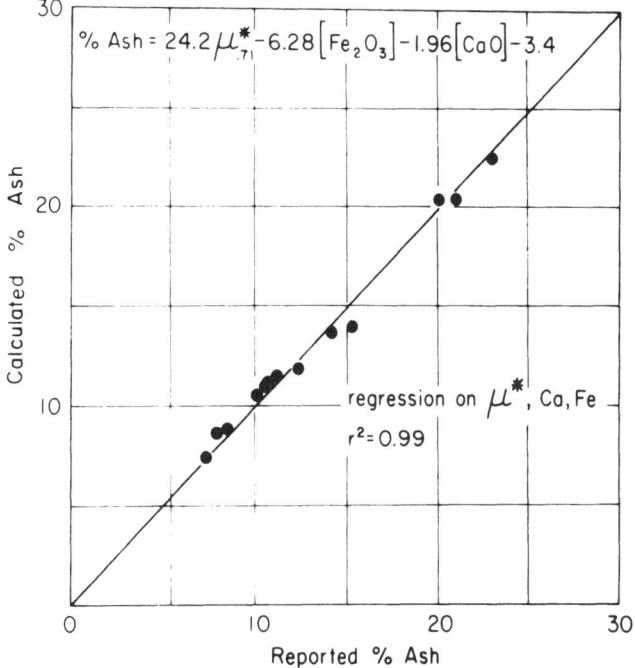

Figure 3. Reported ash versus ash calculated from multiple linear regression on mass absorption coefficient and concentration of iron and calcium.

μ^* is the mass absorption coefficient of the coal calculated at 0.71Å and Fe_2O_3 and CaO are the concentrations of those components in the coal. The correlation is very good. If the iron and calcium concentrations and the mass absorption coefficients are well known, the equation given in Fig. 3 estimates ash content to ±0.5% ash.

X-RAY FLUORESCENCE DATA

Fig. 4 shows the sample preparation sequence. The samples were subjected to incidental atmospheric drying. We prepared undiluted briquets with bakelite backing, but this is only necessary for repeated handling of samples. Five grams were necessary for sufficient thickness at 0.7Å with our briquet geometry. The samples were ground at liquid N_2 temperature in a Spex Freezer-Mill for three minutes. We adopted this procedure to avoid differential grinding of the fuel and ash fractions, to speed up grinding, and to avoid loss of volatiles, but it might be an example of "overkill." The briquets were compressed in a die against a polished silicon carbide anvil at 40,000 psi.

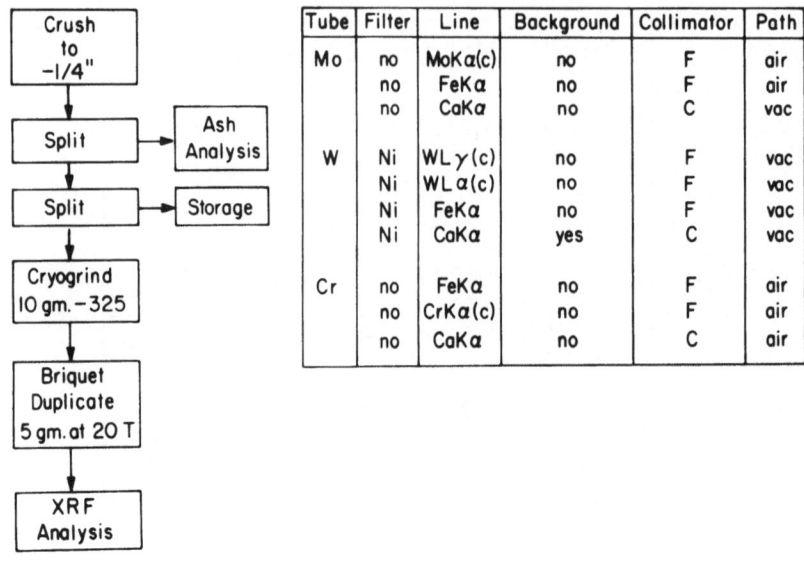

Tube	Filter	Line	Background	Collimator	Path
Mo	no	MoKα(c)	no	F	air
	no	FeKα	no	F	air
	no	CaKα	no	C	vac
W	Ni	WLγ(c)	no	F	vac
	Ni	WLα(c)	no	F	vac
	Ni	FeKα	no	F	vac
	Ni	CaKα	yes	C	vac
Cr	no	FeKα	no	F	air
	no	CrKα(c)	no	F	air
	no	CaKα	no	C	air

Figure 4.   Sample preparation sequence and variable instrumental
            conditions.

We tested two suites of samples using Mo, W, and Cr X-ray tubes.
Each tube has certain advantages. Mo, with a Kα Compton scattering
peak at about 17 Kev is close to the energy that gives the maximum
sensitivity to changes in ash content as shown by Boyce, et al. (3);
furthermore, a large volume of sample is irradiated which reduces
errors due to poor mixing. The W tube provides an estimate of absorption coefficients at two wavelengths of Compton scattering and
maximum excitation of Fe in the face of Ca absorption. The Cr tube
with its Compton peak at a longer wavelength than the Fe absorption
edge estimates the ash content of the coal with less perturbation
by Fe; in addition, it provides the most efficient excitation of Ca.

X-ray intensities were obtained on a full-wave rectified Philips
Universal Vacuum Spectrograph operating at 50 Kv and 40 ma. A minimum of 10,000 counts were obtained on each peak. Counts were
ratioed to a drift briquet kept in one of the eight positions of the
spectrograph.

Experimental conditions used with each X-ray tube were governed
partly by the desire to keep the method as simple as possible and
partly by the way the project evolved. The variable X-ray analytical
conditions are given in Fig. 4. Vacuum was required to obtain sufficient intensity of the Ca line using the Mo and W tubes. We used
a Ni filter to help resolve the WLα Compton peak and this reduced the
Ca intensity so that we had to subract background. The Ca intensity

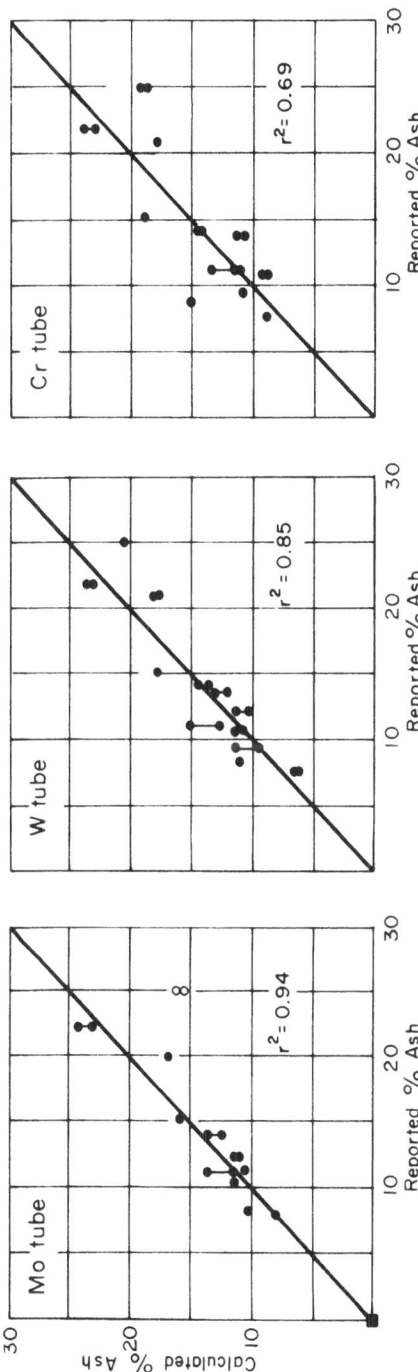

Figure 5. Reported ash versus ash calculated from data obtained using three different X-ray
tubes. Samples represent a diverse suite of New Mexico sub-bituminous coals.

obtained from the Cr tube was marginally high enough to obviate a
background measurement in air path, but results with the Cr tube
would probably have been better had we used a vacuum and subtracted
background.

Data obtained with the Mo tube were ratioed to the intensities
obtained from a spectrographically pure graphite briquet. The graphite count rates at the Fe and Ca wavelengths are low, and ratioing
with the coals degrades their reproducibility, so the graphite briquet was abandoned and a real coal was subsequently used in the
drift position. All curve fitting of X-ray data was done by least
squares linear regression. In all cases, the reciprocals of the
Compton intensity ratios were used in conjunction with the direct
intensity ratios of Fe and Ca. No attempt was made to determine the
true concentrations of Fe and Ca from their intensities. "Reported
% Ash" values are based on dry coal determinations by the U. S.
Geological Survey and by Hazen Research, Inc.

Fig. 5 shows three regressions using different X-ray tubes.
The intensities of Fe and Ca are components of each regression. The
W tube regression contains both WLγ and WLα Compton intensities.
The regression using the Mo tube is the best even if the sample at
24.9% ash is included.

A factor that undoubtedly contributes to the scatter is depth
of penetration. The Mo Kα Compton peak samples the mass absorption
coefficient of a larger volume of coal than the other available
wavelengths, so it is less sensitive to sample inhomogeneities. A
more important factor contributing to the scatter is the intrinsic
variability of the ash from this suite of samples. They are each
from different coal fields and different sampling environments.

Fig. 6 shows the correlation of a more homogeneous group of
samples. These samples are from the Torreon field in the San Juan
basin, and all were obtained from drill core. Here, although there
is a wide range in ash concentration, the correlation is much better.
The data were obtained with the W tube and have an average error of
±2.1% ash.

Although the concentration of Fe and Ca in ash are not closely
correlated, Ca has a conspicuous absorption effect on the Fe intensity. For this reason, the Ca intensity is a relatively more powerful component of the regression equation than the Fe intensity. A
suite of nine samples representing the Ca intensity range was selected for construction of a general purpose calibration curve.
These samples are shown by the solid circles in Fig. 7. The data
points in open circles are for ash concentrations calculated from
this calibration; r-squared for the points of the calibration curve
is 0.97. When the ash of the balance of the data is calculated from
the calibration equation it correlates with the reported ash with an

r-squared of 0.90 and a mean error of ±2.0%. When data for the Cr
tube is processed in the same manner, and Fe intensities are excluded, r-squared for the samples not used in the calibration set
is 0.85 for a correlation coefficient of 0.92.

The fastest time recorded for acquisition of the X-ray data
was 2.5 minutes per sample using the Cr tube. This was on a manually operated spectrograph using an air path and making only two
intensity measurements. Sample preparation after the coal has
been sized to ¼" is less than 10 minutes per sample with only one
worker. With efficient division of labor and an automated spectrometer, it should be possible to obtain ash determinations in less
than 10 minutes per sample.

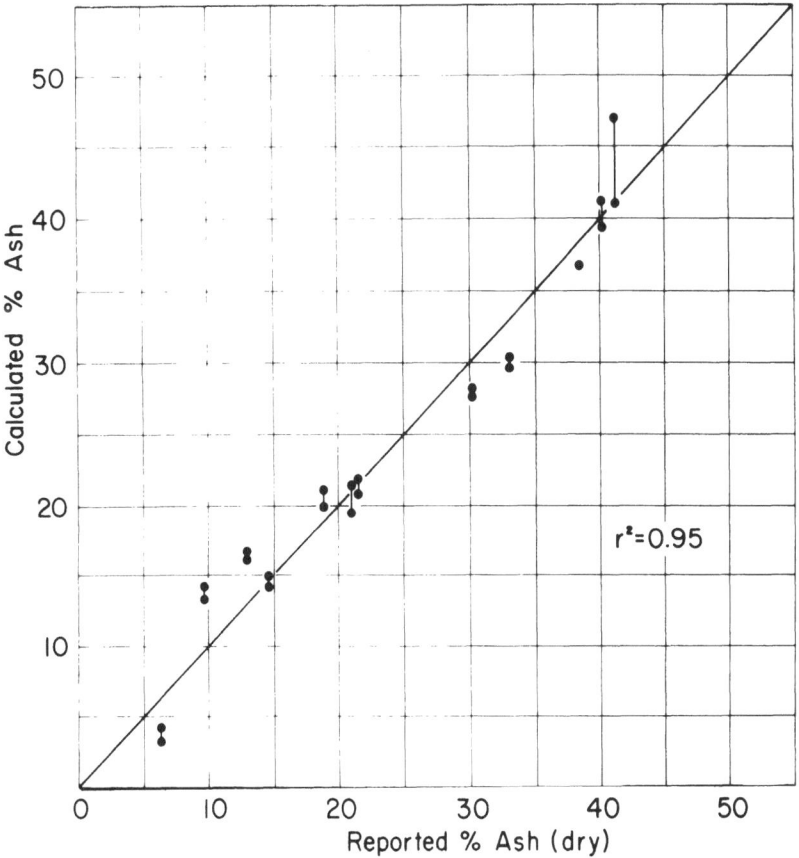

Fig. 6.   Reported ash versus ash calculated from a suite of samples
          from the Torreon field, New Mexico. Regression is on WLγ
          and WLα Compton, and Ca K and Fe K intensities

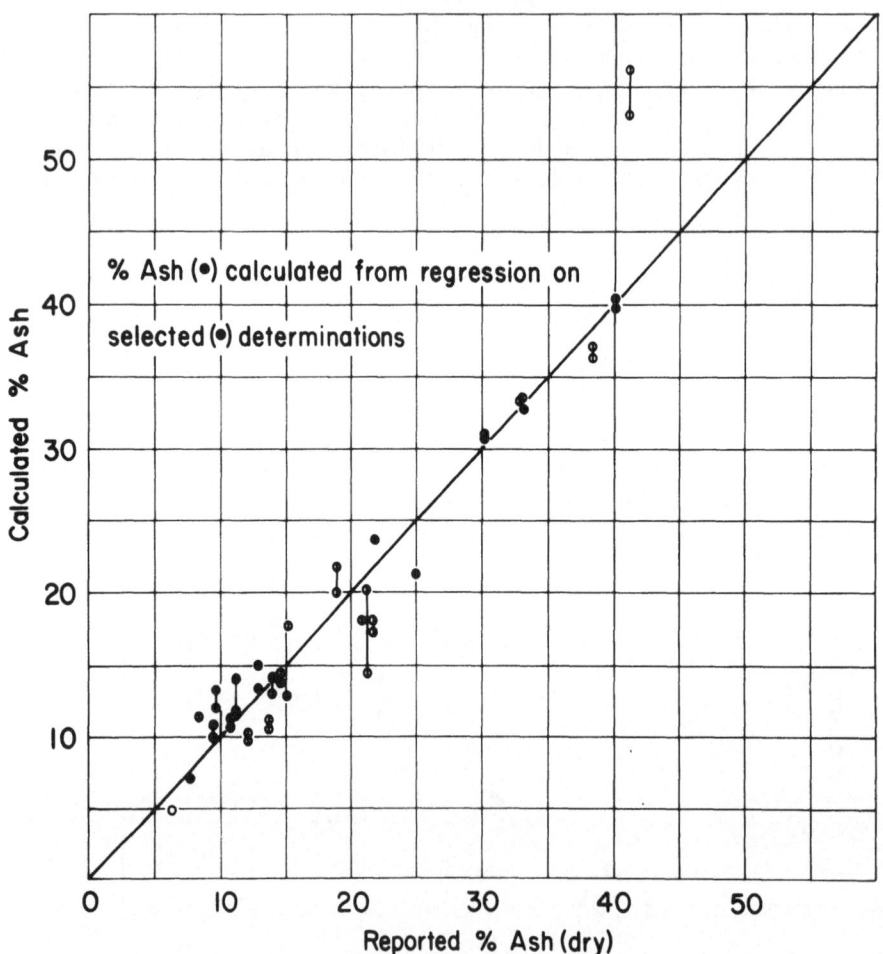

Figure 7.   Variation of reported ash with ash calculated from a
   calibration equation based on a selected group of samples.   Data
   points represent a widely varying group of New Mexico sub-
   bituminous coals.

CONCLUSIONS

We have tried to develop a relatively simple XRF analysis method for the determination of ash in coals. Under the simplest conditions, that is, using a Cr tube, a scintillation counter and an air path and making only two intensity measurements--one on the Cr Kα Compton peak and the other on the Ca Kα line--and using carefully selected standards, one can estimate the ash in a heterogeneous group of coals with a correlation coefficient of 0.92. For better results, a W or Mo tube should be used, and the measurements should be made in a vacuum and background should be subtracted from the Ca peak.

Using calculated mass absorption coefficients of the coals and their analyzed Fe_2O_3 and CaO concentrations, a much better fit is obtained. The mass absorption coefficient can be obtained from Compton scattering with a high accuracy as shown by Reynolds (4,5); this suggests that corrections applied to the Ca and Fe intensities to obtain their concentrations is a viable method for obtaining the ash concentration. Analysis of more components in the coal is required but would be accomplished in conjunction with a complete analysis anyway.

ACKNOWLEDGEMENTS

This work was supported by the New Mexico Bureau of Mines and Mineral Resources, and the author wishes to acknowledge the help of David Tabet and Stephen Frost of the Bureau who made analyzed coal samples available.

REFERENCES

1. American Society for Testing and Materials, 1977 Annual Book of Standards, pt. 26, 370 (1977).
2. R. A. Fookes, V. I. Gravitis, and J. S. Watt, "Determination of Ash Content of Coal by Mass Absorption Coefficient Measurements at Two X-Ray Energies," Nucl. Tech. Miner. Resour., Proc. Int. Symp. 1977, IAEA: Vienna, Austria, 1967-82 (1978).
3. I. S. Boyce, C. G. Clayton, and D. Page, "Some Considerations Relating to the Accuracy of Measuring the Ash Content of Coal by X-Ray Backscattering," Nucl. Tech. Miner. Resour., Proc. Int. Symp. 1977, IAEA: Vienna, Austria, 135-65 (1978).
4. R. C. Reynolds, Jr., "Matrix Corrections in Trace Element Analysis by X-Ray Fluorescence: Estimation of the Mass Absorption Coefficient by Compton Scattering," Amer. Miner. 48:1133-43 (1964).
5. R. C. Reynolds, Jr., "Estimation of Mass Absorption Coefficients by Compton Scattering: Improvements and Extensions of the Method," Amer. Miner., 52:1493-1502 (1967).

ON-SITE DETERMINATION OF ASH IN

COAL UTILIZING A PORTABLE XRF ANALYZER

F. V. Brown and S. A. Jones

Alabama Power Company
600 North 18th Street
Birmingham, Alabama 35202

This paper describes a method for calculating the ash conten[
of coal. In this method a curve is plotted which defines the
relationship between the ash content and corrected backscatter oi
coal. The correction to the backscatter is based upon the iron
content of the sample. The standard deviation of the ash is
calculated to be 1.17 for various particle sizes and moisture
content of the coal.

EXPERIMENTAL

Instrumentation

The x-ray system used was a Columbia Scientific Industries
Corporation Portable X-Ray Fluorescence Analyzer Minilab 700 with
a ten millicurie curium 244 source. This source was chosen
instead of the available 30mCi Pu-238 or the 30mCi Cm-244 source
because it could be supplied under a general radioactivity
materials license. The XRF analyzer could be powered by a
variety of power supplies, either AC or DC. All data were
evaluated and best curves were selected utilizing the Honeywell
Time Sharing System. Programs from the statistical library were
used with minor modifications.

Sample Preparation

Samples were obtained from the coal deliveries by a James A.
Redding automatic sampler. The sampler took approximately fifty
pounds from each load of coal, ground the sample to a top size of

eight mesh and riffled it to provide the eight mesh wet test
sample. The drying of eight mesh wet samples by ASTM methods
gave the eight mesh dry samples. The sixty mesh dry samples were
prepared by grinding eight mesh dry samples. For the sixty mesh
dry coal, the ash range was from 7.62 percent to 18.42 percent.
The eight mesh dry coal's ash varied from 7.56 to 20.82 percent.
The eight mesh wet samples varied from 9.6 to 17.00 percent ash
and from 6.90 to 16.36 percent moisture.

The samples were then weighed to 20 $\pm$ 0.1 grams and hand
packed into a sample cell supplied with the instrument. The
sample cell consisted of a plastic tube covered at one end by
.001 inch polypropylene or mylar.

Calibration and Set Up

In calibrating the XRF analyzer, three sequential steps were
followed: (1) setting the proper probe voltage, (2) setting
the window position and window width controls to correspond to
the iron and backscatter peaks, (3) collecting enough data to
plot a calibration curve.

An arbitrary probe voltage (approximately 4 turns) and a
narrow window width (.4 turns) were selected to begin the
calibration. Fifteen second readings were obtained on an
arbitrarily selected coal sample while increasing the window
position from 0.25 turns through the maximum ten turns at 0.25
turn increments. Counts were plotted versus position to obtain a
coal spectrum. Two peaks should have emerged. The first to
emerge was iron; the second, backscatter. By varying the probe
voltage, the curve was positioned along the x-axis (window
position). As one increased the probe voltage, the peaks were
expanded to the right and *vice versa*. The probe voltage was
adjusted to allow both peaks to fall within six turns of the
window position.

All three controls were set experimentally. It should be
noted that the probe voltage was extremely sensitive and once
adjusted for a given temperature was not tampered with. A
standard was necessary to minimize error associated with
instrument drift (eg. 2% iron in plastic).

With the probe voltage set at 4.0 turns, spectra of several
coal samples were plotted. The iron curve was observed between
positions 0.5 and 1.6 with backscatter falling between 1.6 and
6.0. In order to minimize the effect of drift, the "valleys" of
the curve were used as reference points to determine relative
peak area.

The ash content of numerous samples was determined using the ASTM method. In addition, the iron and backscatter data were collected and analyzed for curve fit.

RESULTS AND DISCUSSION

Attempts were made to establish a satisfactory correlation between the ash content of coal and the backscatter count. Six different equations failed to give an acceptable correlation. Table I gives the standard deviations for these equations for coal samples of various particle size and moisture content. The equations are numbered in Table I for reference throughout the report.

The addition of the iron count to the backscatter count failed to decrease appreciably the standard deviations. Since the iron content of the coals is the major variable which affects the backscatter count, steps were taken to lower the standard deviation to an acceptable level based on iron content. To accomplish this, seven samples were selected which had a low iron content (less than 2000 counts using the parameters listed in the experimental section).

If these seven samples could produce an ideal curve, one in which the ash content varied predictably with backscatter, then the backscatter count for other samples could be corrected by applying a correction factor based on iron content and the ash could be accurately predicted. These seven samples ranged in ash from 4.12 to 13.48 percent. The standard deviations for the six curves are given in Table II. Curves five and six were eliminated as the ideal curve because of their high standard deviations. Curves two and three were eliminated because corrections based upon these curves gave large errors in the higher ash ranges. Curve one was not used because some calculated correction factors were negative. Such negative correction factors implied that the iron content was the cause for the backscatter count's being high when in fact the reverse was true. Thus, curve four was used as the "ideal" theoretical curve.

To calculate the iron correction factor (ICF), the theoretical backscatter was first calculated using the experimental ash. That is

$$\text{Theoretical Backscatter} = B/(\text{Ash} - A)$$

where A and B were the constants of the theoretical curve. Since the actual backscatter was less than the theoretical backscatter and since this decrease was dependent upon the iron content, the relationship

$$ICF = \frac{\text{Theoretical Backscatter} - \text{Actual Backscatter}}{\text{Iron Count}}$$

should have been valid.

The relationship between ICF and the iron content was found to be hyperbolic (curve four). The ICF values were plotted versus the different coal "states" with the results presented in Figure 1. The graph for all samples showed that little error was introduced if the ICF values were calculated independent of moisture content and particle size.

Table I. Standard Deviation of Calculated Ash Using Backscatter Counts with no Correction for Various Curves

Equation	Curve	State-Dry 60 Mesh	State-Dry 8 Mesh	State-Wet 8 Mesh	All Samples
$y=A+(B)(X)$	1	2.55	1.51	2.13	2.02
$y=A\ \text{Exp}(B)(X)$	2	2.66	1.70	2.14	2.08
$y=AX^B$	3	2.74	1.82	2.13	2.12
$y=A+(B/X)$	4	2.71	1.65	2.11	2.11
$y=1/(A+(B)(X))$	5	2.86	3.05	2.22	2.19
$y=X/(A+(B)(X))$	6	3.06	6.63	2.16	2.26

(1) Standard deviation is in ash percent.
(2) X is backscatter, y is ash, A and B are constant.

Table II. Standard Deviations for Selected Samples for Six Curves

Curve	Standard Deviations	Constants A	Constants B
1	0.95	42.83	-1.323×10^{-4}
2	0.91	616.1	-1.680×10^{-5}
3	0.91	2.162×10^{24}	-4.330
4	0.92	-25.82	8.839×10^{6}
5	1.22	$-.4720$	2.354×10^{-6}
6	1.54	-154400	$.7379$

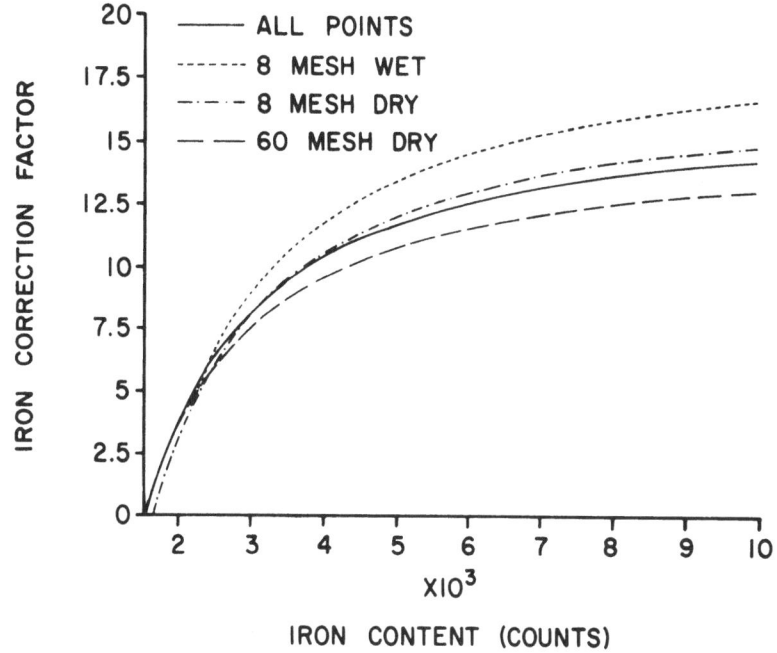

Fig. 1.   Iron content vs. iron connection factor

When the ICF was multiplied by the iron count and added to
the backscatter, the corrected backscatter was obtained. The
relationship between the ash and the corrected backscatter was
found to have the same correlation for both the linear curve
(curve one) and the hyperbolic curve (curve four). These curves
are reproduced in Figure 2 for all samples. Table III shows the
improvement in standard deviations for curve one and curve four.

To determine if the original assumption was reasonable, a
statistical program which plots the best curve based on three
variables was run. This program calculated the constants which
would give the best curve fit for the equation

$$Ash = 1/((K_1)(Iron) + (K_2) (Backscatter) + (K_3)).$$

This equation corresponds to the form of curve five in Table I.
This method gave a standard deviation of 1.34. The ICF method
also gave a standard deviation of 1.34 for curve five.
Unfortunately, the program could not be modified to give a
similar comparison for curve four.

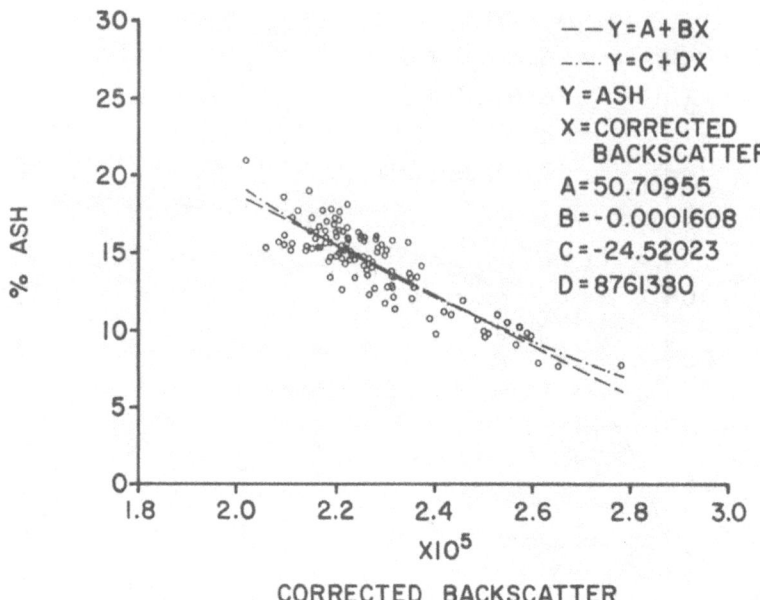

Fig. 2.   Ash vs. corrected backscatter

Table III.   Standard Deviation of Calculated Ash Using
Corrected Backscatter for Various Curves*

Curve	State - Dry 60 Mesh	State - Dry 8 Mesh	State - Wet 8 Mesh	All Samples
1	0.94	1.08	1.04	1.16
4	0.95	1.09	1.08	1.17

*Standard deviation is in ash percent.

To determine if any error resulted because of sample
variations[1], twenty-eight of the dry samples were compared with
the same samples in the wet state. The standard deviations
compared very favorably with the wet coal producing the lowest
standard deviation. (See Table IV).

These comparable values would tend to confirm that the
variations in moisture produced negligible error. In fact, the
lower standard deviation implied that less error was produced in

Table IV.   Standard Deviation for Selected Wet Eight Mesh and Dry
                       Eight Mesh Samples

Sample State	Curve	Standard Deviation	Curve	Standard Deviation
Wet 8 Mesh	1	0.75	4	0.76
Dry 8 Mesh	1	0.83	4	0.83

wet coal. This was perhaps due to the ease of packing for the
wet samples which gave higher densities. However, since only
twenty-eight samples were compared, statistical uncertainty would
call for caution in any interpretations.

The comparison of the sixty mesh and eight mesh samples could
not be made directly since samples were not carried through from
the eight mesh dry stage to the sixty mesh dry stage. However,
both classes of samples had about the same ash range, and since
all samples came from the Warrior coal field, comparisons between
the two states should be valid. The standard deviations for
these two states are given in Table III. The difference in the
standard deviations suggests possible density variations.

The standard deviation of 1.17 could possibly be improved by
(1) selecting a longer count time, and (2) eliminating the error
associated with determining the ash of the standards used in the
calibration of the XRF analyzer. For the Company's particular
application, the short count time was necessary. Other
applications might not be as restrictive. The elimination of the
error associated with the standards did not appear to be
practical.[2]

It was the conclusion of this laboratory that any instrument
with the capabilities of the Columbia   Scientific Industries'
Minilab 700 XRF Analyzer could produce acceptable accuracies on
eight mesh wet coal samples.

REFERENCES

(1) J. R. Rhodes, J. C. Daglish, and C. G. Clayton, Radioisotope
    Instruments in Industry and Geophysics. 1966, I, 457.

(2) 1978 Annual Book of ASTM Standards, Part 26, Gaseous Fuels;
    Coal and Coke; Atmospheric Analyses. Philadelphia,
    Pennsylvania. Designation D 3174-73, 1978, p. 385.

ROUTINE ENERGY DISPERSIVE ANALYSIS OF SULFUR IN COAL

R. Auermann*, J. C. Russ**, and R. B. Shen**

* Florida Mining and Materials Corporation

**EDAX International, Inc.

INTRODUCTION

The analysis of sulfur in coal has been brought to the attention of the general public in recent years primarily as it concerns
the pollution problem with coal-fired electric generation. Although
our application is different, some of the results would be equally
applicable to that circumstance. The Cement Division of Florida
Mining and Materials Corporation produces annually about 500,000
tons of both Portland and Pozzolanic cements, using a coal-fired
rotary kiln. The sulfur content of coal used in the manufacture of
clinker ranges from traces to as much as 6%. Sulfur content in a
dry process plant such as ours (this is the modern energy-efficient
type of cement plant) is very critical, and in our case must be
limited to 1.2% max.

The current interest in coal analysis is reflected in the
activities of ASTM Committee D-5, whose purpose is to formulate
methods of sampling, analysis, testing and specification of coals
and coke. The principal author is an active member of this committee, which is presently considering X-ray fluorescence methods
for analysis of coal and coal ash. Since we are already using an
EDAX energy-dispersive X-ray spectrometer as the principal means
of analytical control for the cement raw mix, it was a natural
decision to use it for the coal analysis as well, and this has
proved to be totally successful.

SAMPLE PREPARATION AND STANDARDIZATION

Incoming coal samples are dried and placed in a shatterbox for

4 minutes. The sample must be thoroughly ground to a fine powder
to avoid particle size problems from the sulfur present as pyrites,
which are much harder than the coal matrix. Nevertheless, it is our
belief that most of the remaining scatter in the standards data is
due to particle size effects, particularly for light elements such
as silicon. The powder is then pressed to pellets at 20 tons, and
the pellets are analyzed for 100-200 seconds, using continuum
excitation, from a Rh target X-ray tube operated at 10kV, 2.56µA.

Analytical standards were obtained by selecting lots of normal
Kentucky coal and analyzing them by conventional means. The sulfur
was determined using an oxygen bomb. Other elements present in the
coal (Mg, Al, Si, K, Ca, Ti and Fe) were also determined in the
standards and are routinely measured in the unknowns. Typical
calibration data are shown in Fig. 1. In the concentration ranges
of interest, acceptable linear calibration-curve relationships
exist between concentration and intensity. In actual practice,
some interelement coefficients are also used to adjust slope and
intercept values; this further improves the accuracy. The one-
sigma standard deviations obtained are better than 0.02% absolute
for all elements except Al and Si, and better than 0.1% for those.

Once the calibration relationships are established, they are
preserved for long periods of time (many months) by the use of
reference standards. These are simply plastic disks in which we
have embedded finely ground fly ash with additions of pure element
compounds such as magnesium sulfate. The exact composition of the
reference standards is not known; however, when analyzed against
the coal calibration curves, the calculated concentrations for
each reference disk should remain constant. Any deviations are
considered to reflect long term gradual drift in the spectrometer
sensitivity, and the calibration curves are automatically compensated.

PRACTICAL APPLICATION

One of the programs in the computer attached to our spectrometer handles coal analysis (there are others for analysis of as-
ground raw mix, laboratory fused raw mix, clinker, finished cement,
and various raw materials as well as calculation of predicted
cement phases, kiln operating parameters, and so on). Typical
output is shown in Fig. 2, corresponding to the measured spectrum
shown with it. The elements are reported in weight percent along
with the predicted amount of their oxides expected in the coal ash.

The sulfur concentration is used as a criterion to accept or
reject the coal shipment. During routine operation, the other
element oxides, particularly the SiO_2 which is often quite

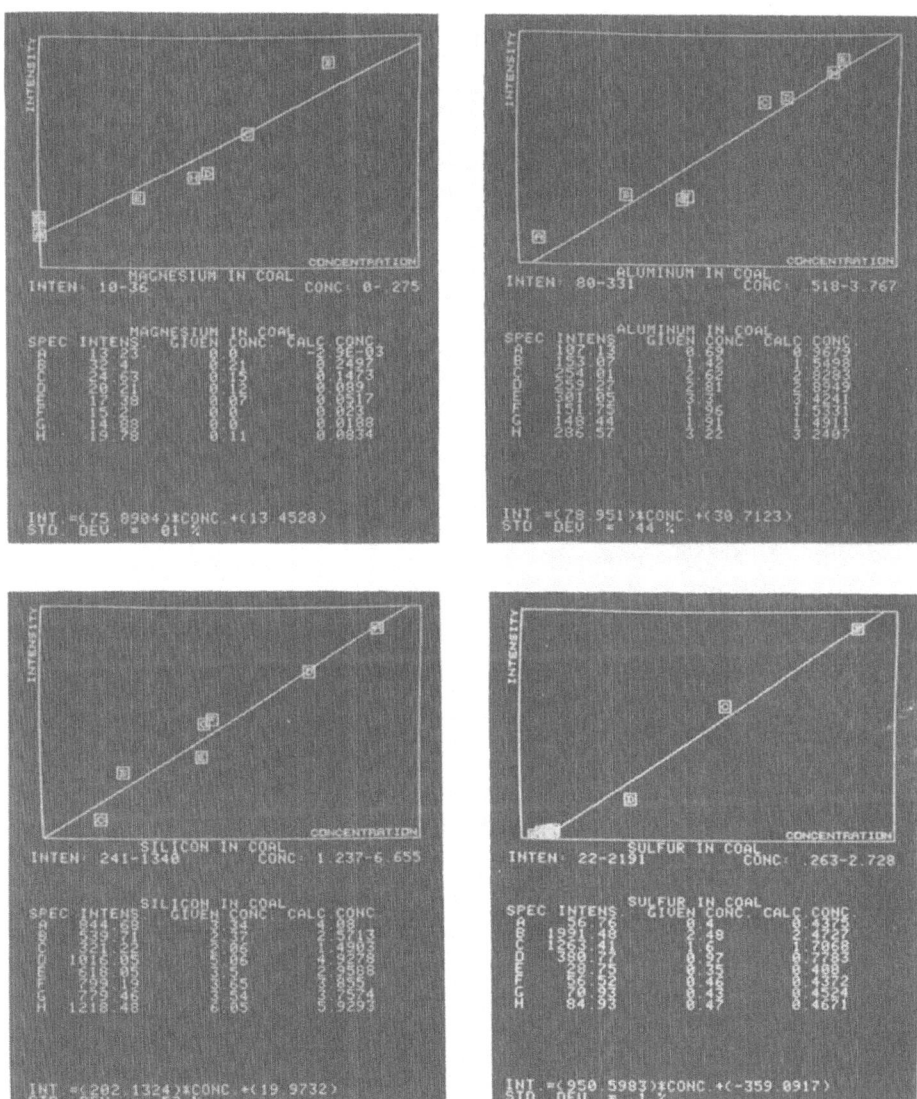

Figure 1: Concentration-intensity relationships for elements in coal.

```
PROGRAM # ? 6
SAMPLE IDENTIFICATION ? B:O 13302

ELEMENTS IN COAL               PROBABLE OXIDES

   MG      0.07 %      19 C/S    MGO     0.12 %
   AL      2.14 %     223 C/S    AL2O3   4.05 %
   SI      3.71 %     774 C/S    SIO2    7.93 %
   S       1.08 %     496 C/S    SO3     2.70 %
   K       0.91 %     261 C/S    K2O     1.10 %
   CA      0.37 %     330 C/S    CAO     0.52 %
   TI      0.11 %     119 C/S    TIO2    0.18 %
   FE      1.45 %     973 C/S    FE2O3   2.07 %

   TOTAL   9.84 %          PROBABLE ASH    15.97 %
```

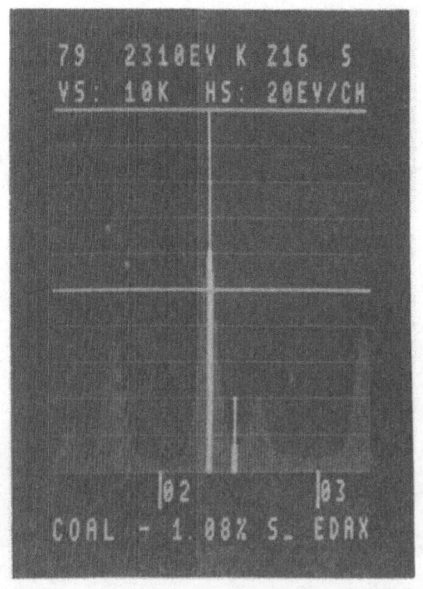

Figure 2: Typical measured spectrum on routine coal sample, with computer output.

substantial,are used to adjust the target concentration of the raw mix. Since the coal ash (5-35% of the coal) can represent as much as 1% of the clinker, this is necessary to maintain the desired final clinker concentration. The target raw mix concentration is thus adjusted by the computer to allow for the coal ash (it is also adjusted to allow for loss of superfine $CaCO_3$ from the mix in the preheater counter air flow). Since the computer also averages (by tonnage) each hourly raw mix analysis to maintain current running weighted average concentration in the blend silos for the raw mix, the inclusion of the coal analysis provides a tight control over the complete kiln operation.

SUMMARY AND CONCLUSION

Routine determination of sulfur in coal is carried out using conventional sample preparation and energy-dispersive X-ray fluorescence analysis. Concentration of sulfur can range from traces to 6%, but concentrations above 1.2% are rejected for this operation. The analysis of the coal is also taken into account by adjusting the cement raw mix to compensate for the coal ash incorporated in the clinker. The entire ED system has been in routine use, 3 shifts per day, for over 3 years, and is the primary analytical control for the entire operation.

APPLICATION OF THE FUNDAMENTAL PARAMETERS MODEL TO ENERGY-DISPERSIVE X-RAY FLUORESCENCE ANALYSIS OF COMPLEX SILICATES

Peggy Dalheim

Colorado School of Mines Research Institute
Golden, Colorado 80401

The elemental analysis of geologic samples such as rocks, minerals and coal ash is a complicated task because of their wide, complex compositional range. Energy dispersive x-ray fluorescence (EDXRF) can provide a rapid, accurate and precise way of analyzing geologic samples. Two approaches to reducing EDXRF intensity data to elemental concentrations are the empirical approach and the fundamental parameters (theoretical) approach. Empirical methods require numerous standards within restricted compositional ranges so can become complex, time consuming and, therefore, expensive if diverse suites of samples are to be analyzed for many elements. Fundamental parameters, on the other hand, requires knowledge of physical constants such as mass absorption coefficients, jump ratios and fluorescent yields, and only one matrix independent standard to calculate a calibration constant for each element making it an ideal approach to the analysis of diverse geologic samples.

The fundamental parameters model based on Shiraiwa's[1] equation as used in EXACT,[2] a mini-computer based program has been successfully applied by Harmon, et al.,[3] to the EDXRF analysis of stainless steels. They noted that calibration constants varied regularly with atomic number for several different exciting energies. In this study, functions were derived to describe the variation of calibration constants with atomic number for the elements Mg through Fe for iron, zirconium and potassium targets. Calibration constants calculated from these functions were successfully applied to the analysis of a variety of geologic samples.

The general form of the fundamental parameters equation given monochromatic excitation is:

$$W_{is} = \left[(I_{is}/T)/K_i\right] \times M_{is,t} \tag{1}$$

where

W_{is} — weight fraction of element i in sample s
I_{is} — x-ray intensity in counts for element i in sample s
T — time in seconds
$M_{is,t}$ — a matrix correction for element i in sample s given target t including enhancement and absorption terms, and a term incorporating the cosecants of the x-ray entrance and exit angles.
K_i — calibration constant for element i and target t based on count rate (I_{is}/T)

Because of severe peak overlap among the elements studied, a deconvoluting technique was employed to calculate peak intensities. This technique applies a digital filter to the sample peaks and elemental reference peaks (R) removing the background and enhancing the peaks. Sample peaks are then compared to these reference peaks resulting in peak ratios for each element i rather than count rates. If the reference peak intensity, R_i is for the same time, T as the sample peak intensity, I_i, then the peak ratio may be substituted for count rate in the fundamental parameters equation resulting in the working equation:

$$W_{is} = \left[(I_{is}/R_i)/K_i'\right] \times M_{is,t} \quad \text{at constant T} \tag{2}$$

where K_i' is a calibration constant based on the peak ratio rather than count rate.

It was observed by Harmon, et al.[3] that K is a function of atomic number, Z. K may be related to K' by combining equations (1) and (2) so that $K_i = K_i' \times R_i$ at a constant T. Therefore, this product is a function of atomic number.

Equation 1 and Equation 2 may be used to solve for K and K', respectively. In theory, one standard should be sufficient to calculate the calibration constant for element i. In practice, various standards for element i give a range of values for K_i'. The problem arises: which value of K_i' is the best value. In addition, it is sometimes very difficult to obtain standards for particular elements, so a reasonable method of choosing K_i' is to calculate it from the experimentally derived function of atomic number, F(Z).

The elements Mg through Fe were chosen to calculate F(Z) since this range includes the major elements in geologic samples. All materials, unless otherwise noted were analyzed as pressed pellets

backed with boric acid and standards were of reagent grade or
better. The equipment used consisted of a Philips x-ray generator,
a Kevex model 0810 with gold tube and secondary targets, and an NS
880 system including a Cal Data computer with 16K memory, the latter
controlling the generator, target changer and sample changer. Three
targets - potassium oxalate, iron and zirconium - were used as secondary x-ray sources. To accommodate the semi-automatic analysis of
many samples with a minimum of operator interference and a minimum
of computer memory, the EXACT program was modified and divided into
two sub-programs, ACQDA1 and XACT2. ACQDA1 set up the analytical
conditions, controlled the generator, acquired intensity data and
stored data on cassette tape. XACT2 read the data from tape and
calculated calibration constants or composition, as required. With
this system, a maximum of 13 elements including known and unknown
elements and a normalizing element or two-element compound could be
analyzed, the number of samples included in one run being limited by
the storage capacity of the cassette tape.

A function, $F(Z) = K' \times R$ was calculated for each target using
a least squares fitting routine. A power function of the form,
$K' \times R = AZ^B$ best described the variation of $K' \times R$ with Z for each
target. For more convenient illustration, this equation becomes a
straight line when plotted on a log-log scale. The potassium target
was used to excite the elements Mg through Cl; the iron target was
used for Mg through V; and the zirconium target for K through Fe.
Figures 1-3 illustrate the functions calculated for each target and
the fit of the experimental data to the curves. Experimental points
for Mg fell far below the potassium target curve and experimental
points for Mg and Al fell far below the iron target curve so they
were excluded from the curve fitting routine resulting in fits to
power functions of r>0.99. Perhaps these elements do not fit the
experimental curves because the secondary excitation is not
strictly monochromatic as required by this fundamental parameters
model. That is, there is probably a bremstrahlung component from
the gold tube adding to the excitation. The result is that there
appears to be a light element limit - Al for the potassium target
and Si for the iron target for the strict application of this model.

Since the value of a calibration constant can vary from instrument to instrument and for different excitation conditions, the
curves developed here are not generally applicable. The useful
information that can be gained from these curves is the set of substances that best determined them. Table 1 lists readily available
materials of reagent grade or better than can provide an efficient
and inexpensive means of calculating curves describing the calibration constant as a function of atomic number for the elements Mg
through Fe.

As a test of the usefulness of this procedure in the analysis
of geologic samples, several USGS and NBS standards were analyzed

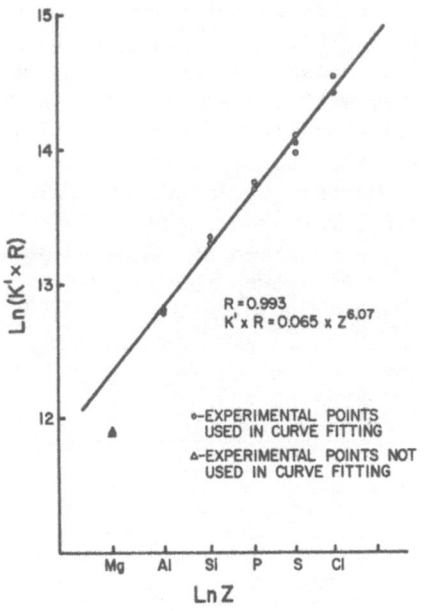

Fig. 1.　Potassium Target

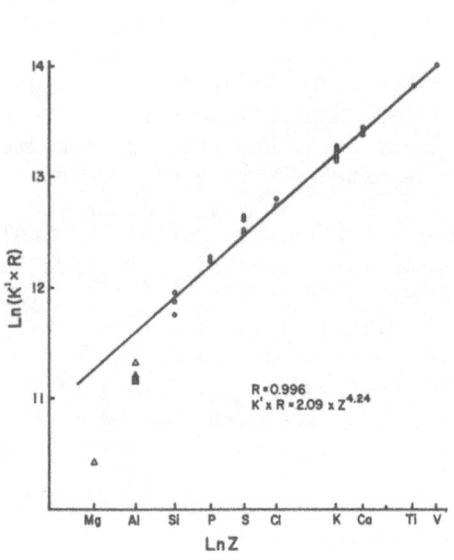

Fig. 2.　Iron Target

Fig. 3.　Zirconium Target

Table 1.　Reliable Standards

Element	Substances
Mg	MgF_2, $MgCO_3$, Mg oxalate
Al	Al metal, NBS 99a
Si	NBS 99a, SiO_2-quartz
P	$LiPO_3$, KH_2PO_4
S	K_2SO_4, NH_2SO_2OH, Na_2SO_3
Cl	KCl, NaCl
K	KH_2PO_4, K_2SO_4, K-Ti oxalate
Ca	$CaCO_3$, CaF_2, $Ca(OH)_2$
Ti	TiO_2, K-Ti oxalate
V	NH_4VO_3
Cr	$K_2Cr_2O_7$, NBS 103a
Mn	$MnCO_3$, $KMnO_4$
Fe	NBS 103a, $Fe(NH_4)_2(SO_4)_2\ 6H_2O$

using calibration constants calculated from the three curves. Mg, Al, Si, P and S were analyzed using the potassium target, counting for 150 seconds; K, Ca and Ti were analyzed using the iron target, counting for 80 seconds; Mn and Fe were analyzed using the zirconium target, counting for 100 seconds. Since experimental points for Mg fell off the potassium target curve, the average of the three experimental points was used for K'_{Mg}. This approximation proved adequate for concentrations of Mg greater than ~3%. Perhaps a lighter element target would have lowered this limit. Table 2 lists the results of these analyses.

Energy dispersive x-ray fluorescence is a useful tool in the multi-element analysis of geologic samples. If a sufficient number of well characterized standards close in composition to unknown samples cannot be obtained, the empirical data reduction methods can become exceedingly difficult. The fundamental parameters method using simple, reagent grade materials to calculate matrix independent calibration constants for each element has proved to be an efficient, inexpensive method for the EDXRF multi-element analysis of such diverse samples as siliceous igneous rocks, ores, coal ash, carbonates and phosphates.

TABLE 2. Analyses of Geologic Samples

					WEIGHT PERCENT OXIDE					
SAMPLE	SiO_2	Al_2O_3	Fe_2O_3	MgO	CaO	K_2O	TiO_2	P_2O_5	MnO	SO_3
G-2[1] A[4]	72.5	15.8	2.43	--	1.99	4.61	0.54	0.14	0.03	0.02
B	69.2	15.4	2.62	0.76	1.99	4.52	0.50	0.14	0.04	0
AGV-1[1] A	65.7	17.5	6.46	--	5.03	3.02	1.13	0.50	0.10	0.02
B	59.7	17.2	6.88	1.54	5.01	2.93	1.05	0.50	0.10	0.02
PCC-1[1] A	41.3	1.07	7.64	40.9	0.52	0	0.005	0	0.14	0.018
B	42.1	0.74	8.24	43.6	0.53	0	0.010	0	0.12	0.025
69a[2] A	5.22	57.0	5.85	--	0.28	0.009	2.90	0.05	0.003	0.10
B	6.02	55.0	5.82	0.02	0.29	<0.01	2.78	0.08	<0.01	0.04
88a[2] A	0.87	0.06	0.31	21.5	31.5	0.09	0.03	0.05	0.02	--
B	1.20	0.19	0.28	21.3	30.1	0.12	0.02	0.01	0.03	--
120b[2] A	4.30	1.47	1.14	--	46.2	0.08	0.20	36.4	0.02	0.55
B	4.69	1.06[3]	1.10[3]	0.28	49.4	0.09	0.15	34.6	0.03	--

[1] USGS Standard Rocks.
[2] NBS Standards.
[3] Listed as "Soluble".
[4] A - EDXRF value; B - Reported value.

REFERENCES

1. T. Shiraiwa and N. Fujino, "Theoretical Calculation of Fluorescent X-Ray Intensities in Fluorescent X-Ray Spectrochemical
   Analysis," Jap. J. Appl. Phys. 5:886 (1966).
2. J. W. Otvos, G. Wyld, and T. C. Yao, "Fundamental Parameter
   Method for Quantitative Elemental Analysis with Monochromatic
   X-Ray Sources," presented at 25th Annual Denver X-Ray Conference, 1976. (not published).
3. J. C. Harmon, G. Wyld, and T. C. Yao, "X-Ray Fluorescence Analysis of Stainless Steels and Low Alloy Steels Using Secondary
   Targets and the EXACT Program," *Advances in X-Ray Analysis*,
   22:325 (1978).

ELEMENTAL ANALYSIS OF URANIFEROUS ROCKS AND ORES

BY X-RAY SPECTROMETRY

Gerard W. James

Kansas Geological Survey

Lawrence, Kansas   66044

INTRODUCTION

Large-scale uranium exploration programs and industrial mine development require rapid and accurate analytical determinations of uranium and associated elements, at concentration levels ranging from a few parts-per-million to the per-cent range. The analytical requirements for associated suites of elements vary with application, but quality control and production practices in U.S. uranium mines and mills commonly require determinations of calcium, iron, copper, molybdenum, and vanadium, in addition to uranium.[1] Previous work[2,3] has demonstrated the capabilities of wavelength-dispersive x-ray spectrometric methods for many exploration applications. The purpose of this investigation is to assess the accuracy of x-ray determinations of uranium at ore-grade levels, utilizing both wavelength-dispersive and energy-dispersive spectrometers, and to discuss the acquistion of multi-element data.

ACCURACY EVALUATION

In order to evaluate the accuracy of x-ray determinations, a suite of thirty-six ore samples was obtained from the *Uranium Analytical Comparison Program* sponsored by the U.S. Department of Energy through the Bendix Field Engineering Corporation. These samples represent ore bodies mined throughout the United States, and contain trace to minor amounts of uranium, vanadium, copper, zinc, arsenic, selenium, molybdenum, and lead. Uranium concentrations, as determined by forty participating laboratories utilizing volumetric, colorimetric, and fluorimetric techniques, ranged from 0.043% to 0.654% U_3O_8.

Two commercially available x-ray spectrometers were optimized
for the determination of uranium: a wavelength-dispersive system
(Philips PW 1410), operated with a molybdenum tube (50 kV, 50 mA),
a LiF$_{220}$ analyzing crystal, and a scintillation detector; and an
energy-dispersive system *(Philips-Edax Exam Six)*, operated with a
pulsed rhodium tube (50 kV, 1 uA) and a rhodium beam filter. The
samples were analyzed directly as rock powders in 2 inch diameter
sample holders; data reduction consisted only of ratioing the U Lα_1
peak response to the Compton-scattered tube radiation. Analytical
times of less than one minute included duplicate ten-second fixed
time measurements at two 2θ positions on the wavelength system, and
duplicate ten-second spectra accumulation times for the energy-
dispersive system, which required 27 second clock times per spectra.

A comparison of the U Lα_1-to-Compton-scatter ratios, obtained
with the wavelength system, with the recommended U$_3$O$_8$ values is shown
in Fig. 1. A linear calibration curve, with a correlation coefficient of 0.997 and a standard error of 0.006% U$_3$O$_8$, can be fitted
to 31 of the 36 spectral response ratios. The standard errors associated with the wet-chemical analyses, expressed as relative deviations, averaged 4.4%. The average relative deviation for the x-ray
values from the chemical values was 2.7% for the 31 calibration
samples; however, the relative deviations for the five outlying
samples from the chemical values ranged from 10% to 24%.

The five samples with outlying x-ray spectral responses contained unusually high amounts of associated metals (Fe, Cu, Pb, Se,
and/or Mo). The ratio of Compton-scatter of these samples to Ottawa
sand ranged from 0.50 to 0.65; similar ratios of the calibration
samples ranged from 0.75 to 0.92.

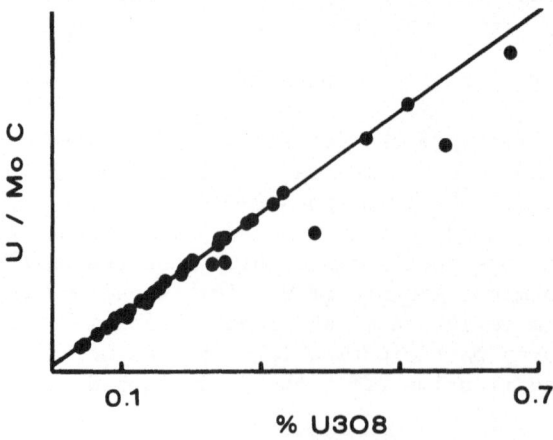

Fig. 1.   X-ray spectral response ratios versus % U$_3$O$_8$.

Additional unpublished results by the author on sample suites
from seven individual ore bodies substantiate the results observed
in this investigation. Despite highly variable lithologic matrices,
uranium determinations by x-ray methods utilizing simple sample
preparation and data reduction techniques can be extremely accurate
for most types of ore bodies. However, the Compton-scatter ratio
technique does not fully compensate for matrix effects due to the
addition of minor amounts of metals (0.1 to 1.0%, or greater) with
high mass-absorption coefficients for the emitted U $L\alpha_1$ radiation.
Consequently, accurate analyses of certain types of ores (e.g.,
U-Cu or U-Mo-Se associations) would require more extensive data
acquisition and the use of inter-element correction programs for
data reduction.

EDX and WDX PERFORMANCE COMPARISON

Both types of x-ray spectrometers provided linear calibration
curves with identical correlation coefficients (0.997) and standard
errors (0.0057% U_3O_8). However, the average spread between duplicate
determinations was lower by a factor of two for the wavelength system
(14 vs. 33 ppm), and the theoretical three-sigma detection limits,
based on background count times of 20 seconds, was also lower by a
factor of two for the wavelength system (5 vs. 11 ppm U_3O_8).

Before considering the merits of each system for additional
data acquisition, it should be emphasized that for many applications
either type of system can perform rapid and accurate determinations
of uranium. However, high levels of performance for the energy-
dispersive unit could most effectively be achieved by (1) utilizing
an air path to attenuate the detection and processing of emitted
light element (major matrix components) radiation, and by (2) utilizing a transmission filter to excite the U energy region with
essentially monochromatic radiation while suppressing the Bremsstrahlung background.

These two factors obviously hinder the energy-dispersive system
from obtaining useful data for elements with emitted energies in the
low keV range (e.g., Ca, V) from the same spectrum utilized for the
uranium determinations. An indication of the loss of sensitivity
can be illustrated by comparing detection limits (95% confidence
level) based on *10 second* background count times for the *1410* (air
path, LiF_{200} crystal) with detection limits based on *100 second*
spectra accumulation (270 seconds clock time) for the *Exam Six* (air
path, Rh filter) for silicate matrices: Ca, *0.008% vs. 0.3%*; V,
0.001% vs. 0.06%; Cu, *0.001% vs. 0.003%*; and U, *0.0009% vs. 0.0004%*.

MULTI-ELEMENT DATA ACQUISITION

Clearly, the energy-dispersive system, when optimized for the
determination of uranium, can accumulate useful data for elements

with emitted energies in the 6 keV to 18 keV range (e.g., Fe, Cu, As, Se, Pb, and Mo) for inter-element correction programs for ore-grade levels of uranium much more rapidly than the wavelength system. However, additional spectra would have to be accumulated under different conditions for elements such as Ca and V. In addition, if these spectra are to be acquired in rapid succession, compromise conditions utilizing a constant tube voltage in combination with beam filter or secondary target changes must be employed to avoid stability problems.

The choice of instrumentation becomes even more difficult if the laboratory also faces trace determinations for exploration programs. In order for the energy-dispersive system to match the levels of sensitivity obtained by the wavelength system, the time required for spectra accumulation for useful trace determinations becomes relatively large, and a wavelength system may prove to be the more efficient choice.

In conclusion, the choice of instrumentation required for the acquisition of multi-element data for geologic samples in the uranium industry should be carefully considered.

ACKNOWLEDGMENTS

The author gratefully acknowledges Philips Electronic Instruments, Inc. for the loan of an Exam Six for evaluation work, the many mineral companies who supplied sample suites of uranium ore bodies, and the Bendix Field Engineering Corporation for allowing participation in the *Uranium Analytical Comparison Program*. Publication authorized by the Director of the Kansas Geological Survey.

REFERENCES

1. R.C. Merritt, "The Extractive Metallurgy of Uranium," Colorado School of Mines Research Institute, Golden (1971).
2. G.W. James and L.R. Hathaway, Recent Advances in Analytical Methods for Determining Uranium in Natural Waters and Geological Samples, in: "Exploration for Uranium Ore Deposits," International Atomic Energy Agency, Vienna (1974).
3. G.W. James, Parts-per-Million Determinations of Uranium and Thorium in Geologic Samples by X-ray Spectrometry, Anal. Chem. 49:967 (1977).

THE FAST ANALYSIS OF URANIUM ORE BY EDXRF

Ronald A. Vane,[1] William D. Stewart[1] and Mike Barker[2]

[1]United Scientific Corporation, Mountain View, CA

[2]Exxon Minerals Company, USA, Three Rivers, TX

ABSTRACT

Energy dispersive X-ray fluorescence (EDXRF) spectrometry has
been applied to the high speed analysis of uranium ore for the purpose of grading and sorting truckloads of ore at the mine. The
system is able to analyze uranium ore in less than 60 seconds with
uranium levels ranging from < 0.005% to > 5% U_3O_8. Precision at
the 2σ level for 200 ppm ore is better than 10% relative. Good
agreement is obtained with wet chemical results in a large variety
of rock matrices.

INTRODUCTION

The sorting of truckloads by grade in a uranium mine demands
high performance from an energy dispersive XRF system. The mined
material is sorted into waste, protore (held for possible future
milling), and ore of various mill feed grade ranges by analyzing a
sample of the truckload and then dumping the material in the appropriate pile. To minimize truck idle time, the analysis time must be
less than 60 seconds including spectrum acquisition, calculation,
and printout. A direct method of analysis is needed because the
uranium in these ores is not found in equilibrium with its daughter
elements, preventing the rapid and accurate use of beta and gamma
radiation counting techniques. Ranges of uranium content cause
count rate variations from 5,000 to 40,000 cps. The uranium $L\alpha$
peak must be separated from the nearby Rb $K\alpha$ and Sr $K\alpha$ peaks, and a
matrix correction applied to provide needed precision.

The mobile laboratory in which the system is mounted is frequently moved between two different mine sites. The system must
be able to withstand moves on unpaved roads and able to operate on
the highly variable electric power supply typical of mining sites.

OPERATION

In actual operation a sample of each truckload is taken by a
technician and then brought inside the mobile laboratory. The sample is dried on a hot plate, pulverized to approximately 100 mesh
and blended. Approximately 3 grams of the blended material is
placed in a standard 1¼" diameter sample cup with a mylar window.
The sample is tapped in order to remove any voids in the powder and
placed in the X-ray spectrometer.

The X-ray spectrometer system has a Mo transmission target
X-ray tube, pulsed tube control, a $30mm^2$ Si(Li) detector with a
resolution of 154 eV at 5.9 KeV, a high count rate amplifier, and
a computer-based multichannel analyzer.   Operating conditions are
30KV, .2 ma, pulsed mode, .05mm Mo filter, and 30 sec. livetime
maximum. The count on each sample is stopped either at 30 seconds
livetime or at 50,000 counts in the uranium peak, whichever comes
first. These operating conditions are compromise conditions which
allow the analysis of uranium ore with levels of uranium ranging
from .005% to 5% or greater. High grade ores have very high count
rates and the limitation to 50,000 counts in the peak allows fast
reporting of the results without delays due to excessive amounts of
deadtime at these levels. For high sensitivity analysis of low
grade ore, the count rate and the sensitivity of the instrument can
be improved by increasing the tube power.

The Mo X-ray tube was chosen because the Mo K line excites the
L_{III} absorption edge of uranium but not the other L edges of uranium,
which greatly simplifies the spectrum. The Mo filter effectively
filters out the bremmstrahlung scatter in the region of the uranium
Lα peak, keeping the background low.

Pulsed tube operation is needed because the high grade ores
produce very high count rates under these operating conditions. The
pulsed X-ray tube[1] significantly reduces system deadtime at high
count rates by preventing pulse pileup by turning off the X-ray production while the amplifier is processing each pulse. The pulsed
tube, in conjunction with the high count rate amplifier, allows a
system throughput (counts stored in the analyzer in real time) of
nearly 40,000 cps.

DATA REDUCTION

The uranium peak in the ores is commonly overlapped with the
Rb Kα peak and is quite close to the Sr Kα peak. In order to obtain the true intensity of uranium, especially in low grade ores,
a multiple least squares fitting program (Super ML) is used.[2]
This program employs reference fitted spectra of pure element peaks
in conjunction with a digital filter to remove background and to
determine the true net peak in the spectrum. The effectiveness of
the fitting technique is shown in Fig. 1. This figure shows the
raw X-ray spectrum of a low grade ore, and the insets show the results of Super ML fitting with first the fitted amount of uranium
subtracted out and then with Rb, U, and Sr subtracted out. The
smooth background that remains is an indication of the method's
effectiveness for determining the true intensity in the presence
of background and overlapping peaks.

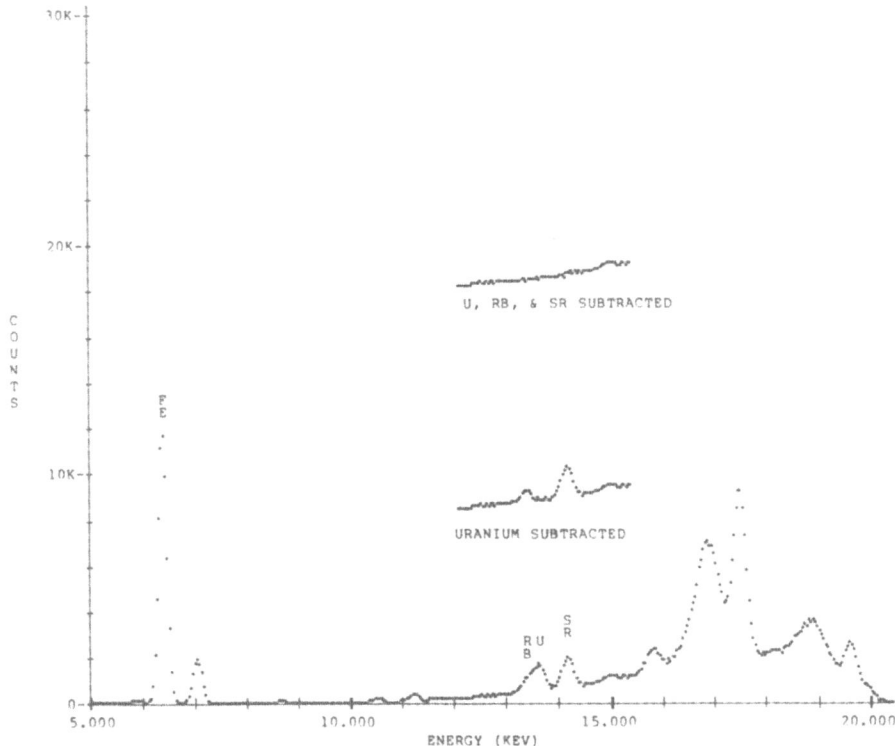

Fig. 1. Spectrum of .0534% uranium ore showing the overlap of the
         U Lα, Rb Kα and Sr Kα. The insets show spectra after
         subtracting out intensities of U, Rb, Sr as determined by
         the multiple least square fitting program.

Due to the variety of rock matrices present in the ore, a matrix absorption correction must be made on the uranium intensities. This is done using the method first proposed by Reynolds.[3] With this method, the peak intensity is ratioed to the peak intensity of the Compton scattered Mo radiation from the X-ray tube. This peak intensity-Compton ratio is then applied to the calibration curve to calculate the concentration of uranium. Use of a ratio technique also corrects for the differences in counting time for the higher grade samples.

The calculation program is also used to keep track of the daily mine output. This program is fed a daily moisture factor and the tare weights of each truck at the beginning of each day. The program determines the U_3O_8 concentration for each truckload of ore and is given the truck number and the gross weight of the load. The program then calculates the net wet tons of ore, dry tons, and pounds of U_3O_8 in the truckload. At the end of each shift a cumulative total by mine grids and stockpiles is printed out.

TIME OF ANALYSIS

In the worst case situation, with a sample producing about 30% deadtime with a livetime of 30 seconds, the real time of analysis is about 45 seconds. Data reduction and printout take approximately 15 seconds. Thus, the maximum analysis time is 60 seconds. For low grade ore, deadtime is less and the analysis quicker. For high grade ores, the maximum count limitation to 50,000 counts in the uranium Lα peak reduces the analysis time to less than 10 seconds in some cases.

SYSTEM CALIBRATION

To calibrate the system, a suite of 45 ore samples, representative of the range of concentrations and ore types, was assembled. These samples were assayed for uranium in three independent laboratories and the results averaged. The samples were analyzed on the EDXRF system and the calibration curve established. It was found that the analytical curve for uranium needed to be broken into three segments in order to obtain the best results. The three analytical curves are characterized by a unique set of coefficients. In calculating the concentration of unknown samples the computer tests the intensity-Compton ratio to determine which set of coefficients should be used to calculate the concentration of uranium.

The high grade ore calibration curve has a quadratic shape, caused by the fact that uranium in these samples is no longer a trace element and is contributing significantly to the matrix

Table 1. Uranium Ore Test Data

Sample Number	% U$_3$O$_8$ Assay	% U$_3$O$_8$ EDXRF
30	1.954	1.84
31	2.225	2.14
32	1.80	1.85
37	0.129	0.128
40	0.125	0.118
44	0.184	0.185
47	0.195	0.191
49	0.230	0.221
50	0.259	0.253
52	0.115	0.108
53	0.169	0.166
54	0.203	0.192
55	0.235	0.238
56	0.267	0.262

absorption coefficient. For the Compton absorption correction
method to be linear there must not be any major element absorption
edges between the analyte line and the Compton scatter. The UL$_{III}$
absorption edge which lies below the Mo Kα line causes the
calibration curve to become nonlinear as the concentration of
uranium increases.

RESULTS

The accuracy of the analysis was tested by analyzing a set of
samples which had known assays. The results of these analyses with
the assay value and the X-ray value are given in Table 1. These
results are all within the precision needed for this analysis. Better precision for the low grade samples can be obtained by extending
the analysis time to 100 seconds livetime, to improve the counting
statistics. In addition, the use of pressed pellets instead of loose
powders would improve precision.

CONCLUSION

The analysis of uranium ores has been applied over a wide range
of concentrations with good accuracy and minimum time. The mine site
environment, while not ideal, presents no significant problems for
the EDXRF system.

REFERENCES

1. J. E. Stewart, H. R. Zulliger, and W. E. Drummond in <u>Advances in X-Ray Analysis</u>, Vol. 19, R. W. Gould et al., eds. (Dubuque, Iowa: Kendall/Hunt Publishing Co., 1976) pp. 153-160.
2. F. H. Shamber in <u>X-Ray Fluorescence Analysis of Environmental</u> Samples, Thomas W. Dzubay, ed. (Ann Arbor, Michigan: Ann Arbor Science Publishers, Inc. 1977) pp. 241-258.
3. R. C. Reynolds, Jr., Am. Mineral., 52, 1493 (1967).

A COMPREHENSIVE ALPHA COEFFICIENT ALGORITHM

G.R. Lachance and F. Claisse

Geological Survey of Canada, Ottawa, Ont. and Dept. Mining and Metallurgy, Université Laval, Québec, Qué. Canada

ABSTRACT

An algorithm is proposed as a model that describes the intensity-concentration relation over a wide concentration range in binary and more complex systems involving absorption and enhancement. All the influence coefficients can be determined from fundamental parameter equations. Terms containing product of two weight fractions are always negligible in pure absorption cases, are often negligible in absorption/weak enhancement cases and are necessary but small in strong absorption/strong enhancement situations.

Once the magnitude of the coefficients is known, the range of application of the one- or two-coefficient models can be established with certitude.

INTRODUCTION

The increasing pressure for analysis of higher accuracy, the advent of automated X-ray spectrometers for multi-element analysis, and the increasing use of mini-computers are good reasons for further developments in mathematical corrections for interelement effects.

Corrections using the fundamental equations need computers with large capacities in routine analysis where they are required to make calculations for corrections that are often negligible or that can be made more easily using simpler algorithms with little loss in accuracy.

The desirable correction equation should be based on theoretical equations and be accurate over the whole range of compositions. One should aim for the smallest number of parameters having specific physical meaning. The model should have enough flexibility for adaptation to cases which do not need sophisticated corrections, and it should be easy to use even without a computer.

An equation that meets most of these requirements was written by Lachance & Traill[1].

$$W_i = R_i \ (1 + \sum_j \alpha_j W_j) \tag{1}$$

where W = weight fraction
       R = relative intensity
       α = a constant which represents the magnitude of the inter-element influence. This relation is strictly correct when a monochromatic radiation is used and when enhancement is absent. Because enhancement is nearly equivalent to negative absorption and because a primary beam of polychromatic radiations behaves very nearly as monochromatic radiation in practice, the ideal algorithm for matrix corrections should be a refinement of equation 1, but it should not be significantly different. One such refined expression was derived by Claisse & Quintin[2].

$$W_i = R_i \ (1 + \sum_j \alpha_j W_j \ + \ \sum_j \alpha_{jj} \ W_j^2 \ + \ \sum_{jk} \alpha_{jk} \ W_j W_k \ ) \tag{2}$$

This relation differs from the L-T relation by the second-order terms which represent minor additional corrections only, but the increased number of parameters increases the accuracy considerably.

In actual situations, the choice is between the oversimplified L-T relation with its limits in range and/or accuracy and the rather inconvenient C-Q relation with its higher accuracy but at the price of too many coefficients for routine use. It is the purpose of this paper to present a comprehensive algorithm which combines a high degree of physical meaning and flexibility.

THE BASIC EQUATION

A significant observation is made by writing the C-Q relation (equation 2) for a binary system in the following form:

$$W_i = R_i \left[1 + (\alpha_j + \alpha_{jj} \ W_j) \ W_j \right] \tag{3}$$

which indicates that the C-Q relation is identical to the L-T relation except that the effective α coefficient is no longer a constant but a value which varies slightly with composition. A first consequence is that a basic equation for a binary system which engulfs the L-T and the C-Q relations and others still to come, can be written as

$$W_i = R_i \ (1 + \alpha_j \ W_j)$$ (4)

where α_j is a parameter that varies with composition.

    In binary systems when i represents the analyte and j is the matrix element, equation 4 states that the weight fraction of i is roughly equal to the relative intensity R_i. This intensity must be multiplied by a matrix correction factor $1 + \alpha_j \ W_j$ to yield the correct weight fraction W_i. This correction is represented by QJ in figure 1. If element i were combined with element k instead of j, the correction would be represented by QK for the same weight fraction W_i. If the matrix contains both elements j and k at the same weight fraction of i, the matrix correction should be represented by QM where the position of M is between J and K and depends on the proportions of j and k.

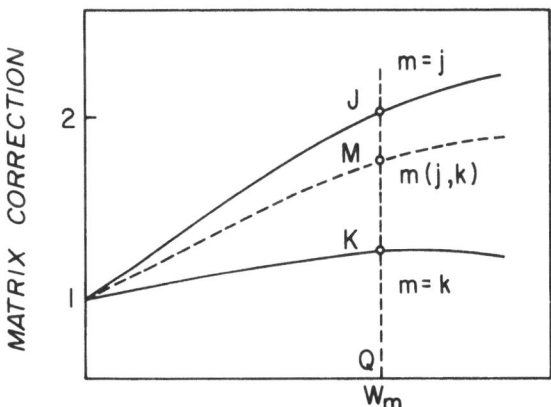

Figure 1. Matrix correction as a function of weight fraction of matrix.

    Since α coefficients are linear functions of absorption coefficients[1] except when strong enhancement is present, and since the absorption coefficient of a complex system varies linearly with composition, as a first approximation the position of M should move

linearly with composition between J and K. Then, in ternary and
higher systems, the basic equation becomes

$$W_i = R_i \left[(1 + \alpha_j W_m) \frac{W_j}{W_m} + (1 + \alpha_k W_m) \frac{W_k}{W_m} + \ldots \right] \qquad (5)$$

$$W_i = R_i \ (1 + \Sigma \alpha_j \ W_j) \qquad (6)$$

where m is the sum of all the elements except i,

which is still the L-T relation with the additional condition that
the α *coefficients must be those at the level Wm* and not at the
individual levels W_j, W_k, etc. That condition was mentioned earlier
by Tertian[3].

WORKING ALGORITHM

The application of equation 6 requires that the α_j coefficients
for binary systems must be expressed as functions of concentrations.
Claisse and Quintin model[2] meets this requirement but for better
accuracy and for reasons that will become clear later, the addition
of a third term is desirable; it has been found that a term in W_m^3
is better than a term in W_m^2:

$$\alpha_j = (\alpha_1 + \alpha_2 \ W_m + \alpha_3 \ W_m^3)_j \qquad (7)$$

where m = j in a binary system.

The three α constants are calculated from equations 4 and 7
using three fluorescence intensities at three binary compositions.

Then, the proposed algorithm is a combination of εqns 6 and 7.

VERIFICATION

Relative intensities were computed[4] from which the α_1, α_2 and
α_3 coefficients were obtained using the data at the W_j= 0.2, 0.8 and
0.95 levels. Typical coefficients are given in Table I for strong
absorption/enhancement which is always associated with significant
values of α_3 . Thus, in the majority of cases, α_3 is very small and
often negligible. The recalculated weight percents obtained by
iteration[1] are an indication of the validity of the model.

Table II shows recalculated concentrations on ternary systems
using α coefficients from binaries at the proper W_m level, i.e. W_m
or its equivalent $1-W_i$. In all cases the analyte is Fe. In the
first case, Fe is combined with Ni and Cu which are both strong

Table I

Fe and Ti in binary systems

Nominal W_i/W_j	2/98	10/90	50/50	90/10	α_1	α_2	α_3
i-j							
Fe-Cr	2.01	10.00	50.12	89.83	2.23	-.21	+.06
Fe-Ni	2.00	9.99	50.08	89.98	-.16	-.19	-.10
Fe-Pb	2.01	10.00	50.10	89.85	1.90	-.27	+.07
Ti-Cr	2.00	9.99	50.10	89.98	-.01	-.20	+.16

Table II

Fe in termary systems

Nominal $W_i/W_j/W_k$	5/80/15	20/40/40	50/20/30	80/10/10	α_{jk}
i-j-k					
Fe-Ni-Cu	4.99	19.92	49.87	79.96	0
Fe-Si-Cr	5.00	20.00	49.95	79.89	0
Fe-Cr-Ni	5.05	20.49	50.48	80.04	0
Fe-Cr-Ni	4.99	20.02	49.94	79.88	-0.24

enhancers for Fe; in the second, it is combined with Si and Cr, a
weak and a strong absorber. In both cases the deviations are
smaller than the usual experimental errors. The third example refers
to the Fe-Cr-Ni system, a combination of a strong absorber and a
strong enhancer for Fe. Results in line 3 indicate that the relative
error is significant. These situations are few, predictable and can
be dealt with by the introduction of a cross-correction term as in
the C-Q relation (equation 2):

$$W_i = R_i (1 + \sum_j \alpha_j W_j + \sum_{jk} \alpha_{jk} W_j W_k) \qquad (8)$$

The last line of Table II shows the improvement that can be
expected when such a correction is added.

Figure 1 illustrates when and why cross-coefficients are expected.
An error is associated with the non-linearity of the interpolation,
especially when the two lines are far apart. It is minimal when the
elements are absorbers and is maximal when the interelement effect
is the result of a combined strong absorption/enhancement. In practice, it is only when dealing with the latter case involving major
constituents that the cross-corrections will be significant. In
other cases, the crossed terms can be dropped (i.e. the coefficients

are set to zero) and compensated in part by a minor shift in the
binary coefficients.

CONCLUSION

     The proposed algorithm is more accurate and simpler to use than
the C-Q relation. The apparent complexity because one more coefficient is used to represent each binary effect is more than compensated
by the internal concordancy of the coefficients, by the fact that
most cross-coefficients are negligible and that it is only in the
most demanding situations that three coefficients are needed. At
the practical level, the model should prove useful for its versatility to be condensed to the two- or even one-coefficient model. When
such simplifications are made, accuracy is retained by adjusting the
coefficients to the range of composition of interest.

REFERENCES

1. G.R. Lachance and R.J. Traill, "A Practical Solution to the Matrix
    Problem in X-Ray Analysis", Can. Spectry, 11: 43-48 (1966).

2. F. Claisse and M. Quintin, "Generalization of the Lachance-Traill
    Method for the Correction of the Matrix Effect in X-Ray Fluores-
    cence Analysis", Can. Spectry 12: 129-134 (1967).

3. R. Tertian, "An Accurate Coefficient Method for X-Ray Fluores-
    cence Analysis", in "Advances in X-Ray Analysis": 85-109 (1976).

4. G.R. Lachance, "Fundamental Coefficients for X-Ray Spectrochemical
    Analysis", Can. Spectry 15: 64-71 (1970).

FUNDAMENTAL-PARAMETERS CALCULATIONS

ON A LABORATORY MICROCOMPUTER

J. W. Criss

Criss Software, Inc.
12204 Blaketon Street
Largo, Maryland 20870

Fundamental-parameters calculations can be made on a laboratory microcomputer for automatic treatment of interelement absorption and enhancement effects in x-ray fluorescence analysis. A new software package, called XRF-11, uses an efficient combination of fundamental parameters and alpha factors to compensate for any lack of measured reference materials, while taking full advantage of whatever standards are available, even just pure elements. In many cases, one multi-element standard is enough for accurate analysis.

The new XRF-11 software uses the same data base of absorption coefficients, fluorescence yields, etc. as the big-computer program NRLXRF[1], and combines theory with experiment in a consistent way that is similar to, but more efficient than, the treatment used in NRLXRF.

There are three major distinctions between NRLXRF and the new XRF-11: (1) NRLXRF runs only on large computers such as the PDP-10, IBM 370, CDC 7600, Univac 1108, etc. -- XRF-11 runs on a PDP-11 with 32K 16-bit words plus floppy disks (and on several other mini-computers that are based on the LSI-11 microcomputer), (2) XRF-11 is not just a modification of NRLXRF -- except for the main data base and a few routines for accessing it, XRF-11 is a totally new software package designed specifically for the LSI-11, (3) NRLXRF was developed by the U.S. government, and is in the public domain -- XRF-11 was developed entirely in the private sector, and is a commercial product available through several different vendors.

This short paper will outline the basic design considerations that made such an elaborate calculation practical for a mini-computer, and will list some of the special features of XRF-11.

DESIGN AND DEVELOPMENT OF THE NEW SOFTWARE

A tape copy of NRLXRF was purchased from COSMIC[1] and loaded onto a PDP-11/70 computer (which has the same Fortran address space as the other PDP-11's). The 11/70 was used mainly for its large disk space and editing capabilities. Final program development and testing was done on a PDP-11/03 system.

NRLXRF could not be made to run on a 32K minicomputer merely by reducing the sizes of arrays, because the instructions themselves required more memory than was available. Moreover, the seventy or so subroutines in NRLXRF were connected to each other in such an intricate way, and there was so much memory required for COMMON storage, that it was not feasible to run NRLXRF by an overlay scheme (swapping subroutines into main memory from disk only when needed).

It was necessary to start from scratch and design the new software specifically for a minicomputer, and to take full advantage of the overlay capabilities. The only subroutines from NRLXRF that could be adapted were a few that read parameters from the master data file -- which could be used, without change, on a PDP-11.

The design of the minicomputer software was guided by several goals: avoid making any approximations to the physical theory or the fundamental parameters, treat as many elements at a time (20) as NRLXRF, make the calculations as fast as the data collection, and permit easy modification of the input/output interfaces with the user and the instrument. These goals were met, as described below.

PHYSICAL THEORIES AND PARAMETERS USED IN XRF-11

The theoretical formulas and physical parameters used in XRF-11 are the same as used in NRLXRF for homogeneous samples of any thickness. The formula of Gillam and Heal[2] expresses the theoretical x-ray intensity, including absorption and secondary fluorescence effects, for any given composition. A simple factor to account for sample thickness is taken from the work of Compton.[3] The intensity formula is integrated numerically (as in NRLXRF) to account for incident radiation from an x-ray tube.

Wavelengths of emission lines and absorption edges are from the tabulations of Bearden and Burr.[4,5] Mass absorption coefficients are obtained from the parametric expressions of McMaster et al.[6] The efficiency of fluorescence production is computed with allowance for Auger and Coster-Kronig transitions, as well as the usual subshell yields, using formulas and data from Bambynek et al.[7]

The software can treat monochromatic incident radiation (as from an isotope or fluorescer), or it can treat the full spectrum

from an x-ray tube. A tube spectrum can be calculated automatically for any anode element, operating voltage, and beryllium
window thickness, using the Kramers formula[8] for the shape of the
continuum, as generated, modifications of the formulas of Green and
Cosslett[9] for line intensities, an anode absorption factor from the
electron microprobe work of Philibert[10] and Heinrich,[11] and the
usual absorption factor for the tube window and any filter.

ORGANIZATION AND USE OF THE MINICOMPUTER SOFTWARE

There are two major programs, called XRF11A and XRF11B. The
first one, XRF11A, is used for the initial problem set-up. The
second part, XRF11B, is the principal tool for analysis of unknowns.

XRF 11A is used to define the operating conditions and the lists
of lines measured and elements present (whether measured or not).
It automatically calculates the x-ray tube spectrum (if appropriate),
and writes disk files containing only those fundamental parameters
that pertain to the specified situation.

XRF11A can also be used to calculate theoretical x-ray intensities for any specified composition, and to calculate alpha factors
that match the fundamental-parameters calculations exactly, for
several different compositions simultaneously: at trace levels of
each measured element, near 100 per cent for each element, and at
some arbitrary nominal composition of greatest interest. The alpha-
factor model used is a limited Claisse-Quintin[12] form,

$$C_i = R_i(1 + \alpha_{i1}C_1 + \alpha_{i2}C_2 + \ldots + \alpha_{i11}C_1^2 + \alpha_{i22}C_2^2 + \ldots)$$

where C and R are the mass fraction and relative x-ray intensity,
and i represents the measured element. The alpha's calculated by
XRF11A are stored on disk for later use by XRF11B.

XRF11B reads the disk files produced by XRF11A, and also reads
other data, which may be typed in by the user or accessed automatically from disk files or from the instrument. The main purpose
of XRF11B is to treat measured x-ray intensities. Data from standards are used to rescale all theoretical intensities (and alpha
factors) to conform to experimental reality. This adjustment
process is essential in using any kind of theory for accurate
analysis, because no theory alone will ever be able to model all
instrumental responses as accurately as needed in all cases.

XRF11B automatically considers all measured standards, and
places greatest emphasis on those that are most similar to each
sample that is being treated. A single multi-element standard is
sufficient for calibrating the calculation process, but it is better
to use whatever standards are available, well characterized, and
reasonably similar to the unknowns. In extreme situations, one may

use only pure elements, or any standards that contain the measured elements (even if they contain elements not in the unknowns).

The composition of an unknown is calculated by XRF11B in the following way. First, the alpha factors (properly adjusted) are used to calculate the composition. Then, a fundamental-parameters calculation is made to predict the intensities from a sample with that composition. The predicted intensities are adjusted to conform to the standards, and compared with the measurements of the unknown. If there is significant disagreement (considering the random errors in measurement) then XRF11B makes a refined estimate of composition, and predicts the intensities again. The iteration stops when there is no significant change in calculated composition. Usually, one or two cycles using fundamental-parameters predictions are enough to refine the concentrations calculated from alpha factors.

Several other options are available. For example, if it were necessary to process unknowns very rapidly, XRF11B could output only the results of the alpha-factor method, and not make fundamental-parameters refinements for each unknown. Also, XRF11B could be run only for the purpose of calibrating the alpha factors, which could then be written to disk for use by other programs of the analyst's choosing. Usually, however, XRF11B will be able to process unknowns as fast as they are measured.

SPECIAL FEATURES

XRF11B can treat compounds (oxides, chlorides, etc.) as the independent constituents of interest. The analyst may fix the amounts of any components, such as internal standards or a flux. An unmeasured component can be calculated by difference. Besides the options (mentioned above) for bypassing the theoretical iterations, there are other ways to balance the two calculation methods, to emphasize either speed or precision.

Because of the modular design of the software, and the complete separation of input/output routines from calculation routines, data handling can be modified to suit the application--on-line process control, interactive data analysis, deferred (batch) processing of accumulated data, different formats for reporting results, etc.

ACCURACY AND SPEED

XRF-11 was run on a PDP-11/03 for a set of published measurements[13] of NBS alloys containing nine elements (P, S, Si, Cr, Mn, Fe, Ni, Cu, and Mo). Using only one multi-element standard and eight pure elements, XRF-11 gave the same results as NRLXRF: concentrations above the 0.5 weight per cent level were calculated with an average accuracy of 1.2 per cent relative (1.2% of the amount present). Trace-amount determinations also agreed with NRLXRF.

The calculations for the nine-component alloys required an average time of 83 seconds per sample on the PDP-11/03, using the full fundamental-parameters iteration. Samples with five elements could be processed about three times as fast, still using the full treatment. In an extreme case having twenty elements, the theoretical predictions for twenty lines took only two and one half minutes.

ACKNOWLEDGEMENT

Valuable suggestions were made by analysts at Edax International, Rigaku/USA, Siemens Corporation, and Telsec Instruments Limited.

1. J. W. Criss, "NRLXRF", COSMIC Program and Documentation DOD-65, Computer Software Management and Information Center, University of Georgia, Athens Ga 30602 (1977 and updates)
2. E. Gillam and H. T. Heal, Some Problems in the Analysis of Steels by X-Ray Fluorescence, Brit J Appl Phys, 3:353 (1952)
3. A. H. Compton, The Efficiency of Production of Fluorescent X-Rays, Phil Mag, 8:961 (1929)
4. J. A. Bearden, X-Ray Wavelengths, Rev Mod Phys 39:78 (1967)
5. J. A. Bearden and A. F. Burr, Reevaluation of X-Ray Atomic Energy Levels, Rev Mod Phys, 39:125 (1967)
6. W. H. McMaster, N. K. del Grande, J. H. Mallet, J. H. Hubbell, "Compilation of X-Ray Cross Sections", Lawrence Radiation Laboratory report UCRL-50174 sec II (1969 and revisions)
7. W. Bambynek, B. Crasemann, R. W. Fink, H. U. Freund, H. Mark, C. D. Swift, R. E. Price, P. V. Rao, X-Ray Fluorescence Yields, Auger, and Coster-Kronig Transition Probabilities, Rev Mod Phys 44:716 (1972)
8. H. A. Kramers, On the Theory of X-Ray Absorption and of the Continuous X-Ray Spectrum, Phil Mag 46:836 (1923)
9. M. Green and V. E. Cosslett, The Efficiency of Production of Characteristic X-Radiation in Thick Targets of a Pure Element, Proc Phys Soc 78:1206 (1961)
10. J. Philibert, pp 379-392 in "X-Ray Optics and X-Ray Microanalysis", H. H. Pattee, V. E. Cosslett, and A. Engstrom, eds, Academic Press, New York (1963)
11. K. F. J. Heinrich, "Present State of the Classical Theory of Quantitative Electron Probe Microanalysis", Nat Bur Stand (U.S.) Tech Note 521 (1970)
12. F. Claisse and M. Quintin, Can Spectrosc 12:129 (1967)
13. J. W. Criss, L. S. Birks, and J. V. Gilfrich, A Versatile X-Ray Analysis Program Combining Fundamental Parameters and Empirical Coefficients, Anal Chem 50:33 (1978)

Modified NRLXRF Program for Energy Dispersive X-Ray Fluorescence

Analysis

R.B. Shen, J. Criss*, J.C. Russ, A.O. Sandborg

EDAX International, Inc., P.O. Box 135, Prairie View, IL 60069

*12204 Blaketon St. Largo, MD 20870

Introduction

An X-ray fluorescence program, similar to NRLXRF[2] has been written by John Criss to fit into the DEC LSI-11 32K computer with floppy disk system. EDAX has combined the new version with our own software to create a set of FORTRAN programs, called "XRAY 95" to use with the new EDAX EXAM 9500 energy dispersive X-ray fluorescence system. The XRAY 95 program employs both fundamental-parameter equations and influence-coefficient equations to optimize the matrix corrections for multi-component samples. It is an extremely versatile program, using whatever reference standards the user provides and supplementing them with physical theory. The result is a practical approach to fast, accurate analysis.

The main key to the success of the program, is its ability to calculate accurate intensities for any simulated composition, scaling the results to match one or more multi-component real standards with (preferred) or without a similar matrix.

A Brief Description of the XRAY95 Program

The program flow chart is shown in Figure 1. There are six independent sub-programs chained together with a small program, called "XRAY95". It takes 5 to 15 seconds to swap a program depending on the program size. The overlay structure has been heavily used due to the very large size of most programs. Within each program, the analyzer function keys will guide the user through the analytical procedures step by step. Choices are given by labelling different keys with simple names, for the user to select. All data pass from one program to another through the diskette files which speeds the process and minimizes human error.

Flow Chart   Description

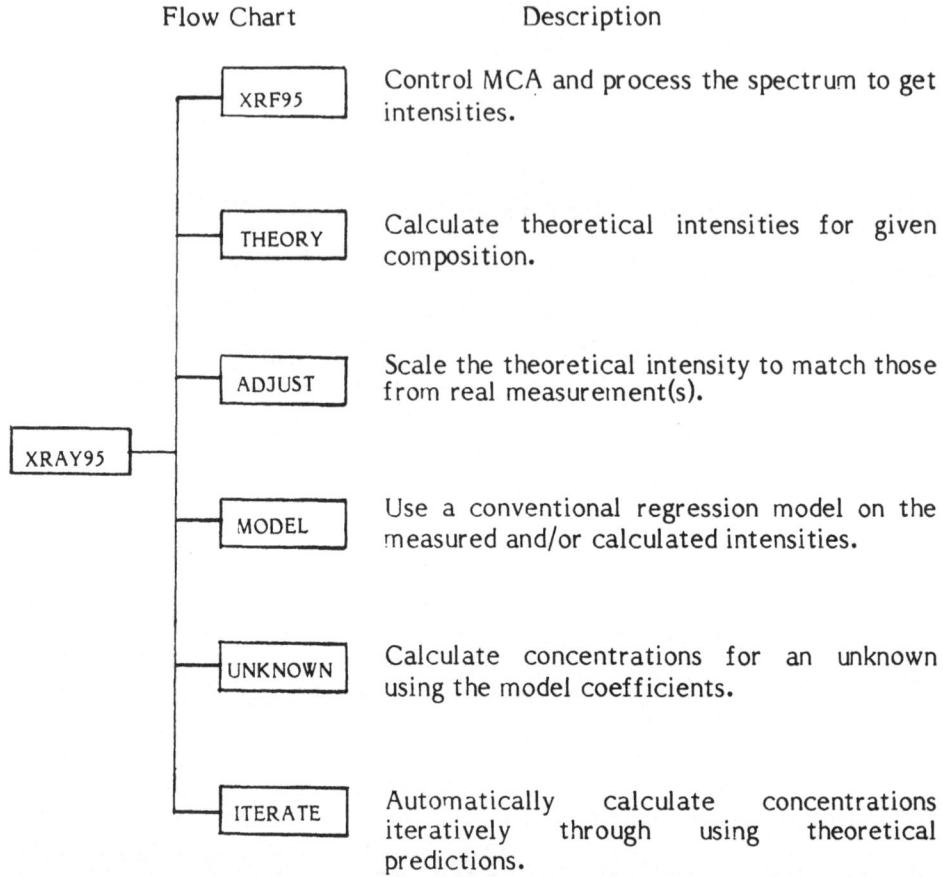

Control MCA and process the spectrum to get intensities.

Calculate theoretical intensities for given composition.

Scale the theoretical intensity to match those from real measurement(s).

Use a conventional regression model on the measured and/or calculated intensities.

Calculate concentrations for an unknown using the model coefficients.

Automatically calculate concentrations iteratively through using theoretical predictions.

Fig. 1.   XRAY 95 Programs Flow Chart

1.   XRF95:   This is a multichannel analyzer control and spectrum processing program.   Besides displaying the spectrum, it outputs net intensities (and concentration results if the coefficients of a regression model have been stored on diskette beforehand).   The results are displayed on the CRT screen or printed out as hard copy.   To get net intensities. the spectrum background is removed first by subtracting a polynomial curve fit to selected points, and then a modified gaussian shape is used to fit and strip peaks one by one.

2.   THEORY: This is the theoretical intensity calculation program.   The program inputs tube target, kV, filter information and element list from disk file which was created by XRF95 program when the samples were analyzed. The X-ray tube spectrum is calculated first. Mass absorption coefficients, fluorescence yield, line energies are read in from a master data file and sub-set of parameter data file is created for later use.   Then the program calculates the theoretical intensities for any real or synthetic samples when their concentrations are typed in. The method of calculation is the same as

NRLXRF[3] program for homogeneous thick and thin samples. In the XRAY95 program, the particle size effect is not included at this time.

3. ADJUST: This program reads in the concentration and theoretical intensity data for all real and synthetic samples. The user selects a few or all real standards and inputs their measured intensities from a disk file. At least one standard is required for each element, however, the more standards used, the more accurate the result. Then the program will adjust all theoretical intensities to match the measured intensities of the real samples. This way, the user could create as many standards as he wishes. This is the foundation of the program.

4. ITERATE: This is a part of combination of the THEORY and ADJUST programs. Under this arrangement, an unknown concentration can be iteratively calculated through the theoretical and adjust intensity cycles after each estimated concentration. This is the best way to use the theory for a unique unknown. For a normal seven element sample, it may take up to three minutes to do the iteration.

5. MODEL: For most quality control types of analyses, a well defined concentration range is known and standards covering it are available. In this case, the conventional type of regression model could be used. In this program, both the modified Lucas-Tooth[4] and the Lachance-Traill models are available. The calculated coefficients can be read back in XRF95 program. Then a "on" line type analysis can be processed in XRF95. Weighted fitting, as in the original NRLXRF program, is provided.

6. UNKNOWN: This program calculated concentrations for an unknown sample from the net intensity and the proper regression model coefficients. This is a small program. If the user has any special output format, additional calculations, etc., this is the program for him to modify or expand.

All of the above programs can handle 16 elements and 20 standards except the ITERATION program in which 12 elements and 12 standards are the maximum due to the large size of the program.

A Comparison of XRAY95 and NRLXRF Calculation Results

It is interesting to compare the results of these two programs with the same input values. The data of NBS stainless steel SRM 1151-1185 in Table II and Table III of J. Criss' paper[2] have been used. The THEORY, ADJUST, MODEL and UNKNOWN programs were used in calculation. The modified Lucas-Tooth model (Δ-I) in MODEL was selected. To make a fair comparison, sample no's SRM 1159, 1160 were excluded because they are Nickel based alloys. One sample, # 1154 was used as the only real standard. Ten synthetic samples were created by the computer after the user input the desired concentration range. The theoretical intensities of all samples were calculated and re-scaled to match the real measured intensities of # 1154. The Δ-I model was used to calculate coefficients. Then measured intensities of all other samples and the coefficients were used in UNKNOWN to calculate their compositions.

The results are listed in Table 1. For the minor elements (Si, Mn, Cu, and Mo), the calculated compositions from the XRAY95 program are almost identical to those from NRLXRF program. And for the major elements (Cr, Fe and Ni), the relative average deviation is 2.0% from the XRAY95, instead of 1.2% from NRLXRF. The difference may come from the different regression model.

Table I - A Comparison of the Calculated Composition (%) Results from XRAY95 and NRLXRF Programs

NBS SRM No.	Si			Cr		
	NBS	NRLXRF	XRAY95	NBS	NRLXRF	XRAY95
1151	0.37	0.53	0.54	22.13	22.38	22.53
1152	0.65	1.17	1.17	18.49	18.80	19.20
1153	0.82	1.13	1.13	16.61	16.75	17.13
1154	1.09	STD	STD	19.58	STD	STD
1155	0.50	0.83	0.84	18.45	18.81	18.60
1171	1.54	1.07	1.06	17.40	17.60	18.07
1184	0.70	1.09	1.12	19.44	19.53	19.29
1185	0.40	1.01	1.00	17.09	17.42	17.40

NBS SRM No.	Mn			Fe		
	NBS	NRLXRF	XRAY95	NBS	NRLXRF	XRAY95
1151	2.17	2.18	2.18	67.15	66.77	67.73
1152	1.19	1.15	1.18	67.81	67.89	69.33
1153	0.61	0.62	0.65	69.03	68.34	69.60
1154	1.74	STD	STD	65.09	STD	STD
1155	1.63	1.64	1.64	64.45	64.21	64.11
1171	1.80	1.80	1.86	68.20	67.70	69.18
1184	1.04	1.04	1.04	65.64	64.77	64.85
1185	1.22	1.24	1.33*	65.88	65.47	65.65

NBS SRM No.	Ni			Cu		
	NBS	NRLXRF	XRAY95	NBS	NRLXRF	XRAY95
1151	7.03	7.03	7.11	0.25	0.25	0.25
1152	10.21	10.34	10.63	0.50	0.49	0.50
1153	12.02	12.07	12.47	0.26	0.26	0.26
1154	10.25	STD	STD	0.56	STD	STD
1155	12.18	12.32	12.51	0.17	0.17	0.18
1171	11.20	11.30	11.62	0.12	0.36	0.37
1184	9.47	9.46	9.61		0.078	0.083
1185	13.18	13.29	13.60	.067	0.072	0.078

* The measured intensity value, 348 instead of 34.8, is used here.

Table I - Continued

NBS SRM No.	NBS	Mo NRLXRF	XRAY95
1151	0.76	0.76	0.78
1152	0.37	0.37	0.37
1153	0.21	0.21	0.21
1154	0.46	STD	STD
1155	2.38	2.38	2.39
1171	0.16	0.15	0.15
1184	1.46	1.48	1.49
1185	2.01	2.10	2.10

We also compared the adjusted theoretical intensities with the real measured ones when the exact compositions were input in THEORY program. Some results are listed in Table II. They agree very well for most of the elements, except Si values for which the measured values (in cps) are low and hence of doubtful precision.

Table II - A Comparison of the Adjusted Theoretical Intensities with the Measured ones for some Stainless Samples

Sample	*	Si	Cr	Mn	Fe	Ni	Cu	Mo
1151	A	21.7	7504	596	12480	883	31.7	229.3
	R	31.2	7590	600	12500	894	32.0	232.0
1152	A	37.6	6485	333	13400	1309	64.7	110.5
	R	67.9	6640	324	12460	1336	64.1	111
1153	A	46.9	5902	171.0	14010	1543	33.7	62.0
	R	65.6	6010	178	13970	1572	32.6	62.3
1154	A	STD**						
	R	63.7	6780	488	12750	1331	73.4	139.0

* A = adjusted intensity, R = real measured intensity
**SRM # 1154 used as standard.

Tests of XRAY95 Program With Literature Data

The main key of the method is that it can calculate the adjusted theoretical intensities accurately for any synthetic samples. The data for this test are taken from the Rasberry and Heinrich paper[5]. There are two types of samples, Fe-Ni and Fe-Ni-Cr. They were measured at two different voltages, 45kV and 20 kV. To do the test, all real given concentrations were used to calculate the theoretical intensities first. Then different real standards were selected to re-scale all theoretical values. The adjusted values were compared with the measured ones. The results are listed in Table III and IV, with the given concentrations and real measured intensities.

Table III - Test XRAY95 Calculated Adjusted Intensities with Literature Data (Binary sample)

Sample #	Given Conc(%)		Real Int. (45 kV)		Real Int. (20 kV)	
	Fe	Ni	Fe	Ni	Fe	Ni
971	4.62	95.16	0.0789	0.8782	0.0823	0.8891
972	6.59	93.22	0.1104	0.8321	0.1152	0.8505
974	10.18	89.64	0.1621	0.7595	0.1684	0.7767
983	22.63	77.11	0.3172	0.5483	0.3307	0.5695
986	30.67	69.31	0.4007	0.4515	0.4151	0.4718
987	34.31	65.52	0.4373	0.4073	0.4484	0.4287
1159	51.00	48.20	0.5907	0.2553	0.6037	0.2695
126B	63.15	35.99	0.6958	0.1720	0.7086	0.1824
809B	95.49	3.29	0.9659	0.0127	0.9594	0.0133

| | Case 1 (45 kV) | | Case 2 (45 kV) | | Case 3 (45 kV) | |
| | A. Int - R. Int* | | A. Int - R. Int | | A. Int - R. Int | |
Sample #	Fe	Ni	Fe	Ni	Fe	Ni
971	STD		0.0006	0.0514	STD	
972	0.0010	-0.0020	0.0019	0.0468	0.0014	-0.0006
974	-0.0011	-0.0015	0.0022	0.0431	0.0015	0.0001
983	0.0024	-0.0108	0.0047	0.0219	0.0034	-0.0078
986	0.0013	-0.0080	0.0042	0.0189	0.0026	-0.0034
987	0.0011	-0.0117	0.0043	0.0128	0.0026	-0.0064
1159	-0.0027	-0.0105	0.0016	0.0051	-0.0005	-0.0024
126B	-0.0031	-0.0082	0.0020	0.0023	-0.0004	-0.0004
809B	-0.0070	-0.0008	STD		STD	
Ave(%)**	0.57	2.8	.99	3.9	.66	.71

| | Case 4 (45 kV) | | Case 5 (20 kV) | | Case 6 (20 kV) | |
| | A.Int - R. Int* | | A. Int - R. Int | | A. Int - R. Int | |
Sample #	Fe	Ni	Fe	Ni	Fe	Ni
971	STD		STD		STD	
972	0.0011	0.0024	0.0013	0.0026	0.0015	0.0054
974	0.0011	0.0035	0.0003	0.0068	0.0006	0.0034
983	0.0026	-0.0035	0.0022	0.0048	0.0028	-0.0020
986	STD		-0.0004	-0.0002	STD	
987	0.0014	-0.0049	-0.0043	0.0035	-0.0045	-0.0004
1159	-0.0020	-0.0046	-0.0034	0.0018	-0.0065	-0.0021
126B	-0.0021	-0.0023	0.0016	-0.0001	-0.0029	-0.0010
809B	STD		STD		STD	
Ave(%)**	.58	.95	.55	.52	.83	.47

* A. Int - R. Int: means the difference of adjusted intensities and the measured real intensities.

** Relative average deviation.

Table IV - Test XRAY95 Calculated Adjusted Intensities with Literature Data (Ternary System)

Sample #	Given Conc. (%)			Real Int. (45 kV)			Real Int. (20 kV)		
	Fe	Ni	Cr	Fe	Ni	Cr	Fe	Ni	Cr
4184	63.22	0.00	36.58				.3751		.4572
4014	0.00	60.64	38.83					.4319	.4392
5074	68.38	4.98	25.25	.4511	.0203	.3258	.4663	.0216	.3336
5181	69.45	9.96	19.88	.4971	.0416	.2651	.5123	.0443	.2721
5324	52.80	19.27	26.96	.3529	.0821	.3311	.3678	.0933	.3393
5321	59.19	20.02	19.88	.4343	.0898	.2582	.4512	.0949	.2662
7271	71.59	8.29	18.79	.5298	.0343	.2536	.5378	.0362	.2599
161	15.01	64.92	16.88	.1460	.4367	.2072	.1522	.4535	.2129
1189	1.40	72.60	20.30	.0125	.5630	.2263	.0135	.5852	.2348

Sample #	Case 1 (45 kV) A. Int - R. Int			Case 2 (45 kV) A. Int - R. Int			Case 3 (45 kV) A. Int - R. Int		
	Fe	Ni	Cr	Fe	Ni	Cr	Fe	Ni	Cr
5074		STD		.0488	.0002	.0389		STD	
5181	.0006	.0004	.0040	.0543	.0007	.0312	.0015	.0002	.0023
5324	.0003	-.0042	.0057	.0384	-.0035	.0396	.0017	.0045	.0002
5321	-.0010	.0010	.0067	.0461	.0018	.0329		STD	
7271	.0097	.0002	.0015	.0659	.0005	.0277	.0106	.0001	.0001
161	.0014	.0033	.0023	.0170	.0069	.0237	.0153	.0042	.0209
1189	-.0015	-.0048	-.0263		STD			STD	
Ave(%)**2.5	1.6	3.2	11.2	2.0	11.5	3.3	1.8	2.8	

Sample #	Case 4 (20 kV) A. Int - R. Int			Case 5 (20 kV) A. Int - R. Int			Case 6 (20 kV) A. Int - R. Int		
	Fe	Ni	Cr	Fe	Ni	Cr	Fe	Ni	Cr
4184		STD		.0034		.0043	.0057		.0070
4014		STD			.0049	-.0020		.0192	.00393*
5074	-.0066	-.0006	-.0015	-.0023	-.0004	-.0030		STD	
5181	-.0073	-.0007	.0022	-.1126	-.0002	.0010	.0002	.0004	.0020
5324	-.0033	-.0011	.0015		STD		.0022	.0015	.0016
5321	-.0056	-.0022	.0045	-.0015	-.0011	.0031		STD	
7271	-.0060	-.0011	-.0005	-.10011	-.0007	-.0016	-.0019	-.0001	-.0004
161	-.0004	-.0191	-.0028	.0009	-.0137	-.0037	.0111	.0017	.0172
1189	-.0013	-.0321	-.0277	-.0011	-.0251	-.0288		STD	
Ave(%) 2.3	2.9	2.4	1.6	2.0	2.3	2.0	1.5	2.2	

* Excluded in calculating average

** Relative average deviation.

For example, the Case 1 in Table III, the sample # 971 was used as real standard which contains 4.62% Fe and 95.16% Ni. These measured intensities were used to adjust all other theoretical intensities. The two columns underneath are the difference (absolute) of the adjusted theoretical intensities and the measured intensities. The relative average deviation in this case is 0.57% for Fe and 2.8% for Ni.

The results show the relative average deviation of the calculated intensities from the measured ones can be less than 1% with properly selected standards. (see Cases 3, 4, 5, 6 in Table III).

For the ternary system (see Table IV), the relative average deviations are slightly higher, about 2%. An interesting thing (see Case 4 in Table IV) should be pointed out: the standards in this case are two binary samples, yet the calculated results still are quite good.

Test and Illustration of XRAY95 Program With Real Samples

Two NBS stainless steel SRM 849 and 850 are used to illustrate the program operation. First the XRF95 program was used to measure net intensities. A hardware configuration and analysis condition were stored on disk. The elements Si, Cr, Mn, Fe and Ni were identified from the spectrum and labled under their peaks. The background points were set. Then these two samples were analyzed; peaks stripped, and net intensities were output. A few pictures are shown in Figure 2 to indicate the procedure.

Table V - The Given or Assumed Concentrations and Calculated Theoretical and Adjusted Intensities

Conc: Sample #	Si	Cr	Mn	Fe	Ni
SRM849	0.68	5.40	1.63	84.92	6.62
F100	1.00	8.00	0.40	70.60	20.00
F200	1.50	1.30	0.50	82.70	15.00
F300	0.20	18.00	0.20	56.60	25.00
F400	0.70	2.00	0.30	67.00	30.00
Theo. Int:					
SRM849	125.2	6513	1391	53570	1987
F100	177.6	8752	339.8	44860	6479
F200	88.5	1606	443.1	56460	4638
F300	35.6	17390	163.6	32440	8458
F400	119.6	2318	267.8	48150	10190
Adj. Int:					
SRM849	10.64	641.0	175.3	5475	199.6
F100	15.10	861.4	42.8	4585	650.7
F200	7.53	158.0	55.8	5771	465.8
F300	3.03	1712.0	20.6	3315	849.4
F400	10.17	228.1	33.7	4921	1023

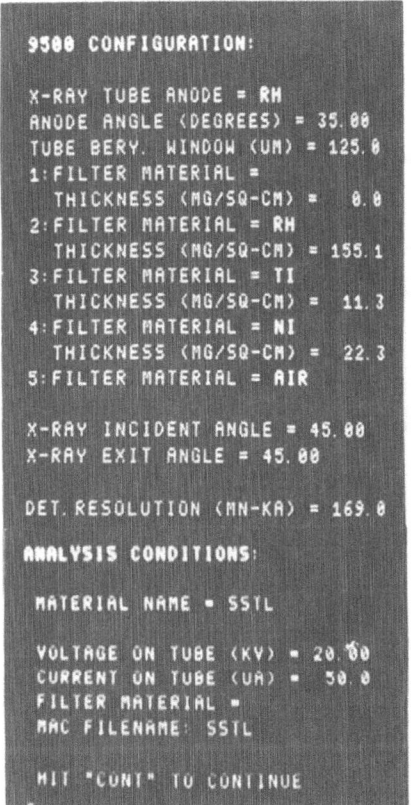

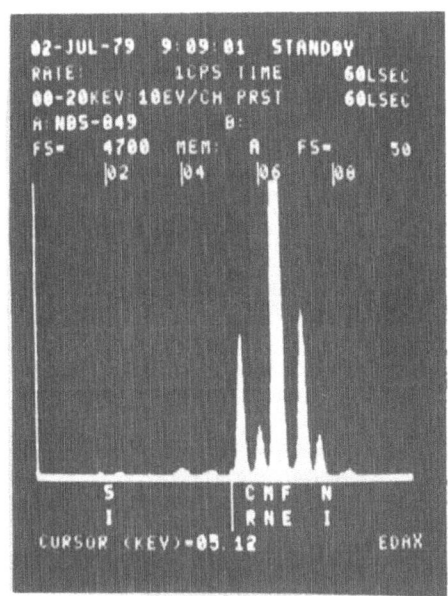

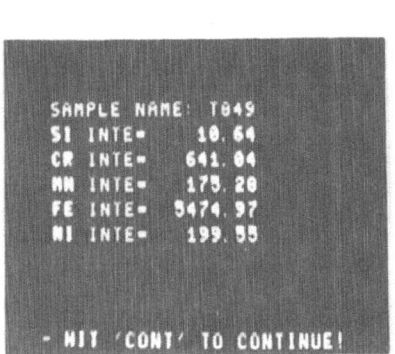

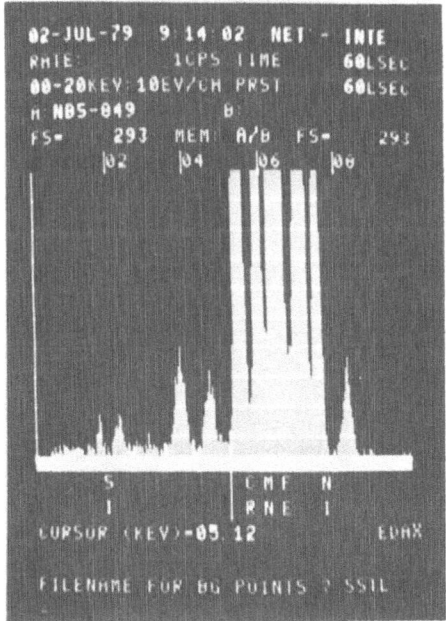

Figure 2: Running the XRF95 program.

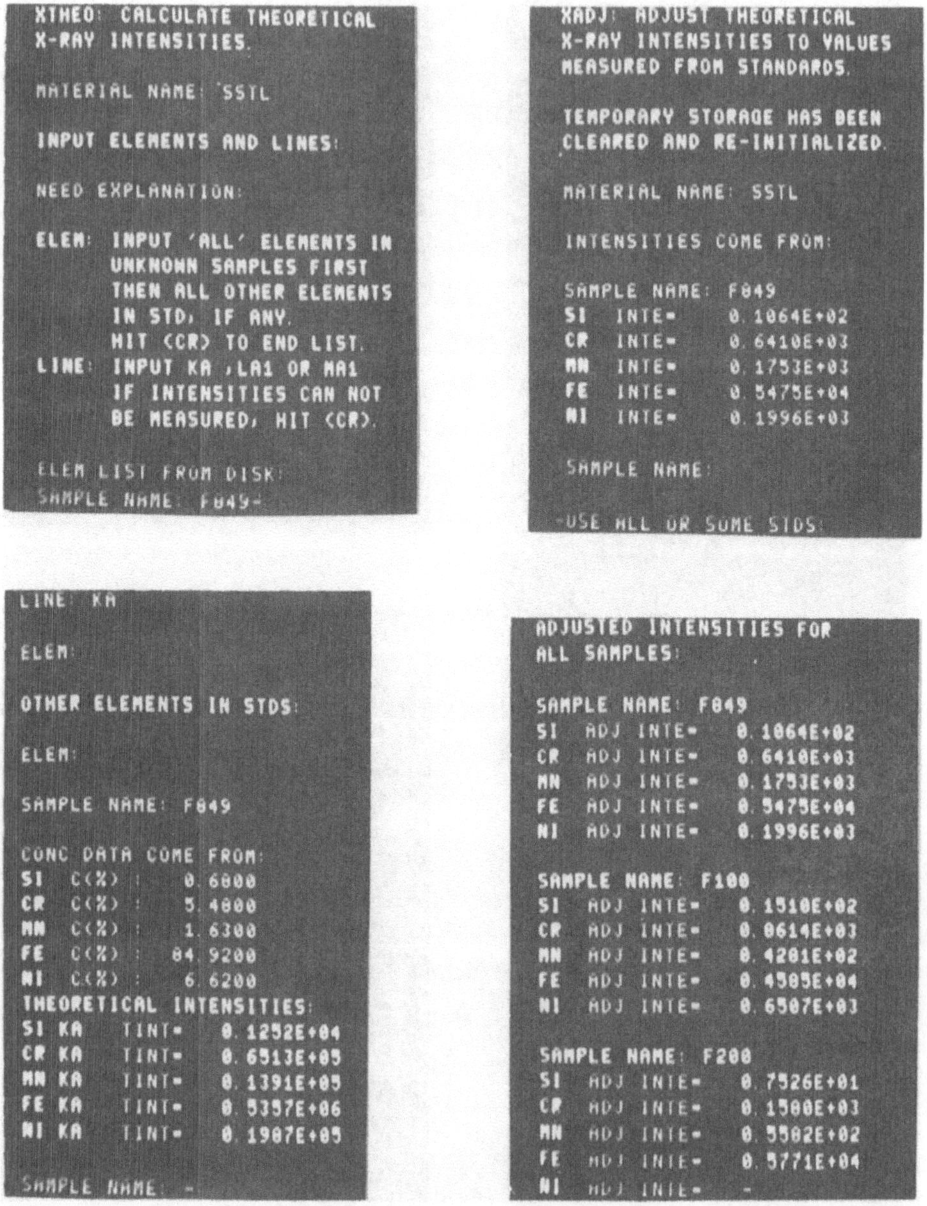

```
XTHEO: CALCULATE THEORETICAL
X-RAY INTENSITIES.

MATERIAL NAME: SSTL

INPUT ELEMENTS AND LINES:

NEED EXPLANATION:

ELEM: INPUT 'ALL' ELEMENTS IN
      UNKNOWN SAMPLES FIRST
      THEN ALL OTHER ELEMENTS
      IN STD, IF ANY.
      HIT <CR> TO END LIST.
LINE: INPUT KA ,LA1 OR MA1
      IF INTENSITIES CAN NOT
      BE MEASURED, HIT <CR>.

ELEM LIST FROM DISK:
SAMPLE NAME: F049-
```

```
XADJ: ADJUST THEORETICAL
X-RAY INTENSITIES TO VALUES
MEASURED FROM STANDARDS.

TEMPORARY STORAGE HAS BEEN
CLEARED AND RE-INITIALIZED.

MATERIAL NAME: SSTL

INTENSITIES COME FROM:

SAMPLE NAME: F049
SI   INTE=    0.1064E+02
CR   INTE=    0.6410E+03
MN   INTE=    0.1753E+03
FE   INTE=    0.5475E+04
NI   INTE=    0.1996E+03

SAMPLE NAME:

-USE ALL OR SOME STDS:
```

```
LINE: KA

ELEM:

OTHER ELEMENTS IN STDS:

ELEM:

SAMPLE NAME: F049

CONC DATA COME FROM:
SI  C(%) :   0.6800
CR  C(%) :   5.4000
MN  C(%) :   1.6300
FE  C(%) :  84.9200
NI  C(%) :   6.6200
THEORETICAL INTENSITIES:
SI KA   TINT=   0.1252E+04
CR KA   TINT=   0.6513E+05
MN KA   TINT=   0.1391E+05
FE KA   TINT=   0.5357E+06
NI KA   TINT=   0.1987E+05

SAMPLE NAME: -
```

```
ADJUSTED INTENSITIES FOR
ALL SAMPLES:

SAMPLE NAME: F049
SI  ADJ INTE=   0.1064E+02
CR  ADJ INTE=   0.6410E+03
MN  ADJ INTE=   0.1753E+03
FE  ADJ INTE=   0.5475E+04
NI  ADJ INTE=   0.1996E+03

SAMPLE NAME: F100
SI  ADJ INTE=   0.1510E+02
CR  ADJ INTE=   0.0614E+03
MN  ADJ INTE=   0.4201E+02
FE  ADJ INTE=   0.4585E+04
NI  ADJ INTE=   0.6507E+03

SAMPLE NAME: F200
SI  ADJ INTE=   0.7526E+01
CR  ADJ INTE=   0.1500E+03
MN  ADJ INTE=   0.5502E+02
FE  ADJ INTE=   0.5771E+04
NI  ADJ INTE=   -
```

Figure 3: Various program output.

Table VI - The Calculated X-Ray Analysis Results using XRAY95 Program

Sample #	SRM950	
	Given Conc. (%)	Calc. Conc. (%)
Si	0.12	0.27
Cr	2.99	2.92
Mn		0.41
Fe	71.68	71.88
Ni	24.80	24.23

Figure 3 continued

In this test, we selected one sample, SRM 849, as a standard and the other, SRM850, as an unknown. Using the THEORY program we calculated theoretical intensities for SRM849, and also for four other arbitrarily selected synthetic samples. The concentrations and their theoretical intensities are listed in Table V. Then ADJUST program was used to adjust the calculated theoretical intensities, (see Figure 3), and the Δ-I model was used to calculate the coefficients. Finally, the UNKNOWN program was used to calculate the concentration for the unknown sample. The results are listed in Table VI. Pictures in Figure 3 show this operation.

The poor agreement for Si result may be due to the small intensity value. A better Ni result should be expected if a higher or pure Ni standard is used. However, the general results are excellent when considering the use of one standard with a quite different matrix.

References

1. J.W. Criss, 28th Annual Conf. on Applications of X-ray Analysis, Denver, Colorado, (1979).
2. J.W. Criss, L.S. Birks, and J.V. Gilfrich, Anal. Chem. 50, 33 (1978).
3. A description of the method is on page 41, COSMIC, DOD-00065, Suite 112 Barrow Hall, Athens, Georgia.
4. R. Shen, J. Russ & W. Stroeve, Adv. in X-ray Anal. Vol. 22, p. 385, (1979).
5. S. Rasberry & K. Heinrich, Anal. Chem. 46, 81 (1974).

UNUSUAL MATRIX FLUORESCENCE EFFECTS

IN X-RAY FLUORESCENCE ANALYSIS

J. W. Criss

Criss Software, Inc.
12204 Blaketon Street
Largo, Maryland 20870

ABSTRACT

On the basis of fundamental-parameters calculations of x-ray fluorescence intensities, it is predicted that relative intensities much greater than one can be observed from thick, homogeneous samples. That is, the intensity of a particular line from a mixture of two or more elements can be much greater than the intensity from the pure element. Similarly, the intensity from a mixture of compounds can be greater than the intensity from a pure sample of the compound containing the emitting element.

INTRODUCTION

For layered samples, relative x-ray intensities greater than one have been calculated previously[1] (and have no doubt been observed experimentally). A thin surface layer can be fluoresced so strongly by a substrate that the intensity is greater than from a thick, pure sample of the surface material. However, a similar effect for thick, homogeneous samples has never before been reported, to this author's knowledge.

The unusually strong matrix fluorescence discussed here is not likely to be encountered in ordinary analysis, but the physical processes that produce the effects operate to some degree in many practical problems. Therefore, it is valuable to recognize the effects, and to account for them appropriately, whenever much reliance is placed on theoretical predictions of fluorescence intensities-- i.e., in analysis based on fundamental parameters.

111

PHENOMENON

The typical shape of the intensity vs. concentration curve for
a binary sample is monotonic--increasing from zero, at zero mass
concentration of the emitting element, up to a relative x-ray intensity of 1.0 for a pure sample of the measured component. The lower
portion of Figure 1 shows the usual forms: the curve lies below the
diagonal if the effect is predominantly matrix absorption, and lies
above the diagonal if there is "negative absorption" (the matrix is
generally less absorbing than the measured component) or if secondary
fluorescence is strong enough to overcome a more absorbing matrix.
Also shown in Figure 1 is an example of unusual matrix fluorescence,
where the relative x-ray intensity reaches a value greater than 2.

The upper curve in Figure 1 represents the relative x-ray intensity of Si K alpha radiation from a mixture of silica and zirconia.
The abscissa is the weight fraction of silica. The calculations
assumed monochromatic incident radiation of 45 keV; the x-ray incidence and take-off angles at the sample were taken to be 45 degrees.
The calculations were made on a minicomputer program[2] that uses the
same formulas and fundamental parameters as the big-computer program
NRLXRF[3] to treat matrix absorption and secondary fluorescence in
homogeneous samples.

Unusually strong matrix fluorescence effects might be observed
for several different combinations of elements, under suitable excitation conditions. The physical processes will be discussed here
in terms of the silica-zirconia mixture.

EXPLANATION

Because the incident radiation is of high energy (45 keV), it
will penetrate rather deep in pure silica (see the schematic Figure
2), and so the Si K x rays will suffer considerable absorption by
the silica in escaping the sample (the mass absorption coefficients[4]
of silica for the incident and measured x rays are 0.162 and 726,
respectively). Thus, the intensity from pure silica will be less
than if the same total excitation energy had been applied at lower
photon energies.

When some zirconia is present, there will be less direct excitation of silicon, but there will be secondary fluorescence of silicon
by Zr K and L lines. Although the total amount of Si K production
will be less than in pure silica, the secondary Si K radiation will
be produced closer to the sample surface, and will escape relatively
easily (the mass absorption coefficient of zirconia for the incident
radiation is 5.84). As a result, the measured Si K intensity can be
much higher for the mixture than for pure silica.

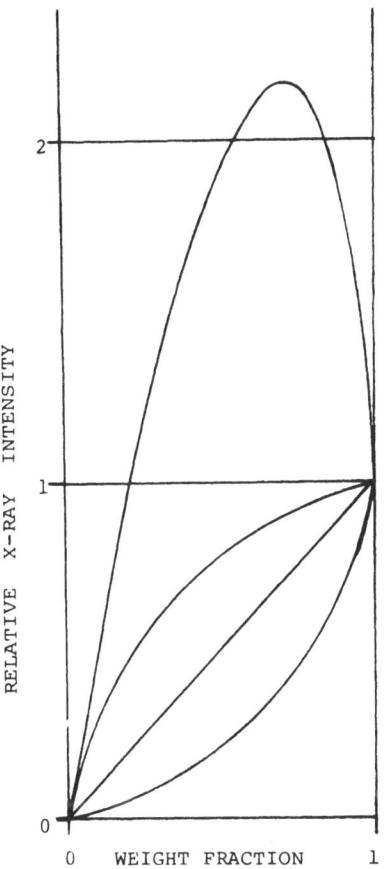

Fig. 1. General shapes of calibration curves. The upper curve is
        Si K intensity vs. wt fraction of silica in silica-zirconia.

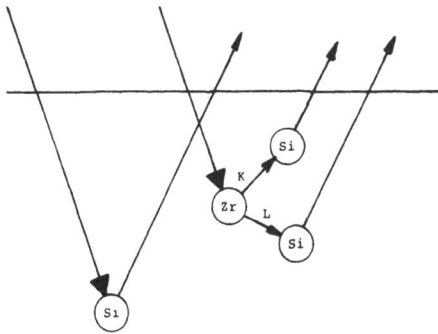

Fig. 2. Direct and indirect fluorescence of Si K x rays.

These same arguments apply to a mixture of the elements Si and Zr, as well as the oxides. Under suitable excitation conditions, similar effects should be observed for certain other combinations of elements (and their oxides, carbonates, etc.)--for example, Ba-Ca Unusual fluorescence would be expected whenever the incident radiation has an energy higher than the K edge of an element (such as Zr) whose K lines (and perhaps L lines) fluoresce a measured element (such as Si) that has a much lower absorption coefficient for the incident radiation.

The calculations for the Si curve in Figure 1 used absorption coefficients from McMaster et al.,[4] and subshell fluorescence yields from Bambynek et al.,[5] and included the effects of Auger and Coster-Kronig transitions[5] on the redistribution of primary vacancies in Zr atoms. Matrix absorption effects are treated by the Gillam and Heal formula.[6]

If the Coster-Kronig and Auger transitions were ignored, the peak of the curve would be about 11 per cent lower. If secondary fluorescence were ignored altogether, the calibration curve would be a familiar absorption-type curve, lying slightly below the diagonal line in Figure 1.

Such dramatic effects as shown in Figure 1 would not occur if the incident radiation included many photons closer to the absorption edge of the measured element, because there would be much more direct fluorescence close to the sample surface.

ACCURACY OF THE PREDICTIONS

An experiment would be expected to provide general qualitative agreement with the calculation methods just described. Precise agreement would not be expected, for several reasons: uncertainties in the values of fundamental parameters, neglect of tertiary fluorescence effects, and neglect of scattering effects.

The values of fluorescence yields and radiationless transition probabilities are more critical for such strong fluorescence effects than for the usual analysis situations. More important, perhaps, is the fact that the case illustrated--fluorescence by both K and L lines--involves a tertiary fluorescence effect (even for a binary mixture of Si and Zr). For example, Zr K photons from one atom can fluoresce Zr L photons from another atom, which can then fluoresce Si. Tertiary fluorescence is not treated in the program used here, because the small size of the effect in analysis hardly ever justifies the much greater computation time required.

Another source of error, especially for very high-energy incident radiation, is the neglect of scattering. The program assumes that the penetration of radiation is governed by the photoelectric

mass absorption coefficient, and does not consider changes in the
directions of photons. To a first approximation, that treatment is
good: photons scattered slightly away from their initial direction
tend to be compensated by other photons scattered into that direction
from slightly different directions.

It remains to be seen just how much disagreement there might be
between experiment and the calculation method used here. A Monte
Carlo calculation could avoid problems caused by neglecting tertiary
fluorescence and scattering, but errors in values of fundamental
parameters would still exist.

SUMMARY

Relative x-ray intensities greater than one are predicted for
thick, homogeneous samples under certain conditions. The unusual
effects are expected when the incident radiation has a high photon
energy, and when there is an element present that absorbs the incident radiation very strongly and also fluoresces the measured
element.

REFERENCES

1. J. W. Criss, Multiple Indirect X-Ray Fluorescence, presented at
    the 21st Mid-America Symposium on Spectroscopy, Chicago,
    Ill, June 2-5, 1970.
2. J. W. Criss, Fundamental-Parameters Calculations on a Laboratory
    Microcomputer, in Advances in X-Ray Analysis, vol. 23, 1980.
3. J. W. Criss, "NRLXRF", COSMIC Program and Documentation DOD-65,
    Computer Software Management and Information Center,
    University of Georgia, Athens Ga 30602 (1977 and updates).
4. W. H. McMaster, N. K. del Grande, J. H. Mallet, J. H. Hubbell,
    "Compilation of X-Ray Cross Sections," Lawrence Radiation
    Laboratory report UCRL-50174 sec II (1969 and revisions).
5. W. Bambynek, B. Crasemann, R. W. Fink, H. U. Freund, H. Mark,
    C. D. Swift, R. E. Price, P. V. Rao, X-Ray Fluorescence
    Yields, Auger, and Coster-Kronig Transition Probabilities,
    Rev Mod Phys 44:716 (1972).
6. E. Gillam and H. T. Heal, Some Problems in the Analysis of
    Steels by X-Ray Fluorescence, Brit J Appl Phys, 3:353 (1952).

MONTE CARLO SIMULATION OF SAMPLE SCATTERING EFFECTS FROM HOMOGENEOUS

SAMPLES EXCITED BY MONOENERGETIC PHOTONS

J. M. Doster and R. P. Gardner

Center for Engineering Applications of Radioisotopes
Nuclear Engineering Department
North Carolina State University
Raleigh, North Carolina 27650

ABSTRACT

The error introduced by sample scattering in EDXRF analysis is
evaluated by Monte Carlo simulation. This is accomplished by deriving
a Monte Carlo model capable of simulating single Compton and Rayleigh
scatters from the exciting photon source and from fluorescent X rays
in homogeneous samples. The model also includes primary, secondary,
and tertiary fluorescence events. (1) Results are given for Ni-Fe-Cr
ternary samples for various exciting energies with and without
scattering and indicate that errors as large as 2% can be attributed
to this effect.

INTRODUCTION

The neglect of scattering in the fundamental parameters approach
to EDXRF analysis is a recognized source of error. Until now no
rigorous mathematical treatment for scattering in multicomponent
systems has been available. Keith and Loomis (2) have developed
analytic expressions for coherent scattering corrections in pure
element samples, however these expressions are only approximations
for modifying attenuation coefficients and do not characterize the
actual scattering processes.

The simulation of scattering events involving photon energies
in the range of fluorescence X rays requires considering the influence
of orbital electron binding energies. This effect is dependent upon
the atomic number of the constituent elements and, therefore, the

composition of the sample. It becomes significant in determining
incoherent energy transfers and is particularly important in calculating coherent and incoherent scattering cross sections. These
bound electron effects are incorporated in the present model.

In this paper, the Monte Carlo simulation is used to examine
the scattering contribution to the calculated intensities and
intensity ratios for several Ni-Fe-Cr ternary samples for various
exciting source energies.

SCATTERING MODELS

The system to be modeled is an annular radioisotope source
mounted coaxially with a circular detector as shown in Figure 1.
Photons interacting in the sample may undergo either a photoelectric
or scattering interaction. Scattering events are characterized by
two angles, a scattering angle θ and rotational angle ϕ. A photon
traveling in a direction given by the vector $\bar{u}$ undergoes a scatter
at point P and leaves the interaction in the direction $\bar{v}$ (see Figure
2) given explicitly by the vector $\bar{u}$ and the angles θ and ϕ. Scattering angles are determined by examining the probability of scattering
into a differential solid angle $d\Omega$ (see Figure 3). These probabilities are expressed in terms of cross sections according to the type
of scattering involved. Scattering events are classified as either:
(1) Rayleigh or Coherent scattering where only the photons direction
is changed and (2) Compton or Incoherent scattering where both the
direction and energy of the photon are changed.

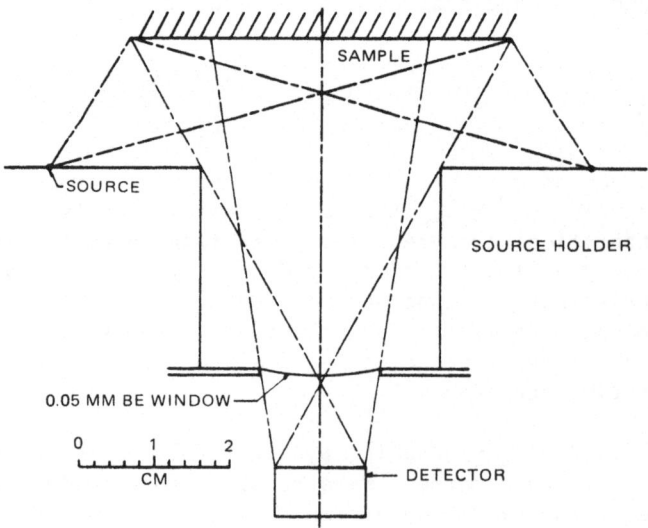

Figure 1.   Drawing of radioisotope annular source X-ray analyzer
            system.

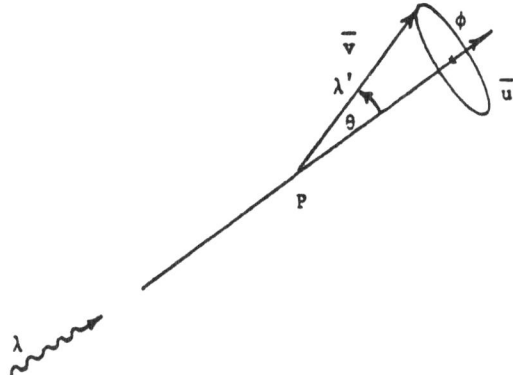

Fig. 2.   Scattering vectors

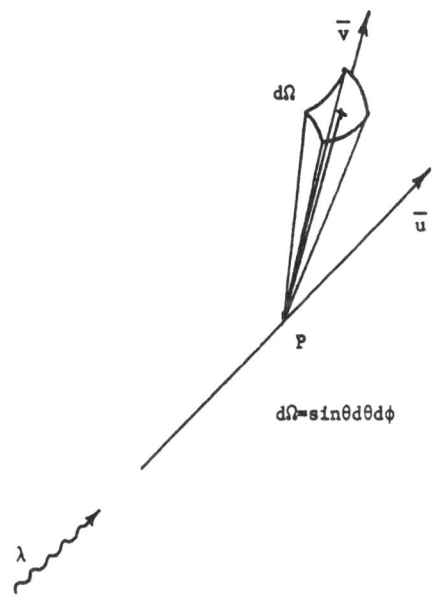

Fig. 3.   Differential solid angle for scattering

For Coherent scattering, the probability of scattering into a differential solid angle $d\Omega$ is given by the expression

$$d\sigma_{Coh}/d\Omega = (r_o^2/2)(1 + \cos^2\theta)[F(q,z)]^2 \qquad (1)$$

where $d\sigma_{Coh}$ is the differential coherent scattering cross section, r_o is the classical electron radius, θ is the scattering angle, and $F(q,z)$ is the atomic form factor. The atomic form factor represents the probability that momentum q is transferred to the z electrons of the element with no energy absorption. For incoherent scattering the probability of scattering into a differential solid angle $d\Omega$ is given by

$$d\sigma_{Inc}/d\Omega = (r_o^2/2)[1 + \alpha(1 - \cos\theta)]^{-2} \left[1 + \cos^2\theta + \alpha^2(1 - \cos\theta)^2/\right.$$

$$\left.[1 + \alpha(1 - \cos\theta)]\right]S(q,z) \tag{2}$$

where $d\sigma_{Inc}$ is the differential incoherent scattering cross section, α is the ratio of the photon energy to the equivalent rest mass energy of an electron, $S(q,z)$ is the incoherent scattering function, and r_o and θ are as previously defined. The incoherent scattering function represents the probability for an atom to be raised to an excited or ionized state when the photon imparts a recoil momentum q to any of the z atomic electrons. Both the incoherent scattering function and the atomic form factor have been calculated and tabulated by several authors (3,4).

In this simulation, the scattering angle at any given scattering event is obtained from the expression

$$\xi = \int_0^\theta (d\sigma/d\Omega)\sin\theta' d\theta'/\int_0^{\pi/2}(d\sigma/d\Omega)\sin\theta' d\theta' \tag{3}$$

where ξ is a rectangularly distributed random number between 0 and 1, θ is the unknown scattering angle, and $d\sigma/d\Omega$ is either the Compton or Rayleigh differential scattering cross section. The scattering angle would be determined ideally by carrying out the integrations and inverting the resulting expression to find an analytic relationship for θ in terms of ξ. As both the Rayleigh and Compton formulas involve the tabulated functions $F(q,z)$ and $S(q,z)$, an analytic solution is impossible. The integrals are therefore evaluated numerically at discrete energies and values of θ. The scattering angle is then obtained by a second order difference interpretation scheme for any randomly selected ξ. The rotational angle is selected uniformly between $-\pi$ and π by the relation

$$\phi = \pi(2\xi - 1) \tag{4}$$

where ξ is again rectangularly distributed.

The change in wavelength of an incoherently scattered photon is given by the expression

$$\lambda - \lambda_o = (h/m_o c)(1 - \cos\theta) + p\lambda*/m_o c \tag{5}$$

$$\lambda* = 2\lambda_o \sin\theta/2$$

where λ_o is the initial wavelength, λ is the wavelength after scatter, h is Planck's constant, m_o is the electron rest mass, c is the speed of light, θ is the scattering angle, and p is the projection of the initial electron momentum on the scattering vector (5). The scattered photon energy is therefore

$$E = E_o[1 + (E_o/m_oc^2)(1 - \cos\theta) + (2p/m_oc)\sin\theta/2]^{-1} \qquad (6)$$

where E_o is the initial photon energy. The electron momentum p is obtained by sampling from the electron momentum distributions J(p) of the constituent elements according to the relation

$$\xi = \int_{-\infty}^{p} J(p')dp'/\int_{-\infty}^{\infty} J(p')dp' \qquad (7)$$

where ξ is again a rectangularly distributed random number. Electron momentum distributions have been calculated and tabulated from Hartree-Fock wave functions (6). The integrals are easily evaluated numerically and p is obtained by interpolation.

RESULTS

The scattering contribution was determined for three Ni-Fe-Cr ternarys with four common exciting sources. The scattering contributions are reported in Table 1 in terms of percent relative change from total intensity without scattering to total intensity with scattering. The results under the heading Scattered I reflect only contributions from source scatter. The results under the heading Scattered II reflect contributions from both source scatter and the scatter of primary fluorescence photons. For the chosen samples the maximum scattering contribution was on the order of 2%. These results, however, are dependent upon such factors as system geometry and sample thickness. No attempt was made to optimize this effect.

CONCLUSIONS

A Monte Carlo program has been developed and demonstrated for examining the scattering effect in multicomponent systems. This effect can be significant and the simulation capability developed is necessary for further development of the complete spectral response approach to analysis. The computer program called XSCAT is available from the authors.

TABLE 1.   Relative Percentage Differences in Calculated Intensity Ratios

Source*	5074 Ternary (%)[a]						Hybrid Ternary (%)[b]						Inconel (%)[c]					
	Scattered I			Scattered II			Scattered I			Scattered II			Scattered I			Scattered II		
	Ni	Fe	Cr	Ni	Fe	Cr	Ni	Fe	Cr	Ni	Fe	Cr	Ni	Fe	Cr	Ni	Fe	Cr
Zn Xrays	0.37	0.41	0.47	0.37	0.42	0.70	0.36	0.41	0.46	0.36	0.45	0.68	0.42	0.44	0.44	0.42	0.81	0.87
Cd-109	0.78	1.01	1.24	0.78	1.02	1.61	0.83	1.00	1.20	0.83	1.05	1.60	1.16	1.14	1.19	1.16	1.77	1.90
Am-241	0.65	0.80	0.99	0.65	0.81	1.31	0.65	0.78	0.96	0.65	0.83	1.26	0.90	0.84	0.91	0.90	1.37	1.49
Gd-153	1.15	1.34	1.65	1.15	1.35	2.04	1.15	1.33	1.59	1.15	1.38	1.96	1.58	1.54	1.60	1.58	2.23	2.38

a   The composition of this sample is 4.98, 68.38, and 25.25% of Ni, Fe, and Cr, respectively.
b   The composition of this sample is 20.5, 55, and 24.5% of Ni, Fe, and Cr, respectively.
c   The composition of this sample is 79, 6.5, and 14.5% of Ni, Fe, and Cr, respectively.

*   The source energies for Zn Xrays, Cd-109, Am-241, and Gd-153 are, respectively:
   (1) 8.63 and 9.57 keV; (2) 22,25, and 88 keV; (3) 14-21, 26.4, and 59.6 keV;
   and (4) 41, 97, and 103 keV.

APPENDIX

Mathematical Treatment of Photon Scattering

A photon with direction cosines α, β, γ (see Figure A-1) travels along a vector $\bar{u} = \alpha\hat{i} + \beta\hat{j} + \gamma\hat{k}$. If the photon scatters at point X_o, Y_o, Z_o with scattering angle θ and rotational angle ϕ relative to $\bar{u}$, the new direction of travel is along a vector $\bar{v} = \alpha'\hat{i} + \beta'\hat{j} + \gamma'\hat{k}$ where α', β', γ' are the new direction cosines (7) given by

$$\alpha' = \alpha\cos\theta + \gamma\alpha\frac{\sin\theta\cos\phi}{\sqrt{1-\gamma^2}} - \beta\frac{\sin\theta\sin\phi}{\sqrt{1-\gamma^2}} \qquad (A-1)$$

$$\beta' = \beta\cos\theta + \gamma\beta\frac{\sin\theta\cos\phi}{\sqrt{1-^2}} + \alpha\frac{\sin\theta\sin\phi}{\sqrt{1-^2}} \qquad (A-2)$$

$$\gamma' = \gamma\cos\theta - \sqrt{1-\gamma^2}\ \sin\theta\cos\phi \qquad (A-3)$$

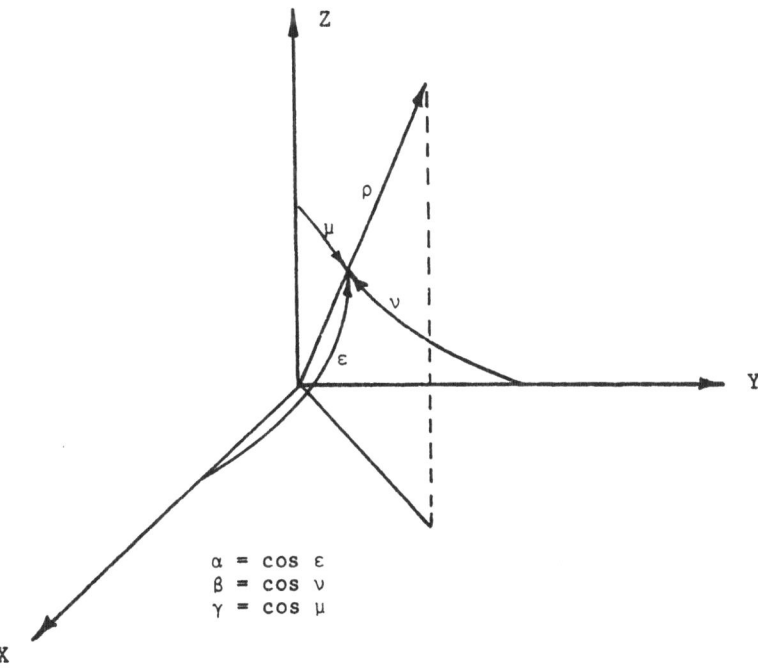

$$\alpha = \cos\ \epsilon$$
$$\beta = \cos\ \nu$$
$$\gamma = \cos\ \mu$$

Figure A-1. Direction angles and direction cosines

If the photon travels a distance ρ to the next interaction point, that point is given by

$$X = Xo + \alpha'\rho \qquad\qquad\qquad\qquad\qquad\qquad (A\text{-}4)$$

$$Y = Yo + \beta'\rho \qquad\qquad\qquad\qquad\qquad\qquad (A\text{-}5)$$

$$Z = Zo + \gamma'\rho \qquad\qquad\qquad\qquad\qquad\qquad (A\text{-}6)$$

and the vector $\bar{v}$ may be expressed as

$$\bar{v} = \frac{X-Xo}{\rho}\,\hat{i} + \frac{Y-Yo}{\rho}\,\hat{j} + \frac{Z-Zo}{\rho}\,\hat{k}. \qquad\qquad (A\text{-}7)$$

REFERENCES

1. Robin P. Gardner and Alan R. Hawthorne, X-Ray Spectrometry, Vol. 4, 138-148, (1975).
2. H. D. Keith and T. C. Loomis, X-Ray Spectrometry, Vol. 7, No. 4, 225-240, (1978).
3. J. H. Hubbell, Wm. J. Veigele, E. A. Briggs, R. T. Brown, D. T. Cromer, R. J. Howerton, J. Phys. Chem. Ref. Data, Vol. 4, No. 3, 471-538, (1975).
4. J. H. Hubbell and I. Overbó, J. Phys. Chem. Ref. Data, Vol. 8, No. 1, 69-105, (1979).
5. Brian Williams, Compton Scattering, P. 105-106, McGraw-Hill (1977).
6. F. Biggs, L. B. Mendelsohn, J. B. Mann, Atomic Data and Nuclear Data Tables, 16, 201-309, (1975).
7. N. M. Schaeffer, Reactor Shielding for Nuclear Engineers, p. 241, TID-25951, (1973).

THE APPLICATION OF DIGITAL FILTERS TO
THE ANALYSIS OF GE AND SI(LI) DETECTOR
X-RAY SPECTRA

L. A. Rayburn

Department of Physics
The University of Texas at Arlington
Arlington, Texas 76019

ABSTRACT

One of the uncertain aspects in the analysis of x-ray spectra
is the determination of the proper background to subtract from the
raw data. In those cases where the background is a smoothly varying
function of the x-ray energy, the application of a digital filter
to the raw data will effectively remove the background leaving only
the filtered peak information. These filtered peaks can then be
fit by using a non-linear least squares method in conjunction with
a suitably chosen mathematical model of the peak structure.

INTRODUCTION

A typical Dy L x-ray spectrum resulting from 0.50 MeV protons
incident on a thin target ($\sim 20 \mu g/cm^2$) of DyF_3 is shown in Fig. 1.
The DyF_3 was vacuum deposited on a $20 \mu g/cm^2$ carbon film. The
detector used had an energy resolution of approximately 195 eV FWHM
at 5.9 keV.

We attempted to computer analyze spectra such as this by using
a program that fit the spectra to Gaussian peaks with high- and low-
energy exponential tails and with provision for selecting a linear
background function or a background function that included a quadratic term (our program forced the background maximum to occur at
the midpoint of the interval being analyzed in this latter case).
We were primarily concerned with the determination of the area under
each of the six L peaks. It is evident that the areas under the
L_ℓ and $L_{\gamma_{2,3}}$ peaks, in particular, will depend quite drastically on

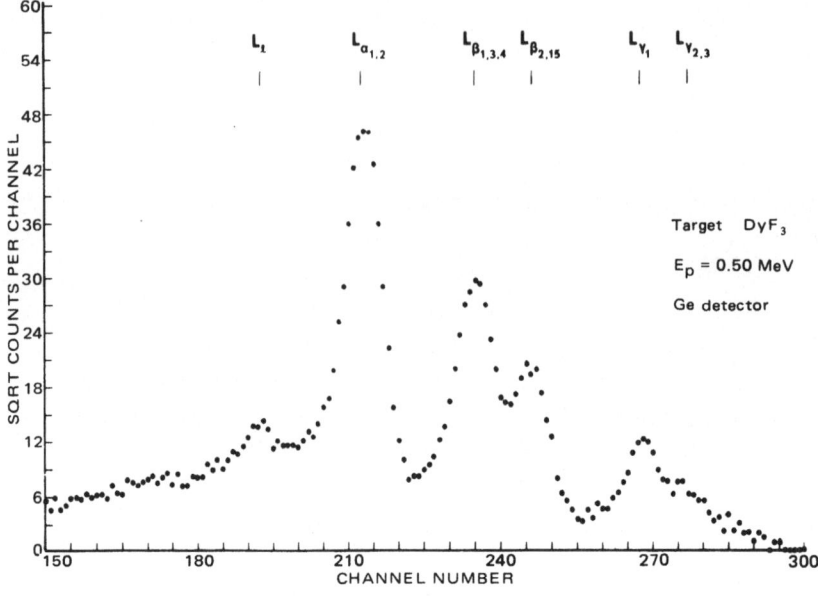

Fig. 1   Dy L x-ray spectrum.

the particular background function that is used. The results that
we obtained from our computer analysis program led us to believe
that a linear or quadratic background function did not correspond
to the background that was present in these spectra. So we decided
to investigate methods of removing the continuum and then fitting
the remainder of the spectrum.

The most promising method appeared to be the application of a
digital filter to the raw data. The filter would strongly attenuate
the slowly varying part of the spectrum while exhibiting a strong
response to any peaks that were present. The method that we finally settled on is based on the work of Schamber (1).

FILTER FUNCTION

Consider a simple rectangular filter that consists of the average of UW channels ($\frac{(UW-1)}{2}$ channels on each side of channel i)

minus the average of 2x(LW) channels (LW channels on each side of
the UW region). The filtered spectrum data points would be given
by:

$$Y_i' = \left(\frac{1}{UW}\right)\sum_{j=-k}^{k} Y_{i+j} - \left[\frac{1}{2(LW)}\right] \sum_{j=1}^{LW} \{Y_{i+j+k} + Y_{i-j-k}\}$$

where k = (UW-1)/2

We found that the best value of UW was approximately equal to the FWHM of our x-ray peaks (7 channels in the present case) and with LW=4 channels.

We applied this filter to various test spectra consisting of Gaussian peaks with low-energy exponential tails (this corresponded to our x-ray peak structure) superimposed on several different background functions (linear, quadratic and broad Gaussian). In each case the background was suppressed but there was a strong response to each of the peaks.

MATHEMATICAL MODEL AND PEAK FITTING

The mathematical model that we used for peak fitting requires the use of four parameters for each peak; these are,

P_j=number of counts at maximum of peak j

PC_j=channel number of centroid of peak j

W_j=width parameter for peak j

L_j=low energy exponential tail parameter for peak j.

The number of counts in channel i due to peak j is then given by:

$$f_{ij} = P_j \; Exp(-(i-PC_j)^2/2W_j^2) \qquad\qquad for \; i \geq (PC_j-L_j)$$

$$f_{ij} = P_j \; Exp(-L_j(2PC_j-2i-L_j)/2W_j^2) \quad for \; i < (PC_j-L_j)$$

Our fit procedure then consisted of the application of the digital filter to an experimental spectrum (Y_i) from which we obtained an experimental filtered spectrum (Y_i'). Then we used initial estimates for P_j, PC_j, W_j and L_j for each peak and obtained a mathematical model filtered spectrum (f_i'). A minimization program (2) was then used to find the minimum of the function:

$$F = \sum_i [Y_i' - f_i']^2/(\sigma_i')^2$$

where

$$(\sigma_i')^2 = Variance \; of \; Y_i' = \sum_m \left[\frac{\partial Y_i}{\partial Y_m}\right]^2 \sigma_m^2$$

The program varies each of the parameters P_j, PC_j, W_j and L_j and finds a minimum F. Values of the parameters are returned for each peak. These values are then used to calculate the area under each peak.

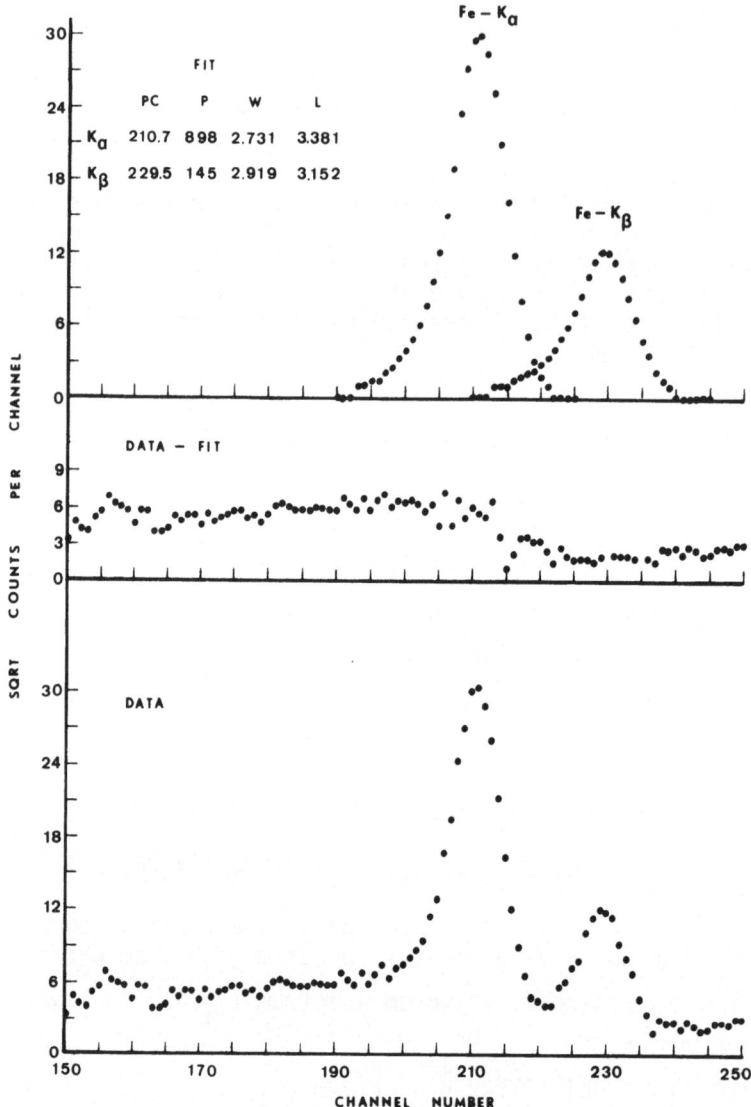

Fig. 2   Results of fit procedure applied to x-ray spectrum from
          ^{57}Co source.

        We applied the fit procedure to selected calibration spectra
which covered the energy region of interest and obtained values of
W_j and L_j for a number of values of PC_j.   The results of the application of this procedure to a ^{57}Co source spectrum is shown in
Fig. 2. The raw data is shown at the bottom of the figure, the
fitted Fe K_α and K_β peaks at the top, and the (Data-Fit) is shown
in the center. Values of PC_j, W_j and L_j determined from several

calibration spectra were used to obtain equations from which values of W_j and L_j could be calculated as a function of PC_j. These equations were used in the computer analysis of the L x-ray spectra. The same minimization procedure was used except that only two parameters, P_j and PC_j, were determined for each peak. We designate this as our modified fit procedure. This reduced CPU time considerably for the L spectra.

RESULTS FOR L X-RAY SPECTRA

The results of the application of our modified fit procedure to the Dy L x-ray spectrum shown in Fig. 1 is shown in Fig. 3. The spectrum shown in the center of Fig. 3 is a plot of the six Dy L x-ray peaks where values of PC_j and P_j determined by our fit procedure have been used. The (Data-Fit) is shown at the top of the figure. This more closely approximates a broad Gaussian than a linear or quadratic background. In each of these cases the square root of the number of counts per channel has been plotted versus channel number. At the bottom of the figure the deviation of the experimental filtered counts per channel is plotted versus channel number. Most of the points fall within ±2 standard deviations.

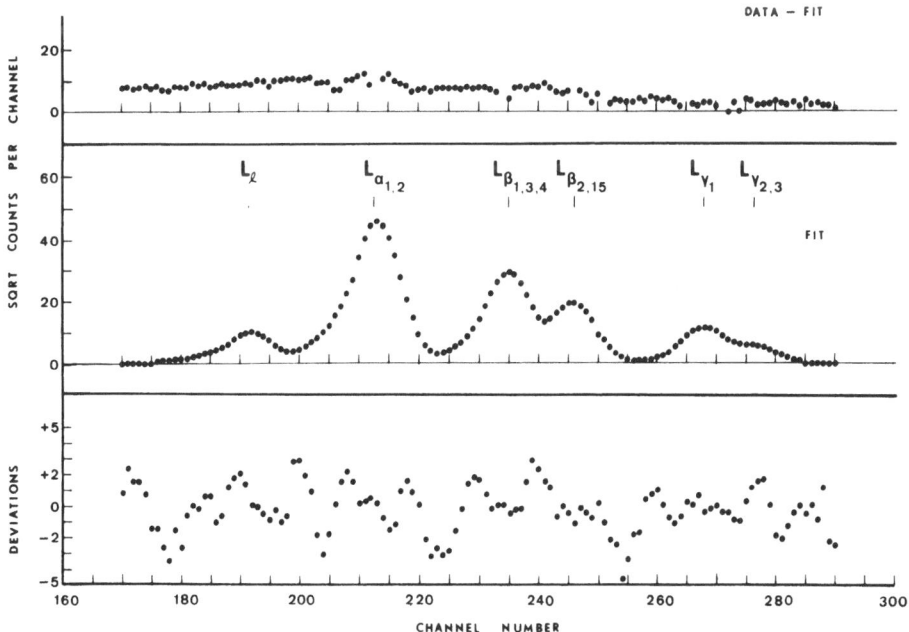

Fig. 3   Results of modified fit procedure applied to Dy L x-ray spectrum shown in Fig. 1.

The values of PC_j, P_j, $Area_j$ and $Energy_j$ determined by our modified fit procedure are shown in Table 1. Also shown for comparison purposes are the energies obtained from ASTM-DS46 (3). The maximum difference of only 16 eV between any experimentally determined energy and the DS46 value indicates that our program was locating the peak centroids accurately. Meaningful comparisons of the relative peak intensities with values given in ASTM-DS46 could not be made since these values are dependent upon the energy and type of ionizing radiation.

Similar results were obtained for L spectra at other incident proton energies and in those cases where a Si(Li) detector was used.

CONCLUSIONS

The filter method does remove slowly varying (with energy) components of experimental L x-ray spectra while preserving the peak information. This makes it possible for the experimental filtered spectra to be fit using suitable mathematical model filtered spectra. It is not necessary to include terms describing the continuum in the mathematical model.

This method does require considerable CPU time. For example, the CPU time on an IBM-370/155 was approximately twelve minutes for the Dy L x-ray spectrum shown in this work. However, we found that the method used here was very valuable in identifying the kind of continuum that was present in our L spectra. We then used a mathematical model that included continuum terms and fit the experimental spectra in the usual non-linear least squares way. This reduced CPU time to approximately 2 to 3 minutes per spectrum.

Table 1. Modified Fit Results for Dy

J	PC_j	P_j	Area	Energy-keV	(ASTM-DS46)
1	192.4	78.2	559	5.752	5.742
2	213.0	2059	14564	6.476	6.491
3	235.0	840	5954	7.250	7.255
4	245.5	383	2723	7.618	7.634
5	268.0	137	984	8.411	8.417
6	276.8	42.1	306	8.722	8.733

REFERENCES

1. F. H. Schamber, A Modification of the Linear Least-Squares
    Fitting Method Which Provides Continuum Suppression, in "X-Ray
    Fluorescence Analysis of Environmental Samples," T. G. Dzubay,
    ed., Ann Arbor Science, Ann Arbor Mich. (1977).

2. Program FMFP, IBM System/360 Scientific Subroutine Package,
    Version III, Programmer's Manual, IBM Corporation, White
    Plains, N.Y.

3. X-Ray Emission Wavelengths and KeV Tables for Nondiffractive
    Analysis, American Society for Testing and Materials Data
    Series DS46.

X-RAY SPECTROMETRIC DETERMINATION OF SULFATE IN NATURAL WATERS

C. A. Seils and G. T. Tisue

Argonne National Laboratory

Argonne, Illinois 60439

A recent surge of interest in sulfur in the environment has revealed the need for improved methods of analysis for sulfate, $SO_4^=$, in rain, freshwater and sediment interstitial fluids. Ion chromatography permits the rapid determination of $SO_4^=$ in the ppm range (1 ppm = 1 mgL^{-1} $\cong$ 10 μmol L^{-1} sulfate) on relatively small samples with good specificity. If a suitable instrument is available, this technique is a good choice for many environmental analyses. Other approaches to sulfate analysis are based on its precipitation with organic or heavy metal cations, usually barium or lead. The amount of precipitate formed may be determined by inter alia gravimetry, turbidimetry, radiometry (using 133-Ba), atomic absorption spectrophotometry (Ba or Pb detection), potentiometry (using a Pb^{++} ion selective electrode), colorimetry, or by x-ray fluorescence spectrometry (Ba, Pb or S detection)(1). Because of our experience with x-ray fluorescence analyses, we chose to develop and test a procedure using that technique.

It was our strategy to determine sulfate by precipitating it as $BaSO_4$, then exciting and detecting the K x-rays of Ba using a Ho secondary source in a non-dispersive x-ray spectrometer. This approach demands that the atom ratio between Ba and S in the precipitate be close to unity, or at least be the same in samples and standards. One impediment to meeting this requirement is the reported difficulty of reproducibly washing excess (i.e., unreacted) Ba from the precipitate. We chose an indirect determination, one

based on heavy metal cation detection, rather than a direct one
based on S K x-rays. This choice rested on knowledge that Pb L-
or Ba K-x-rays could be detected more sensitively and with fewer
spectral interferences than the K x-rays of sulfur. Similar considerations suggested the choice of Ba K- over Pb L-x-rays.

In the cases of rain, snowmelt, or lakewater, one is seldom
limited by sample size. In other instances, here exemplified by
sediment interstitial fluids, only small sample volumes are available, and some modifications of technique become necessary. These
two circumstances are treated separately. It is demonstrable that
the presence of cations other than H^+ or the alkali metals may
interfere with the determination by acting to hold back sulfate in
solution, or by coprecipitating with $BaSO_4$, thereby altering the
stoichiometric relationship between Ba and S on which the indirect
method depends.

For these situations, we describe a pretreatment of the sample
with chelating resin. (The rain samples we analyzed did not
require this conditioning.)

PROCEDURE FOR SAMPLES AND STANDARDS UP TO 0.1 L AND 15 μmol $SO_4^=$

This procedure is applicable to sample volumes of 100 mL containing up to 15 μmol $SO_4^=$. For more concentrated samples an aliquot
portion containing $\leq$15 μmol $SO_4^=$ should be diluted to 100 mL with
10 mM HCl. Reagents: 10 mM HCl; 0.5 M $BaCl_2$; 0.1% by weight
"Tween-20" surfactant (Fisher Scientific Co.); 10 mM HCl/10%
methanol by volume; and 6 M HCl. Note: All reagents are prepared
with Barnstead "Nanopure" deionized water, or equivalent, passed
through a 0.22 μm final filter. Likewise, all samples are filtered
through a 0.4 μm membrane prior to further treatment.

A 100 mL aliquot portion of the filtered sample is pipetted
into a 250 mL beaker. The pH is then adjusted to about 2 by the
dropwise addition of 6 M HCl to eliminate $CO_3^=$ interference; pH
paper is a suitable indicator. While stirring gently with a teflon-
coated magnet, one then adds 5 mL of 0.1% Tween-20, and 1 mL of
0.5 M $BaCl_2$. The resulting mixture is stirred at 80°C for one hour,
then cooled and allowed to stand overnight under a watch glass. We
found it convenient to dislodge the precipitate from the beakers'
walls by ultrasonication for 2–5 min, and then to suspend it by
magnetic stirring. The precipitate is collected by passing the
resulting slurry through a 25 mm diameter, 0.4μm Nuclepore filter

contained in a 5/8" ID glass Millipore vacuum filtration holder.
To provide reliable sealing around the inner circumference of the
funnel's base, one may place a 5 μm Millipore membrane behind the
Nuclepore filter. The beaker and filtration funnel are washed with
several small portions of 0.01 M HCl/10% methanol to complete the
transfer and to remove excess Ba^{+2}. The collected precipitate is
dried in clean air and mounted between 0.5 mil Mylar film in a
2 x 2" cardboard slide holder.

If pretreatment to remove interfering cations is indicated by
preliminary tests, the following technique is applicable. Additional
reagent: AG 50W-X8 cation exchange resin, 100-200 mesh, hydrogen
form (BioRad Laboratories), or equivalent, washed first with 6 M
HCl, then with 10 mM HCl. Note: Our experience has been that the
resin absorbs small amounts of $SO_4^=$ irreversibly. Samples containing ≤3 μmol $SO_4^=$ gave low and erratic recoveries. This difficulty
may be overcome in the main by pre-equilibrating the resin with a
small amount of the sample being analyzed (see Figure 1). However,
in ultratrace analyses using the small spot geometry it is desirable
to substitute Chelex-100 in the H+ form, since that material does
not appear to absorb $SO_4^=$ vide infra.

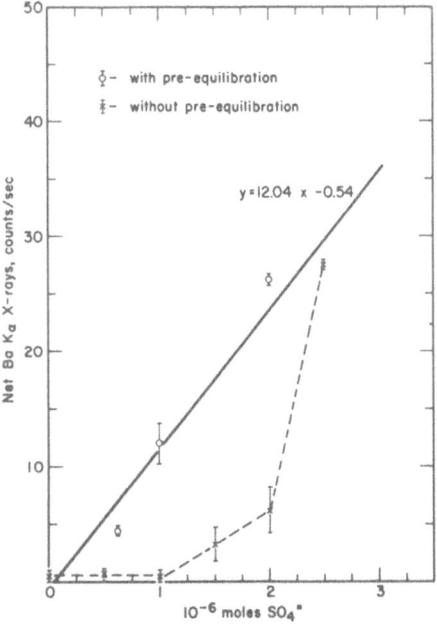

Figure 1. Ion exchange treatment: effect of pre-equilibrating
           resin with sample to be analyzed.

A 47 mm, 0.4 μm Nuclepore membrane is mounted under a vacuum filtration funnel of >100 mL capacity then covered for 5 min with ca. 1 g of prewashed cation exchange resin and ∿25 mL of the sample. Vacuum is applied and the resin drawn dry <u>without washing</u>. After changing the receiver, one adds the sample (about 100 mL) and allows 5-10 min for equilibration with the resin while stirring occasionally. The sample filtrate is then drawn off by vacuum without washing; the used resin is discarded. 100 mL of the filtrate is pipetted into a 250 mL beaker. The $BaSO_4$ precipitate procedure described earlier is followed in subsequent steps.

PROCEDURE FOR SAMPLES UP TO 10mL CONTAINING LESS THAN 1 μmol $SO_4^=$

This procedure is applicable to sample volumes up to ca. 10 mL containing ≦1 μmol $SO_4^=$. The reagents are the same as were described for the large spot procedure.

A portion of the filtered sample is pipetted into a 50mL centrifuge tube and adjusted to a volume of ca. 15 mL with 0.1 M HCl. To it is added ca. 0.1 g of Chelex-100 resin, 100-200 mesh, in the H^+ form. (This reagent may be omitted in the absence of interfering cations.) Resin and sample are allowed to react for ∿ 30 min with occasional stirring, then filtered through a 25 mm diameter, 0.4 μm Nuclepore membrane backed by a 25 mm diameter, 5 μm Millipore membrane. It is convenient to collect the filtrate in a 50 mL centrifuge tube. The collected resin should be washed with amounts of 0.1 M HCl sufficiently small that the combined volume of sample and wash liquid is ≦25 mL.

To the filtrate one then adds 1 mL of 0.1% of 0.5 M $BaCl_2$. The resulting precipitate is digested for 1 hour at $80°$ in a water bath, covered, and allowed to stand overnight. Prior to filtration, it is desirable to dislodge the precipitate by ultrasonication for 2-5 min. It may then be collected on a 25 mm diameter, 0.4 μm Nuclepore membrane backed by a 5 μm Millipore membrane. To confine the precipitate to an area smaller than the x-ray beam in our spectrometer, we employ a 1/4" diameter filter holder and funnel of our own design (see Figure 2). To obtain leak free operations, the upper and lower portions are clamped together tightly by means of a screw clamp; the usual spring-loaded clamps do not prevent loss of precipitate around the gasket seals.

Transfer and washing of the precipitate are completed by washing the centrifuge tube and filter with several small portions of

10^{-2} M HCl containing 10% methanol by volume. After drying in clean air, the filter containing the precipitate is mounted for x-ray fluorescence analysis as before.

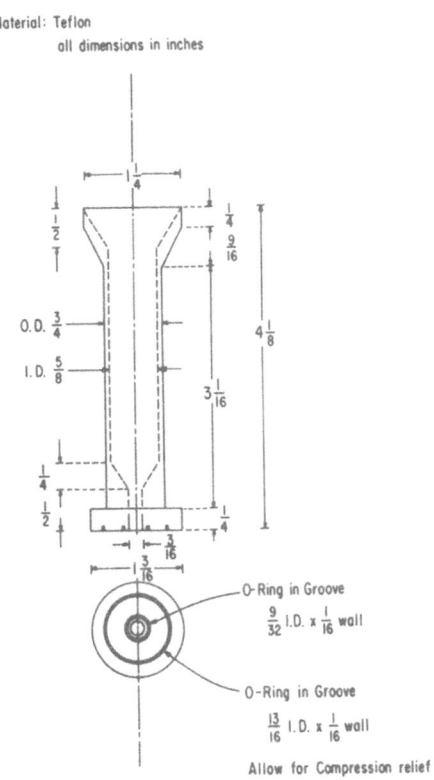

Figure 2. Filter funnel for small spot geometry.

X-RAY FLUORESCENCE ANALYSIS

The spectrometer used for these analyses has been described elsewhere (2). Ba K x-rays are fluoresced in the mounted precipitate with radiation from a Ho foil excited by a W-anode x-ray tube operated at 70 KVP (full-wave rectification) and 5-20 mA. Copper collimators define the optical path. 20 mils of high purity aluminum between the sample and Si(Li) detector serves to absorb the large amount of scattered low energy radiation that would otherwise flood the amplifier. The incompletely resolved Ba $K\alpha_1$-$K\alpha_2$ doublet appears in an otherwise featureless and nearly flat region of the spectrum.

Counts in defined regions lying to either side of this doublet are summed and averaged to provide an estimate of the background in the region containing the Ba Kα net count rate (peak integral minus background integral, per unit time). The usual counting interval was 300 sec live time, or about 350 sec true time. For very low amounts, some increase in precision is realizable in practice by doubling or quadrupling the counting time.

MEASURES OF PERFORMANCE

The average relative error for 29 pairs of replicates analyzed using the large spot geometry was 6% (range 0.8–19.6%). In general, the larger errors are associated with the lowest levels of $SO_4^=$, <5 μmol. Most of this error is probably attributable to mechanical difficulties associated with collecting and handling the small amounts of $BaSO_4$ precipitate – 15 μmol $BaSO_4$ weighs less than 4 mg. Any given sample may be recounted in the x-ray spectrometer with a precision approaching 1% relative error.

Useful analyses are possible down to about 1 μmol of $SO_4^=$ using the large sample procedure. Some sensitivity is lost because the 5/8" diameter circle in which the $BaSO_4$ is confined is larger then the beam of exciting radiation. However, use of small diameter filters often leads to unacceptably slow filtration rates. We found a 5/8" spot size to be a workable compromise for 100 mL samples. We define the minimum detection limit (MDL) as the amount of analyte that will produce a signal equal to or exceeding two times the standard deviation of the background. This definition may be expressed as

$$\text{MDL in μmol} = \frac{2\sigma \text{ (background) in counts/sec}}{\text{sensitivity in counts/sec/μmol}}$$

In a larger number of trials, we found 2σ (background) to average 6.2 counts/sec, while the sensitivity averaged 13.2 counts/sec/μmol. These values lead to a calculated MDL ≅ 0.4 μmol corresponding to a concentration of 4 μM in a 100 mL sample.

Sensitivity is improved with the smaller spot geometry, probably because the precipitate is confined to an area smaller than the effective beam size. This improvement in sensitivity lowers the calculated MDL to 0.08 μmol, which corresponds to a concentration of 8 μM in a 10 mL sample.

Plots of net Ba Kα count rate against the amount of $SO_4^=$ added were linear for aqueous standards (Figures 3 and 4), for rain water (Figure 5) and for standard additions to lake water and sediment interstitial water (Figure 6). (The method of standard additions failed to reveal any serious interferences in the rain samples, but the interstitial water samples required Chelex-100 pretreatment to remove interfering cations.) The Ba Kα count rates doubtlessly would be linear over even a wider range of concentrations, were it possible to collect amounts of $BaSO_4$ greater than ~20 μmol (large spot geometry) or ~2 μmol (small spot geometry) without a high risk of mechanical loss during handling.

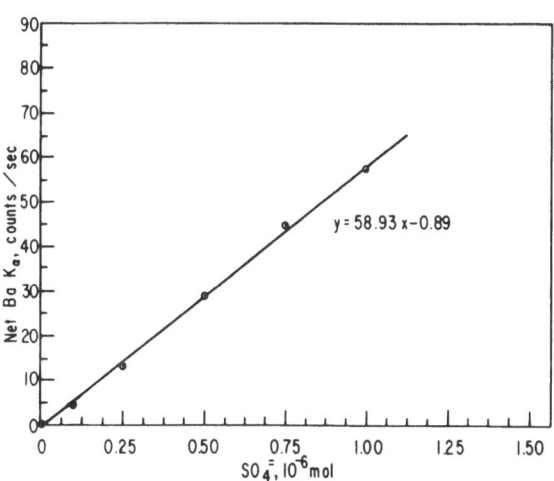

Figure 4. Sulfate in aqueous standards, small spot geometry.

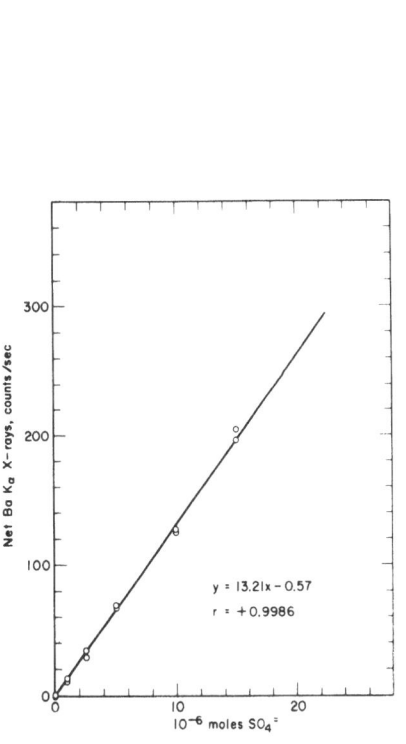

Figure 3. Sulfate in aqueous standards, large spot geometry.

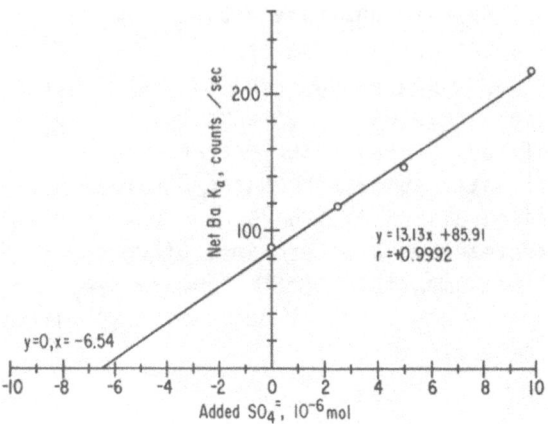

Figure 5.   Standard additions plot, rain samples.

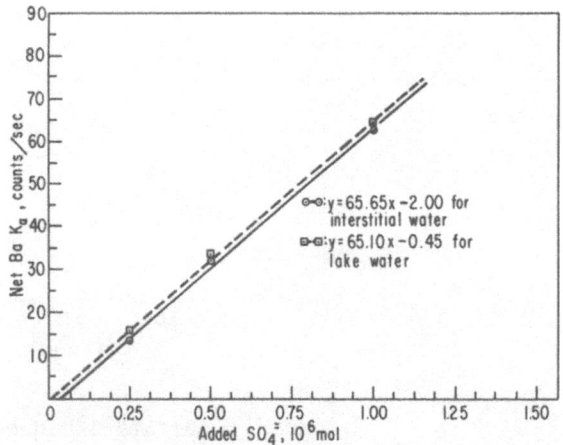

Figure 6.   Standard additions plots for sediment interstitial fluid
and lake water.

At present our only assessment of the accuracy of the method
is the results it yielded on a set of intercomparison samples
circulated by organizers of the MAP3S program. Table 1 compares our
values for $SO_4^=$ in these samples with the means of the analyses
performed by the 23 laboratories who were formal participants in the
intercomparison study. In no case did our results differ in a
statistically significant amount from the reported means.

Table 1. Analysis of MAP3S intercomparison samples.

Sample	Mean (μM)	Our Value (μM)	t
A	118.8±11.06	117.3±2.1	-0.19
B	79.6±5.40	74.9±2.0	-1.20
C	77.5±5.29	77.8±1.1	+0.05
D	94.1±7.09	89.9±0.4	-0.83
E	65.6±3.84	60.1±2.0	-1.94
F	0.36±0.43	0.39±0.0	
G	83.3±8.12	90.3	
H	2.37±2.10	0.0	

It is our conclusion that the method described here has the precision, sensitivity, specificity, and accuracy to be employed routinely on the sample types with which it has been tested.

ACKNOWLEDGEMENTS

We are grateful to Stan Mortimer for his assistance with sample collection chores and in the early stages of method development.

REFERENCES

1. M. J. Fishman and D. E. Erdman, "Water Analysis," Anal. Chem. 49, 139R—158R (1977).

2. R. P. Larsen and G. T. Tisue, "A Secondary-Source Energy-Dispersive X-Ray Spectrometer," Radiological and Environmental Research Division Annual Report, January—December 1975, ANL-75-60, Part III, pp. 51—54.

APPLICATION OF THE PIXE METHOD IN ATMOSPHERIC AEROSOL

INVESTIGATIONS

C. Q. Orsini and L. C. Boueres*

Instituto de Física, Universidade de São Paulo
C.P. 20516, São Paulo, SP, Brazil

INTRODUCTION

Investigations of Atmospheric Aerosol Systems (let us abbreviate AAS) have now spread to many laboratories around the world, mainly because of the important role they assume in the air pollution problem.

In most cases, the studies on AAS attempt to evaluate the so-called "size-distribution function", through which the general behavior of such a system is commonly described.

Basic continuous size-distribution functions represented by $n(v, r, t)$ and $n_D(D, r, t)$ are usually defined from:

$$dN = n(v, r, t)dv$$

and

$$dN = n_D(D, r, t)dD$$

where dN is the number of particles per unit volume of air at a given position r in space and at a given time t , respectively in the particle size volume range v to v + dv and diameter range from D to D + dD (in our case, aerodynamic diameter).

Once the size-distribution is known, it can be used in a general

---

*Present address: Department of Oceanography, Florida State University, Tallahassee, FL 32306.

dynamics equation, which was derived taking into account the physical processes involved with the AAS, for instance, diffusion transport and coagulation[1], as the starting point for predictive modeling of air pollution.

A still more general description of an AAS could be obtained, in principle, if we define a generalized size-composition probability density function where all the chemical species components of the aerosol particle are included[1].

SIZE-DISTRIBUTION AND PIXE

Nowadays a great number of investigations on the AAS's make use of the PIXE method (or any other X-ray analytical method) to measure elemental concentrations in air particulate matter. Particularly, many of them use what we call the Cascade Impactor + PIXE method (which we denote by CI + PIXE) in their measurements. This consists of sampling the AAS by means of cascade impactors (of 6 stages in our case) and in the analysis of the collected samples by the PIXE method.

Usually, the resulting data from the CI + PIXE method are presented as "elemental size-distribution curves". They consist of the measured set of values $\{m_{K,Z}\}$, where $m_{K,Z}$ means the mass concentration of the element with atomic number Z (in general $Z > 13$) detected in the K-stage of the cascade impactor sampler plotted as a function of the particle aerodynamic diameter. Mathematically we might express this mass concentration by:

$$m_{K,Z_i} = \frac{M_{Z_i}}{t} \int_0^t N\infty \int_{V_1}^{V_2} E_K(v) \left[\int \ldots \int g \cdot n_{Z_i} \cdot dn_{Z_2} \cdot \cdot dn_{Z_i} \cdot \cdot dn_{Z_L} \right] \cdot dv \cdot dt'$$

where:

$g(v, n_{Z_i}, \ldots, \ldots n_{Z_i} \ldots n_{Z_L}, \mathbf{r}, t)$ is the size-elemental composition probability density function;

n_{Z_i}, number of gram-atoms of the Z_i element within the sample;

M_{Z_i}, the atomic weight of the Z_i element;

t, the time interval of sampling;

$N\infty$, the total number of particles per unit volume of sampled air;

$E_K(v)$, the K-stage collection efficiency.

Note that the integration variables start with n_{Z_2} (not n_{Z_1}), because there is a relationship between v and the n_{Z_i} variables:

$$v = \sum_{i=z}^{L} n_{Z_i} \bar{v}_i$$

($\bar{v}_i$ is the partial volume of the gram-atom of the Z_i element.)

Finally, there is also a mathematical relation connecting the first defined g size-composition function to the ordinary size-distribution $n(v, r, t)$, that is:

$$n(v, r, t) = N\infty \int_{n_{Z_2}} \cdots \int_{n_{Z_L}} g \, dn_{Z_2} \cdots dn_{Z_L}$$

In spite of this interesting mathematical description, it is not yet feasible in practical cases to determine the n size-distribution from the measured m_{K,Z_i} values, unless under very restrictive and unrealistic approximations[2].

Consequently, the possibility of making the transformation:

$$\{m_{K,Z}\} \longrightarrow \{N_{K,Z}\} \longrightarrow \{n_Z\} \longrightarrow n$$

is still not available and we should explore other ways to utilize the CI + PIXE data for AAS characterization.

AAS CHARACTERIZATION BY CI + PIXE DATA

To characterize an AAS by data generated by CI + PIXE it seems to be more convenient to consider directly the elemental size-distribution curves obtained, if possible together with atmospheric chemistry information and meteorological data. Whenever it is possible to combine information from all these sources, safe conclusions concerning the origin, transport and fate of gaseous and particulate pollutants, involved in the physical and chemical processes related to the measured aerosol samples, can be derived.

To illustrate the above statement we give below a partial interpretation of elemental size-distribution data, obtained by the CI + PIXE technique, which we usually find in ambient aerosol samples from the city of São Paulo, in Brazil[3].

Figures 1a-1c show the elemental size-distribution curves as sampling geometric averages of $m_{K,Z}$ for the elements S, Ca, Fe, Zn, Br, and Pb, for five sampling intervals conducted during 1976-77. These geometric averages are plotted against the impactor K-stage, in inverse order (sites and times of the sampling are also included

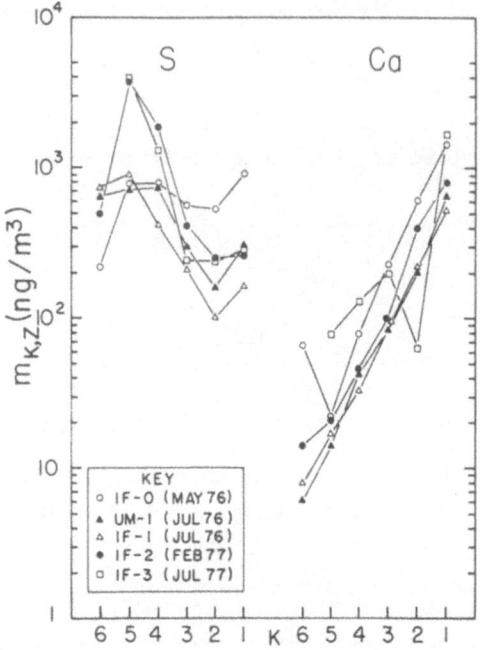

Fíg. 1a.   Sulfur and calcium distributions for São Paulo

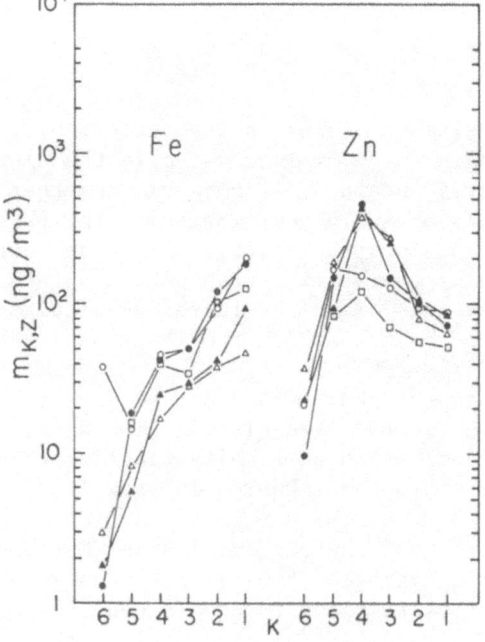

Fig. 1b.   Iron and zinc distributions for São Paulo

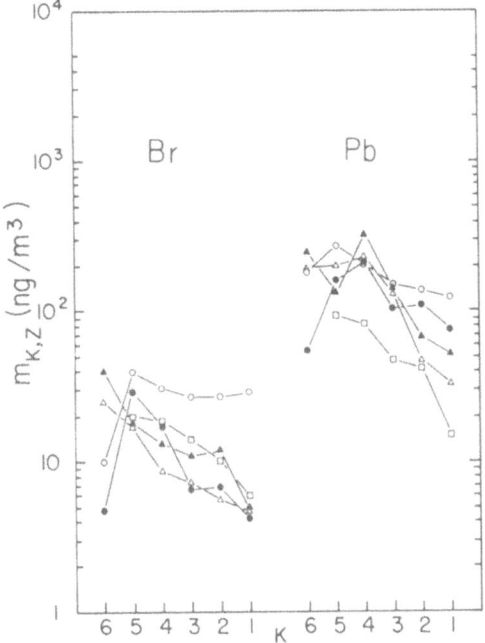

Fig. 1c.   Bromine and lead distributions for São Paulo

in a key table). Roughly speaking, in our case the stages 6 to 1
correspond to the mid-range values 0.2, 0.375, 0.75, 1.5, 3 and
6 μm, respectively, for the particle aerodynamic diameter D, plotted
on a log-scale.

A simple visual examination of Figures 1a-1c reveals that the
S, Zn, Br and Pb distributions peak in the submicron range, a fact
that characterizes a combustion plus gas-to-particle conversion
mechanism as responsible for these airborne particulates.

On the contrary, the Ca and Fe distributions show a concentration of particles in the region of coarse particles (D > 2 μm)
which is compatible with mechanical processes of particle production
(soil blown dust plus industrial abrasion sources)[4,5]. However,
the average slopes of the Ca and Fe distributions are quite different. In fact, our most recent analysis of the Fe distributions[6]
have shown a large fine particle Fe component in the São Paulo
aerosol, also derived from combustion sources as yet not identified.

As for sulfur, observed mainly in fine particles, the particularly well known behavior of the atmospheric sulfur-cycle strongly
supports the hypothesis of an SO_2 origin (in São Paulo produced in
both gasoline and industrial oil combustion). Gross quantitative
evaluation to check this assumption can be done if the total amount
of fuel consumption is known.

The correlation between the Br and the Pb size-distributions is consistent with their supposed origin in gasoline combustion[6].

Many other conclusions referring to the natural or anthropogenic origin of the particles can be obtained from statistical analysis of the data sets $m_{K,Z}$ (e.g. correlations among $m_{K,Z}$ elements for a given site and particular date, or else varying the time and space coordinates). Likewise, the comparison of the $m_{K,Z}$ values with "standard aerosol" values may be used to separate the natural from the anthropogenic components[5,6].

ACKNOWLEDGEMENTS

This work was partially supported by FAPESP - "Fundação de Amparo à Pesquisa de Estado de São Paulo", Brazil. The aerosol data reported was obtained with the PIXE system of the Florida State University through a cooperative program supported in part by the U.S. Environmental Protection Agency (R802132 and R803887), the North Atlantic Treaty Organization (SRG/ES.007) and the U.S. National Institute of Environmental Health Sciences (5T32 ES07011).

REFERENCES

1. S. K. Friedlander,"Smoke, Dust and Haze, Fundamentals of Aerosol Behavior," John Wiley and Sons, New York (1977).
2. C. Q. Orsini and L. C. Bouéres, Atmospheric aerosol characterization by means of impactor samples analyzed by PIXE, to be published by Revista Brasileira de Física (1979).
3. C. Q. Orsini and L. C. Bouéres, A PIXE system for air pollution studies in South America, Nucl. Instr. and Methods 142:27-32 (1977).
4. K. T. Whitby, The physical characteristics of sulfur aerosol, Atm. Environ. 12:135-159 (1978).
5. D. R. Lawson, Chemistry of the natural aerosol: a case study in South America, Ph.D. Thesis, Florida State University, Tallahassee, FL (1978).
6. L. C. S. Bouéres, C. Q. Orsini, D. R. Lawson and J. W. Winchester, Fine particulate sulfur and metals in the atmosphere of São Paulo, Brazil, in preparation for Interciência (1979).

COMPUTER CODE FOR ANALYSING X-RAY FLUORESCENCE

SPECTRA OF AIRBORNE PARTICULATE MATTER

E.A. Drane, D.G. Rickel, W.J. Courtney
T.G. Dzubay*

Northrop Services Inc. -- Environmental Sciences
Research Triangle Park, N.C. 27711

*U.S. Environmental Protection Agency
Research Triangle Park, N.C. 27711

INTRODUCTION

During recent years energy dispersive x-ray fluorescence (EDXRF) has been used[1-3] to measure the elemental content of atmospheric aerosols. The code described here is used to reduce EDXRF data and determine the elemental composition of samples collected on membrane filters. The program has been specifically written for EDXRF analysis of size-fractionated aerosols collected by a dichotomous sampler.[4]

The x-ray fluorescence spectrometer[5] used in our laboratory employs a pulsed-mode x-ray tube[6] and a lithium-drifted silicon detector. Pulse-height spectra are produced for elements ranging in atomic number from $Z = 13$ to $Z = 82$ (corresponding to an energy range from 1.4 to 32.1 keV). Approximately uniform x-ray production is attained by producing independent spectra from three secondary targets (Ti, Mo, Sm).

The code itself is executed under the UNIVAC 1110 operating system and was produced in standard ASCII FORTRAN for maximum transportability. The program consists of a main directing routine which calls 21 subroutines and requires 61K words of core for execution on this system. For batches of 72 filters, a routine run requires about 3 seconds for processing each filter. The procedure has four steps: data collection, least-squares fit, corrections and conversion to useful dimensions, and user output. Figure 1 presents a flow diagram.

Table 1. Elements Stored in Standards Library

NAME	AT NO.	2° FLUO-RESCOR	CHANNEL RANGE STORED IN LIBRARY* FILE	SYSTEM SENSITIVITY counts (ng/cm^2)$^{-1}$
Al	13	Ti	20-40	.93
Si	14		20-45	2.36
P	15		20-55	4.32
S	16		20-70	6.69
Cl	17		20-78	9.92
K	19	Ti/Mo	20-101/ 80-100	21.38/ 1.05
Ca	20		20-118/ 80-118	30.04/ 1.66
Ti	22	Mo	80-147	2.50
V	23		100-160	3.69
Cr	24		100-178	4.89
Mn	25		100-193	6.01
Fe	26		120-212	7.63
Co	27		175-230	8.99
Ni	28		150-250	11.23
Cu	29		170-270	12.72
Zn	30		190-290	14.69
Ga	31		210-315	16.77
Ge	32		230-340	18.83
As	33		240-360	20.95
Se	34		270-385	23.25
Br	35		280-410	25.65
Rb	36		340-460	30.53
Sr	38	Mo/Sm	360-460/400-490	33.95/ 4.41
Cd	48	Sm	660-820	6.35
Sn	50		700-900	6.30
Sb	51		750-950	6.21
Ba	56		940-1000	4.89
W	74	Mo	118-370	11.86
Hg	80		240-460	20.44
Pb	82		240-460	21.41

*SYSTEM ENERGY CALIBRATION DEFINED BY

$$E(keV) = 0.544 + 0.03219 \left(\frac{keV}{ch}\right) * CHANNEL$$

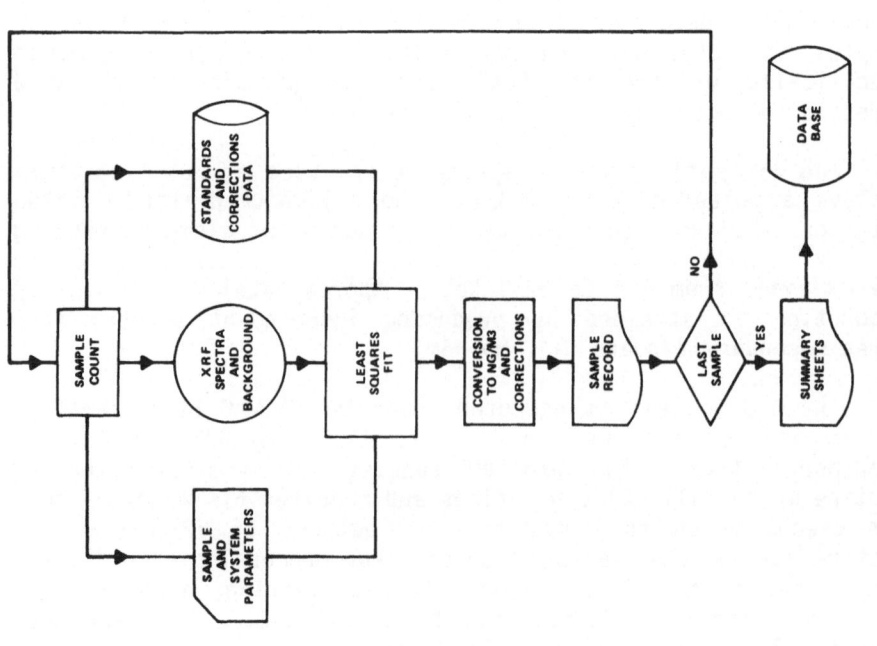

Figure 1. Flow Chart for EDXRF Data Processing

DATA REDUCTION

It can be shown empirically that the x-ray intensities produced from thin specimens are linearly related to mass densities. This relationship can be characterized by the production of pure element spectra from standard films with known deposit mass.[7,8] These calibration values have been measured and are reported in Table 1 and the calibration technique is described by Giauque et al.[8]

The determination of mass densities for sample filters is accomplished by a least-squares fit similar in approach to a method outlined by Arinc et al.[9] By assuming that the observed spectra can be expressed as a linear combination of pure element spectra, a function which provides the basis for a least squares fit is:

$$Y_i = \sum_{j=1}^{m} a_{ij} x_j$$

where:

Y_i is the total counts in the ith channel

a_{ij} is the total counts in the normalized library spectrum for i^{th} channel from the jth element

x_j is the value for the relative amount of element j in the sample to be determined from the fit.

The fitting function (a_{ij}) is acquired by fluorescing the standards used for calibration. After an appropriate background subtraction, these spectra are stored in an on-line disk file. Because storage requirements would be enormous for complete pulse-height spectra for 31 different elements, only a section of each spectrum is stored in the library. Except for W, Hg, and Pb which are determined using L-x-ray lines, the range of channels stored for each element includes its Kα, Kβ, and Si escape peaks. Table 1 lists the elements included in the library and the associated channel ranges.

The fit also requires characterization of the background spectra. If the loading is light compared to the mass of the sample matrix, then the background shape is fairly constant from sample to sample and can be well characterized by the pulse-height spectrum from an unloaded filter.[9,10] By choice of sampling conditions this assumption is nearly always valid; thus, the spectrum for an unloaded filter from the same lot as the loaded sample is temporarily stored in the library and used as another independent unknown in the fit. This is used to account for the intensity differences between blank and loaded background.

By maximizing the probability that any set of elemental coefficients will reproduce the observed spectra, a system of normal equations can be obtained with the relative elemental concentrations as unknowns. This system is most conveniently expressed in matrix form:

$A \cdot x = B$

where:

$$A_{jk} = \sum_{i=1}^{N} \frac{a_{ij}a_{ik}}{y_i} \qquad y_i \quad \text{is the experimentally observed count rate.}$$

$$B_k = \sum_{i=1}^{N} a_{ik} \qquad B_k \quad \text{is the total library element intensity.}$$

This formulation makes use of the validity of Poisson counting statistics for pulse-height spectra (i.e., variance of count rate is approximately equal to the count rate itself).

Because both y_i and a_{ij} represent count rates, the matrix A can be shown to be positive definite and symmetric.[11] These properties suggest solution of the system by factoring A into upper and lower triangular matrices through a Cholesky decomposition.[12]

$A = L \ L^T$

for A an nxn matrix

j=1, ...n

i=j

$$L_{ii} = [A_{jj} - \sum_{k=1}^{j-1} L_{jk}^2]^{\frac{1}{2}}$$

i=j+1, j+2, ...n

$$L_{ij} = \frac{1}{L_{jj}} [A_{ji} - \sum_{k=1}^{j-1} L_{ik}L_{jk}]$$

The vector of coefficients, X, is then found by back substitution. This algorithm is both numerically stable and

computationally economical. The computation is carried out in
double precision mode to reduce the danger of large round-off
errors. Finally, multiplication of these coefficients by the
pure-element intensity and system sensitivity produces a set of
masses per unit substrate area for each of the elemental species.

ADJUSTMENTS TO RESULTS

Once the masses per unit area are determined, they are[13]
corrected for attenuation of the x-rays by the sample itself[13]
and for x-ray line interferences between elements analyzed by
different fluorescors. Correction for line interferences is
accomplished by the simple method used in connection with
wavelength-dispersive x-ray analysis.[14] The true concentration
of element i with interfering lines from elements j=1, m is:

$$C_i = C_o - \sum_{j=1}^{m} K_{ij} C_j$$

where:

C_o is the observed concentration of element i
C_j is the concentration of the interfering element
K_{ij} is an empirically determined interference coefficient.

The interference coefficient is determined by performing the
least squares fit to the shape standard for element j. K_{ij} is
simply the intensity ratio of the observed interference line to
the analyte line of element j. Table 2 shows the values for
these coefficients determined for our analysis system.

LIMITATIONS AND UNCERTAINTIES

Because the shape standard for analysis of an analyte line
is not the best parent function for fitting an interference in
that energy region, there is rather poor reproducibility in the
empirical determination of the interference coefficient. This
results in association of a high uncertainty with the line
interference correction. This uncertainty is set at 10%. We
plan to investigate a slower but perhaps more accurate approach
which involves stripping the interfering peaks from the spectrum
prior to the least-squares fit.

The uncertainty associated with the fit is indicated by the
computed value for the normalized sum of the squares of the
residuals. The uncertainties assigned to each of the fitting
coefficients are just the diagonal elements of the inverse of
matrix A,[15] which are computed using the results of the Cholesky

Table 2.   X-ray Line Interference Coefficients

INTERFERENCE		MEASURED INTERFERENCE K_{ij} COEFFICIENT
ELEMENT j	ELEMENT i	
As	Al	.051
Se	Al	.266
Br	Al	.255
Rb	Si	.408
Sr	Si	.277
Sr	P	.107
Cd	K	.227
Sn	K	.162
Sn	Ca	.179
Sb	Ca	.270
Ba	Ti	.373
Ba	V	.128
W	Si	.216
W	P	.062
Pb	S	.404
Pb	Cl	.054

decomposition. The total uncertainty in the concentrations is comprised of the uncertainties in the fit, system calibration, attenuation corrections, and interference corrections. The values are calculated using the standard propagation of error theory[15] and the root sum square result is reported.   The 5% standard deviation of the system calibration is due primarily to uncertainty in the calibration standards. The uncertainty in the attenuation correction is reported by Dzubay and Nelson.[13] After error analysis is complete, each mass density is tested against the reported uncertainty.   If the mass density is less than 2.3 times the uncertainty, that value is reported to be below the decision limit.[16]

REFERENCES

1.   B.W. Loo, W.R. French, R.C. Gatti, F.S. Goulding, J.M. Jaklevic, J. Liacer, and A.C. Thompson, Large-Scale Measurement of Airborne Particulate Sulfur, Atmos. Environ. 12:759 (1978).

2.   T.G.Dzubay, R.K. Stevens, and L.W. Richards, Composition of Aerosols over Los Angeles Freeways, Atmos. Environ. 12:653 (1978).

3.   R.K. Stevens, T.G. Dzubay, G.M. Russwurm, and D.G. Rickel, Sampling and Analysis of Atmospheric Sulfates and Related Species, Atmos. Environ. 12:55 (1978).

4.   T.G. Dzubay and R.K. Stevens, Ambient Air Analysis with Dichotomous Sampler and X-ray Fluorescence Spectrometer, Environ. Sci. and Tech. 9:663 (1975).

5.   F.S. Goulding and J.M. Jaklevic, X-ray Fluorescence Spectrometer for Airborne Particulate Monitoring, U.S. Environmental Protection Agency, Office of Research and Development, Report EPA-R2-73-182, Research Triangle Park, N.C.

6.   J.M. Jaklevic. D.A. Landis, and F.S. Goulding, Energy Dispersive X-ray Fluorescence Spectrometry Using Pulsed X-ray Excitation, Adv. in X-ray Anal. 19:253 (1975).

7.   T.G. Dzubay and D.G. Rickel, X-ray Fluorescence Analysis of Filter-Collected Aerosol Particles, in: "Electron Microscopy and X-ray Applications," P. Russell, ed., Ann Arbor Science Publishers, Ann Arbor (1977).

8.   R.D. Giauque, R.B. Garrett, and L.Y. Goda, Calibration of Energy Dispersive X-ray Spectrometer for Analysis of Thin Environmental Samples, in: "X-ray Fluorescence Analysis of Environmental Samples," T.G. Dzubay, ed., Ann Arbor Science Publishers, Inc., Ann Arbor (1977).

9.   F. Arinc, L. Wielopolski, and R.P. Gardner, The Linear Least-Squares Analysis of X-ray Fluorescence Spectra of Aerosol Samples using Pure Element Library Standards and Photo Excitation, in: "X-ray Fluorescence Analysis of Environmental Samples," T.G. Dzubay, ed., Ann Arbor Science Publishers, Inc., Ann Arbor (1977).

10. J.F. Harrison and R.A. Eldred, Automatic Data Acquisition and Reduction for Elemental Analysis of Aerosol Samples, Adv. in X-ray Anal. 17:560 (1974).

11. G.S. Strang, "Linear Algebra and Its Applications," Academic Press, N.Y. (1976).

12. Henry Kauffman, private communication (1979) and R.J. Goult, et al., "Computational Methods in Linear Algebra," Halsted Press, N.Y. (1974).

13. T.G. Dzubay and R.O. Nelson, Self-Absorption Corrections for X-ray Fluorescence Analysis of Aerosols, Adv. in X-ray Anal. 18:619 (1975).

14. J.V. Gilfrich, L.S. Birks and J.W. Criss, Corrections for
     Line Interferences in Wavelength-Dispersive X-ray Analysis,
     in: "X-ray Fluorescence Analysis of Environmental Samples,"
     T.G. Dzubay, ed., Ann Arbor Science Publishers, Inc., Ann
     Arbor (1977).

15. P.R. Bevington, "Data Reduction and Error Analysis for the
     Physical Sciences," Mc-Graw Hill, N.Y. (1969).

16. L.A. Currie, Detection and Quantitation in X-ray
     Fluorescence Spectrometry, in: "X-ray Fluroescence Analysis
     of Environmental Samples," T.G. Dzubay, ed., Ann Arbor
     Science Publishers, Inc., Ann Arbor (1977).

ENERGY DISPERSIVE X-RAY FLUORESCENCE (EDXRF) ANALYSIS AS A

RELIABLE NONDESTRUCTIVE INDUSTRIAL TOOL

L. E. Miller and H. J. Abplanalp

BOEING AEROSPACE COMPANY
Seattle, Wa.   98124

For the past several years the Boeing Aerospace Company has been implementing advanced nondestructive chemical analysis methods to improve product reliability and reduce material inspection costs. Previous testing of incoming material for conformance to vendor test reports or of production materials for verification of alloy composition, had consisted of either time-consuming destructive testing or nondestructive chemical spot testing, which often was insensitive to differences between alloys of similar chemical properties. Beginning in 1974, development of EDXRF techniques was initiated to provide a rapid nondestructive analysis capability for both laboratory and factory use. For materials containing elements easily excited by EDXRF methods, costly destructive sampling and testing can be avoided. Generally, chips, wire, barstock, sheet or plate can be analyzed using an annular radioactive source. The uniformity of the X-ray flux diminishes sample geometry and surface roughness effects. A possible exception is the analysis of highly irregular parts such as the threaded area of a steel bolt, which tends to give high results on the elements Nb and Mo. Ti, Cr, Mn, Fe, Co, Ni, Cu, W, Nb, Mo, Pb, and Bi are among the many elements that can be excited with a 109-Cd source. This permits evaluation of most steel, copper, and nickel alloys with one X-ray source except when Al and Si must be included in the analysis.

Some alloys cannot be easily analyzed with radioactive sources. A commonly used titanium alloy, 6Al-4V, requires the use of a 55-Fe source for aluminum and vanadium detection. However, the efficiency of this source in exciting Ti is so great that the detector is swamped before sufficient Al counts can be realized. Aluminum alloys are also difficult to analyze by radioisotope sources due to low sensitivity of Mg, Al, and Si to the 55-Fe source.

One of the first Company applications of EDXRF was in differen-
tiating two similar high strength steel alloys, used in hydrofoil
fabrication. One customer required assurance that the alloys would
not be mixed. A KEVEX 5000/6000 radioisotope source system successfully maintained alloy separation. The modified KEVEX system (Figure
1) illustrates the required features, i.e., portability, maneuverability, and ability to operate in an industrial environment.

For portability the commercial KEVEX system was installed in a
special shock-isolated cabinet and is easily transported and operated
out of a van. For maneuverability, forty foot cables were fabricated
enabling the detector to be manipulated independently of the analyzer.
A special stand was constructed to provide height adjustment from
floor level to four feet and angle adjustment from horizontal to vertical. For environmental conditions (i.e., noise, temperature changes,
humidity, vibration and electrical fields), modifications to the
system or operating conditions were made. Noise effects on resolution
can be minimized by avoiding high noise level areas. Temperature effects on both resolution and calibration cannot be controlled in many
factory locations. They must be minimized by detector design. High
humidity can cause condensation on cool metallic detector parts resulting in arcing. The addition of a dessicant to the preamplifier
unit resolved this problem on the KEVEX system. Special shock isola-
tion was provided to minimize the effects of vibration on detector
resolution. Detector response to electrical fields, such as those
generated by an induction furnace or an electric forklift, was found
to be an individual detector problem. Two supposedly identical detectors had radically different responses to the same electric field;
one was strongly affected, the other not at all.

Since 1975, two portable, small X-ray tube EDXRF systems have
been acquired and evaluated. Although the X-ray tube systems are
less portable, they provide more efficient excitation of the elements
in aluminum alloys. A modified EDAX laboratory system acquired in
1975 was found to be unsatisfactory since it was not easily manipulated and did not function reliably in a factory environment.

A recently acquired system, the DuBois 404 Object Analyzer
(Figure 2), utilizes some significant advances that have been made in
portable X-ray tube systems. The system is capable of operating with
forty foot cables, which allows considerable source-detector movement
and stationary, stable operation of the data analyzer and power supply.
The X-ray tube excitation provides good resolution of the Mg, Al and
Si peaks, as well as exceeding detection limit requirements (0.10%)
for other elements of interest. When the FWHM for the Al peak is
110 eV or less, Mg and Si can be detected down to 1% concentrations.

Automated data handling has been initiated with the acquisition
of an ND6620 computer. Both the KEVEX and DuBois 404 detectors have
been interfaced to this system as has the KEVEX analyzer which batch

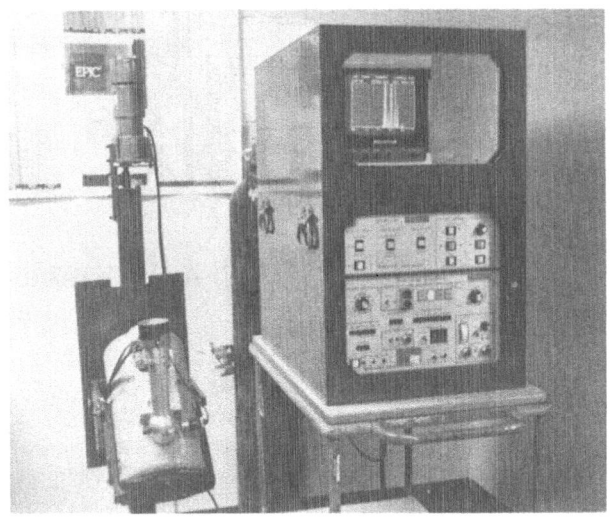

Figure 1   Modified KEVEX EDX System

Figure 2   DuBois 404 Object Analysis

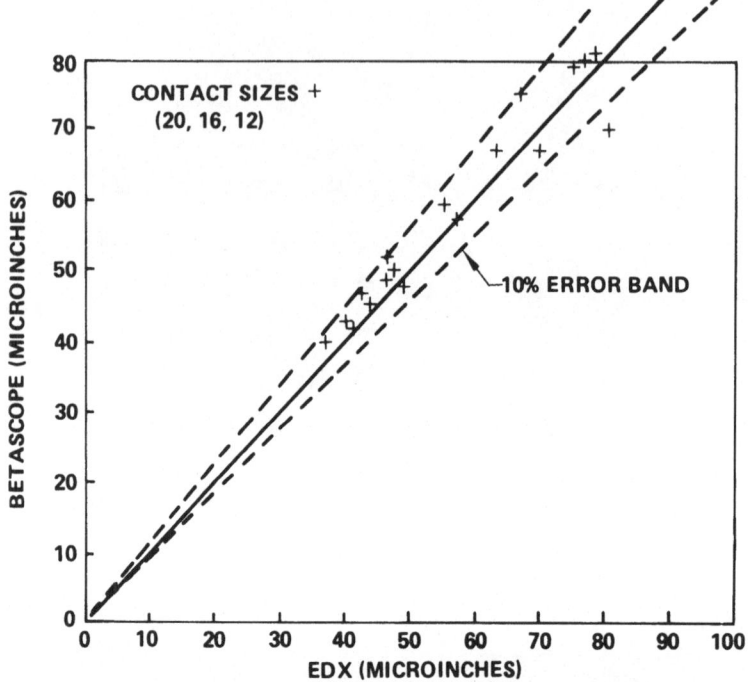

Figure 3   Gold Plating EDX vs Betabackscatter

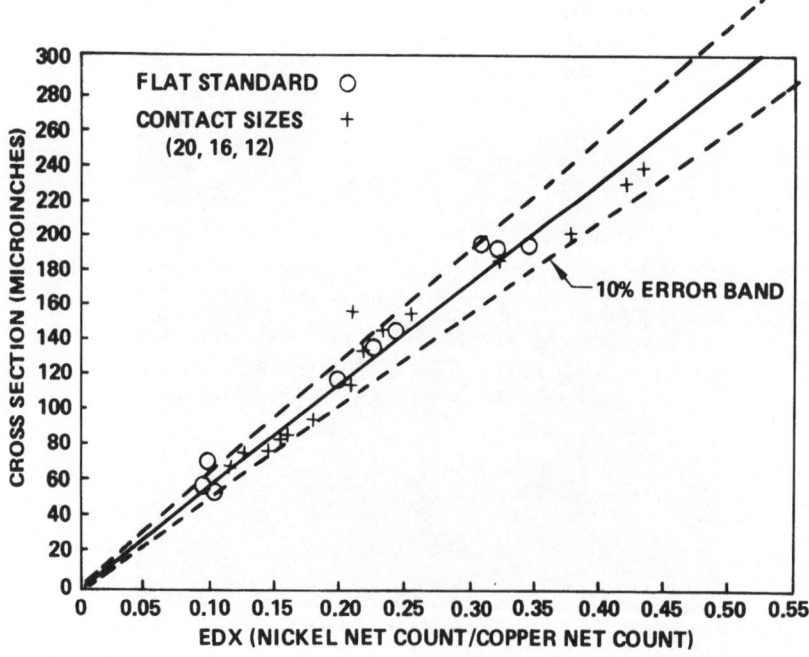

Figure 4   Nickel Plating EDX vs Cross Section

dumps its data through a 400 foot cable. A sample changer has also been added to allow unmanned operation of the KEVEX system.

Generally, we have found that a portable X-ray tube system works best for aluminum and titanium alloys; however, for all other alloys a radioactive source system is better because of (1) insensitivity to sample geometry, (2) long term system stability, (3) low background, (4) better portability, and (5) better sensitivity for 14-17.5 keV X-rays (Sr-Mo). Both systems provide results reproducible to +10% of the analyzed value for elements with "K" or "L" lines between 4.5 and 17 keV and concentrations above 0.20%. However, the reproducibility of Mg, Al, and Si have not yet been fully determined for the X-ray tube source.

Other EDXRF applications at Boeing include:

(1) Circuit boards, made of fiberglass and resin, are treated for fire retardancy by adding a brominated hydrocarbon to the resin. By using a 109-Cd source and a one second count time, fire retardant and nonfire retardant board material are easily separated.

(2) The Company receives two grades of pins for use as connector contacts; a low temperature grade requiring 50 microinches of gold over copper subplating and a high temperature grade requiring 50 microinches of gold over nickel subplating. EDXRF techniques have been developed to measure the gold thickness (Figure 3) and to detect and measure nickel subplating (Figure 4). Other plating combinations including rhodium over silver, rhodium over nickel and gold over silver, have also been investigated.

(3) Quick and accurate identification of metallic particles in fluid-containing systems is necessary for effective preventive maintenance on operating equipment. The EDXRF capability to nondestructively detect and analyze nanogram amounts of material has provided valuable information on the contaminant source, thus shortening equipment downtimes and minimizing unnecessary servicing.

(4) Some unusual surface effects have been observed during testing with EDXRF. These are attributed to: (1) chemical etching which preferentially depletes certain elements, (2) surface cleaning with different types of grit, or (3) chemical treatment of the surface. Removal of about 0.0005 inch of the surface usually elminates the problem.

New applications for EDXRF are constantly being identified and developed, providing the Boeing Aerospace Company with a valuable capability for dependable, rapid, nondestructive materials testing.

NONDESTRUCTIVE, ENERGY-DISPERSIVE, X-RAY FLUORESCENCE ANALYSIS OF ACTINIDE STREAM CONCENTRATIONS FROM REPROCESSED NUCLEAR FUEL*

D. C. Camp and W. D. Ruhter

Lawrence Livermore Laboratory
University of California
Livermore, Calif. 94550

INTRODUCTION

In the event that nuclear fuel from light water reactors (LWR) is reprocessed to reclaim the uranium or plutonium, several analytical techniques will be used for product accountability. Generally, the isotopic content of both the plutonium and uranium in the reprocessed product will have to be accurately determined. One plan for the reprocessing of LWR spent fuel incorporates the following scheme.[1] After separation from both the fission products and transplutonium actinides (including neptunium and americium), part of the uranium and all of the plutonium in a nitrate solution will merge together to form a coprocessed stream. This solution will be concentrated by evaporation and sent to a hold tank for accountability. Input concentrations into the hold tank could be up to 350 g U/ℓ and nearly 50 g Pu/ℓ. The variation to be expected in these concentrations is not known. The remaining uranium fraction will be further purified and sent to a separate storage tank. Its expected stream concentration will be about 60 g U/ℓ. These two relatively high actinide stream concentrations can be monitored rapidly, quantitatively, and nondestructively using the technique of energy-dispersive x-ray fluorescence analysis (XRFA).[2]

---
*This work was performed under the auspices of the U. S. Department of Energy by Lawrence Livermore Laboratory under contract No. W-7405-Eng-48.

EXPERIMENTAL EQUIPMENT

Excitation Source Requirements

    Gamma rays can be used to excite x rays from atoms within a
sample. The binding energies of K electrons in U and Pu are 115.59
and 121.72 keV, respectively. Since the primary gamma ray emitted
by ^{57}Co has an energy of 122.05 keV, it is an optimum exciting radiation for these two actinide elements. The exciting radiation is
usually collimated in some fashion that depends on the geometry of
the sample. This is to reduce the amount of radiation that can scatter off of nonsample materials or that can cause them to fluoresce.

    Lithium-drifted silicon, Si(Li), is an excellent radiation de-
tector for x rays with less than 30 keV of energy, but it becomes
very inefficient for the detection of radiation energies above 60 keV.
Since the K x-ray energies of U and Pu extend from 98 to 120 keV, a
lithium-drifted or high-purity germanium detector, Ge(Li) or HPGe,
is used. For this work a 10-mm-deep, 500-mm^2 HPGe detector was
used. It had an energy resolution of 600 eV FWHM for the 122.05-
keV gamma-ray peak of ^{57}Co.

    The source-detector collimation assembly is shown in Fig. 1.
Two ^{57}Co sources are partially collimated to create two beams. The
radioactivity was electroplated onto a 1.6-mm-diameter spot and en-
cased in a welded stainless steel capsule* 4.8 mm in diameter and
3.2 mm thick. The 0.37-mm-thick stainless steel plate indicated in
Fig. 1 is part of the bottom of the glove box, which was used when
handling all of the solutions. The source-detector collimation assembly and liquid-nitrogen (LN) dewar are separate from and located
below the glove box. The collimator assembly is 7.5 cm in diameter
and 5.0 cm thick.

    Since ^{57}Co also emits 570- and 692-keV gamma rays with branch-
ing intensities of about 0.16%, as well as other weaker gamma rays
above 300 keV, their intensities must be strongly attenuated by introducing shielding between the source and the detector. X rays from
lead and Hevimet (tungsten alloy) can also be excited by the source
gamma rays; hence, graded absorbers of cadmium and copper are used as
liners on the top and bottom surfaces to eliminate these x rays. A
central 12.5-mm hole within this collimator assembly allows part of
the x rays released within the sample to strike the detector.

_____
*Available from Isotopes Products Laboratories, Burbank, Calif.
Reference to a company or product name does not imply approval
or recommendation of the product by the University of California
or the U.S. Department of Energy to the exclusion of others
that may be suitable.

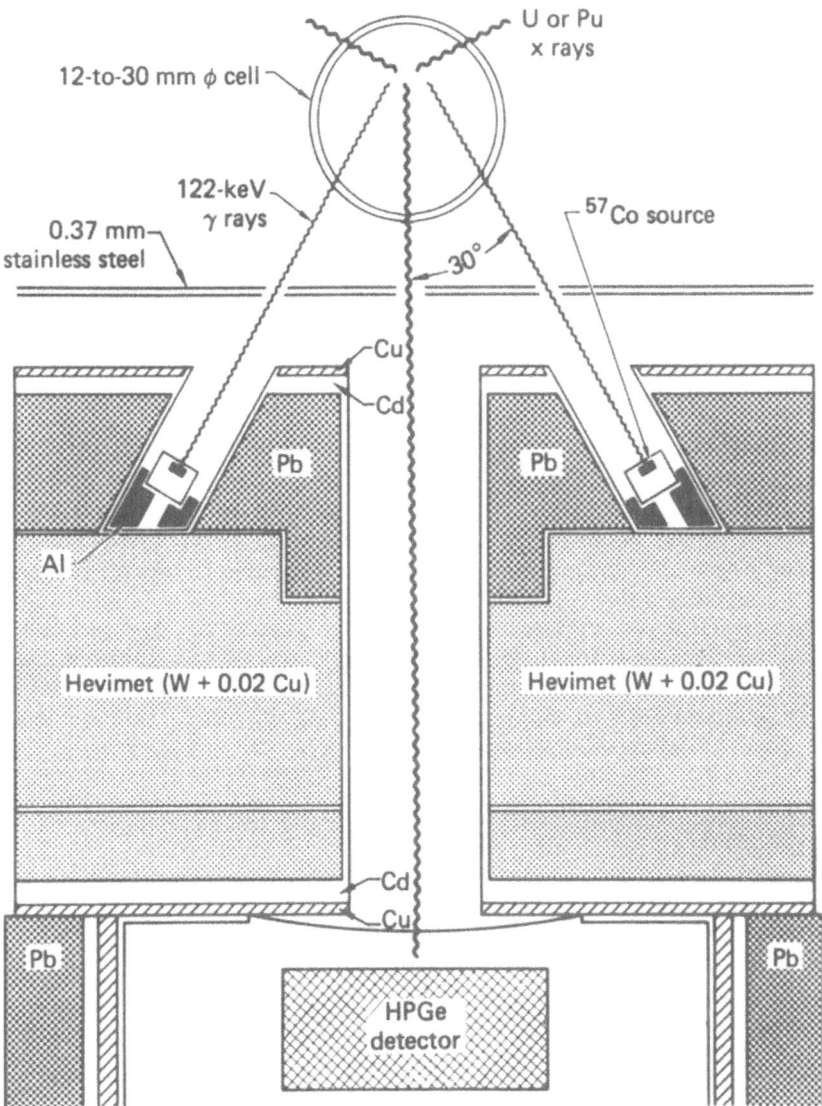

Fig. 1   A cross sectional view of the source exciter and de-
          tector collimator assembly. The 0.37-mm stainless
          steel plate is part of the bottom of a glove box as-
          sembly.

     In the application of interest here the sample is a solution
contained within a cylindrical geometry. A solution cell or pipe
section used for calibration purposes could have any diameter, but
should be larger than the inside diameter of the detector's collimator. The collimated 122-keV gamma rays interact with atoms in the

solution, creating x rays characteristic of those elements dissolved in the solution. A portion of the emitted x rays are collimated to strike the detector, and from the energies and intensities detected the elemental concentrations in the solution can be determined.

The HPGe detector cryostat used in these experiments utilized a beryllium window, but such a window is not necessary if actinide K x rays are to be detected. Clearly, the x-ray intensity recorded by the HPGe detector increases as the sample volume-to-detector distance decreases. However, this distance cannot be decreased indefinitely. As the distance is decreased, less shielding is possible between the intense ^{57}Co sources and the detector. Those higher energy gamma rays, which pass through the Hevimet and interact with the HPGe detector, create a Compton continuum that appears as a constant, energy independent background in the vicinity of U and Pu x rays. This background contribution increases very rapidly with decreasing amounts of shielding, degrading the x-ray signal-to-noise ratio quickly. Some high Z shielding is also required around the detector housing (above the cryostat) to reduce background radiation detected from the local environment and source-air scattering.

The radiation sources and detector collimator are necessarily coupled to the detector and its LN dewar, which was located below the glove box containing the solution cell. A glove box must be used when handling solutions containing Pu so that, in the event of a spill, contamination is confined.

Solution Cells

In an actual reprocessing plant, U and Pu solutions will probably flow through stainless steel pipes. In order to examine the behavior of these solutions when in motion, a flow system was constructed. A variable speed peristaltic pump moves the solution through the Tygon tubing and cell by a cyclical squeezing action. The solution circulates from the separatory funnel through the cell and pump, and then is returned to the funnel. The flow direction can be reversed if desired.

Calibrated, unknown, or wash nitric acid solutions containing only U were transferred from their containers to the separatory funnel by use of a unidirectional air flow hand pump. This avoided any pouring action. Quantitative transfers were usually not necessary. By reversing the dual-channel stopcock, one could empty the flow system using the peristaltic pump. Since 2 to 3% of the solution remained in the tubing, the system was flushed out after each use, and then primed with a solution close in concentration to the next to be measured. The glass cell allowed visual inspection of the flow conditions. Three sizes of Pyrex cells, 12, 25, and 38 mm in outside diameter, were used.

Once it was demonstrated that the x-ray intensities were independent of whether the solution was flowing or static, solution cells
were constructed using stainless steel (SS). Each cell was machined
to have an outside diameter of 18.80 mm and 1.50-mm-thick walls.
They had a nominal solution length of 10.0 cm. SS cells filled with
Pu nitrate or mixed U-Pu nitrate had to be handled in a sealed glove
box environment. To insure repeatability in the measurements, both
the Pyrex cells and SS cells could be positively located in a reproducible position with respect to the detector collimation axis.

Computer-Based Analyzer

The x rays released in the solution samples were detected by the
HPGe detector. Preamplified pulses were routed to a Canberra 1413
amplifier and 1468A pile-up rejector. Valid output pulses were routed to a Nuclear Data ND600 pulse height analyzer (PHA). The PHA,
with its own LSI-11 microprocessor, was coupled to an LSI-11 minicomputer that had a 32K 16-bit-word memory. A dual floppy disk was
coupled to the LSI-11 and each disk had a 216K-byte (108K-word) capacity. Also, a dual hard disk system with a 10M-byte capacity was
coupled to the LSI-11. Other system peripherals include a Hazeltine
video teletypewriter terminal, an LA-180 high-speed line printer,
and a Tektronix digital data plotter.

A computer-based pulse height analyzer added a considerable
amount of versatility to the experimental system. In fact, in an
actual reprocessing installation a computer-based data analysis system would be essential, and probably would be linked to a command
computer center. This particular system allowed successive spectra
to be stored on disk and permitted spectrum analysis to be carried
out simultaneously while data were also being acquired. Whatever
automatic analysis sequence is desired, appropriate software can be
written and stored on floppy disk. Generally, the software programs
can be written in FORTRAN, BASIC or another language familiar to the
user.

EXPERIMENTAL PROCEDURES AND RESULTS

Spectra

The spectrum shown in Fig. 2 is the result of an x-ray fluorescence analysis of pure uranium nitrate solution at 100 g U/ℓ
contained in a 25-mm-diameter cylindrical Pyrex cell. Total analysis time was 334 live time seconds at 19.4% analyzer dead time,
using two 5-mCi ^{57}Co sources. The net Kα1 x-ray intensity is almost 2 × 10^5 counts. The energy region shown extends from 0 to
200 keV. One of the more dominant features in the spectrum is the
broad, intense peak centered at about channel 450. This peak is a
result of the primary exciting radiation (the 122-keV gamma ray of
^{57}Co) incoherently (Compton) scattering through an angle of 140°.

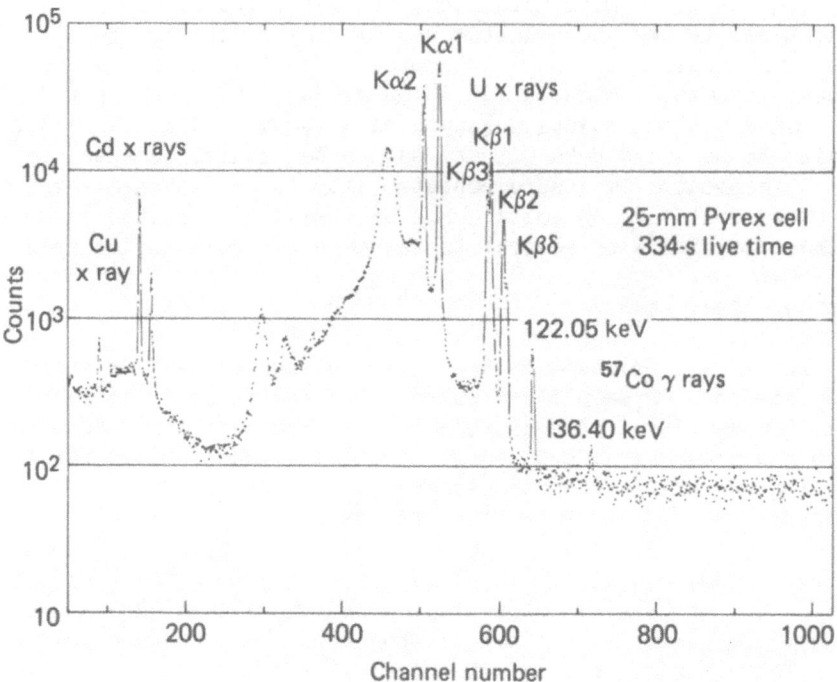

Fig. 2   A 1024-channel XFRA spectrum of 100-g-U/ℓ uranium
          nitrate in the 25-mm cell. The cadmium x rays re-
          sult from a liner on the inner wall of the detector
          collimator. The two broad peaks above channel 300
          are incoherently scattered platinum (source backing
          material) x rays.

The sharp, intense peaks located between channels 500 and 620
are the K x rays of U. The most intense x ray is Kα1 at 98.439
keV. Kα2 is about half as intense and located slightly lower in
energy at 94.665 keV. The Kβ x rays contain more than one component. Hence, they appear as multiplets. The (Kβ1 + Kβ3 + Kβ5)
x ray is located near channel 584, while (Kβ2 + Kβ4 + Kδ) is less
intense and is located near channel 608. The weak 122-keV peak at
about channel 644 is coherently scattered (no energy change) exciting radiation. The very weak peak observed at channel 720 is coherently scattered 136.4-keV gamma ray also from ^{57}Co. Its intensity is one-tenth that of the 122-keV radiation. The short, nearly
flat distribution centered about channel 488 is the 140° incoherently scattered radiation from the 136.4-keV gamma ray.

Figure 3 shows an expanded view of a fluorescence spectrum of
the 80- to 130-keV region for two different concentrations of U in
nitric acid. The most intense spectrum represents a concentration

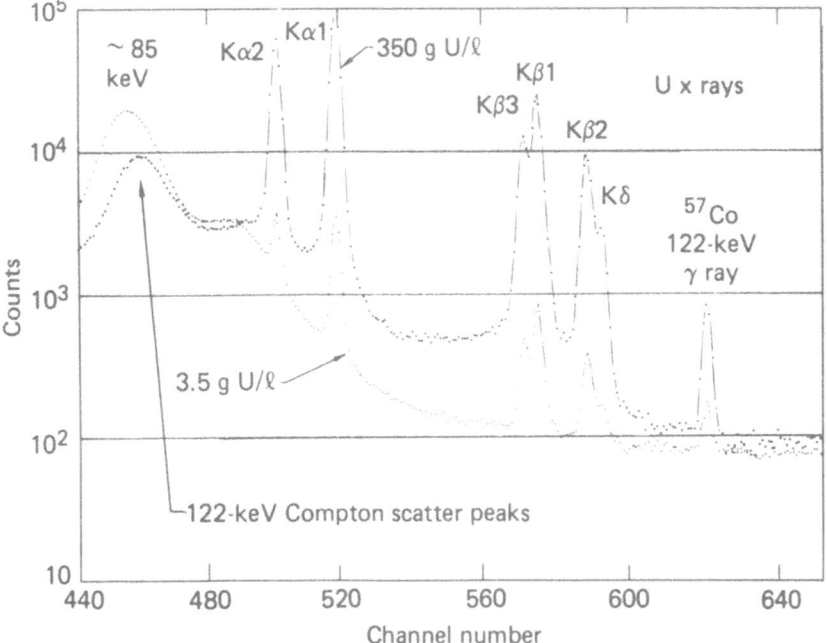

Fig. 3   Two spectra of uranium x rays in the 80-to-130-keV
           region for two concentrations of uranium nitrate
           plotted for equivalent counting times. Note the
           behavior of the incoherent and coherent scattering
           peaks of the 122-keV exciting radiation versus
           concentration.

of 350 g U/ℓ; while the weaker spectrum results from a nitrate solution containing only 3.5 g U/ℓ. These two spectra span a dynamic concentration range of 100. The actual dynamic range offered by
the XRFA technique is in excess of 1000; however, it is most useful
for concentrations above 1 g U/ℓ. Below 1 g U/ℓ the counting
times required become much longer in order to obtain sufficient statistical accuracy.

     Note that as the solution concentration increases the intensity
of the coherently scattered exciting radiation peak at 122 keV increases (because the effective Z of the solution increases). Also,
as the solution concentration increases, the broad incoherently scattered 122-keV peak at about 85 keV decreases in intensity and shifts
slightly toward a higher energy. The increase in energy of this peak
from 3.5 g U/ℓ to 350 g U/ℓ is 0.79 keV, which corresponds to a
change from 143.6° to 138.7° in the backscattering angle. In effect, at higher concentrations the center of the solution volume

exposed moves slightly closer to the detector, thus decreasing the backscattering angle.

The top spectrum in Fig. 4 shows an expanded view of the 80- to 130-keV spectrum region for U and Pu nitrate solution. The U concentration is 350 g U/ℓ; while the Pu concentration is 48 g Pu/ℓ. This

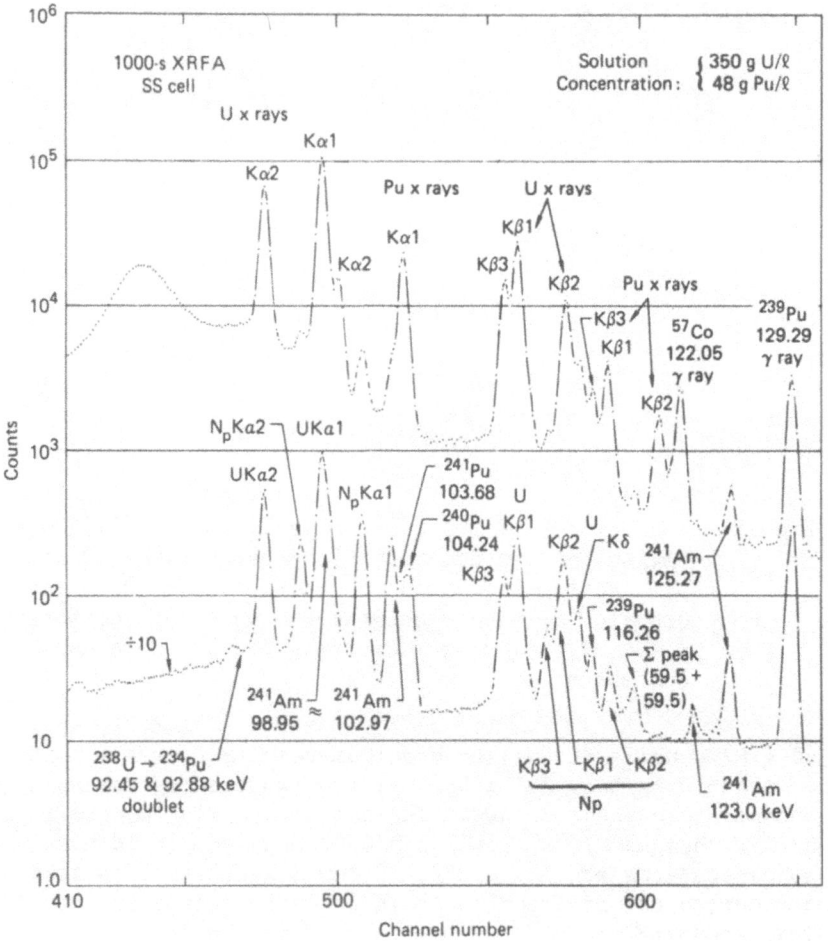

Fig. 4   The top spectrum is an XRFA of a mixed uranium plutonium nitrate solution in a stainless steel cell. Their respective concentrations are shown. The lower spectrum results from the natural radioactivity in the solution. It is plotted one decade lower for clarity: the 129.29-keV peak from ^{239}Pu in the top spectrum results entirely from the solution's natural radioactivity.

mixture corresponds closely to the U-Pu concentration ratio expected
to flow into the final hold tank from a nonspiked, coprocessed product stream. The principal features in this spectrum are the fluoresced x rays of U and Pu and several Pu gamma rays from its natural
radioactivity. The natural radioactivity arising from this U-Pu solution is shown in the lower spectrum, which has been plotted one
decade lower for clarity. All features in the spectrum have been
identified; however, many of these are quite different than those
expected in the XRFA of a freshly reprocessed nitrate stream. These
differences merit some discussion.

The difference in isotope percentages will have the following
influence on the natural radioactivity issuing from a freshly reprocessed U-Pu nitrate stream. The 92.45- and 92.88-keV doublet
gamma ray in ^{234}U from the decay of ^{238}U will still be present, but
will not be significantly troublesome. The neptunium K x-ray intensity, which accompanies the decay of ^{237}U and ^{241}Am, should be less
intense. The ^{241}Am, which has been chemically separated, will not
have its 98.95-, 102.97-, 123.0-, and 125.29-keV gamma rays present.
The (59.53 + 59.53-keV) sum peak at 119.06 keV will be absent. The
^{241}Pu 103.68-keV gamma ray will be much stronger. The ^{240}Pu 104.24-
keV gamma ray will also be stronger. The ^{239}Pu gamma-ray lines at
98.71, 116.26, 124.5, and 129.29 keV will not be as strong.

Finally, it is difficult to say how strong the U K x-ray lines
will be from natural radioactivity. They will grow in strength as
the α-decay of ^{241}Pu approaches equilibrium, after about 42 d. A
small contribution from internal α and γ self-fluorescence may be
present but will depend on solution concentration. Any separated
^{237}U blended back into the coprocessed stream will contribute to
the neptunium K x-ray intensity. Since the Pu isotope percentages
will not remain the same from fuel batch to batch, the natural radioactivity in the coprocessed stream will have to be monitored.
Thus, it appears that in the x-ray spectrum U Kα1 and Pu Kα2 will
be relatively free of interference, but Pu Kα1 will have strong contributions from the ^{240}Pu, 104.24-keV and the ^{241}Pu, 103.68-keV
gamma rays.

Count Rate vs Concentration and Cell Calibration

A set of standard solutions were prepared using ACS-grade natural uranium nitrate. Sufficient HNO_3 acid was used to adjust the
acid concentration to 3.0$\underline{M}$. The standards covered the range from
0.6 g U/ℓ to 350 g U/ℓ and their values were determined by potential
coulometry. The solutions were introduced into the flow system as
described earlier. The 25-mm-diameter Pyrex cell was the first to
be used, and Fig. 4 showed a spectrum obtained from the 100-g-U/ℓ
solution. Generally, to obtain data for each concentration three
separate runs were made with the system under flow (or static) conditions once or twice out of the three runs. Analysis live times

for each run were set to obtain better than 0.5% statistics (40,000
counts) in the gross Kα1 peak, but no run was less than 100 s. Only
one long run was used to obtain data on solution concentrations
≤ 1 g U/ℓ. Subsequently, this procedure was repeated for the
12-mm Pyrex cell. For the 38-mm Pyrex cell and SS cell, data were
obtained with the solution not flowing. Other extensive experiments
had confirmed that concentration measurements made on the solution
under either flow or static conditions were equivalent (see the next
section).

    Figure 5 shows the net counting rate in counts/s (left border)
for the U Kα1 peak as a function of solution concentration (top
border), as measured with the 25-mm Pyrex cell. The two similarly

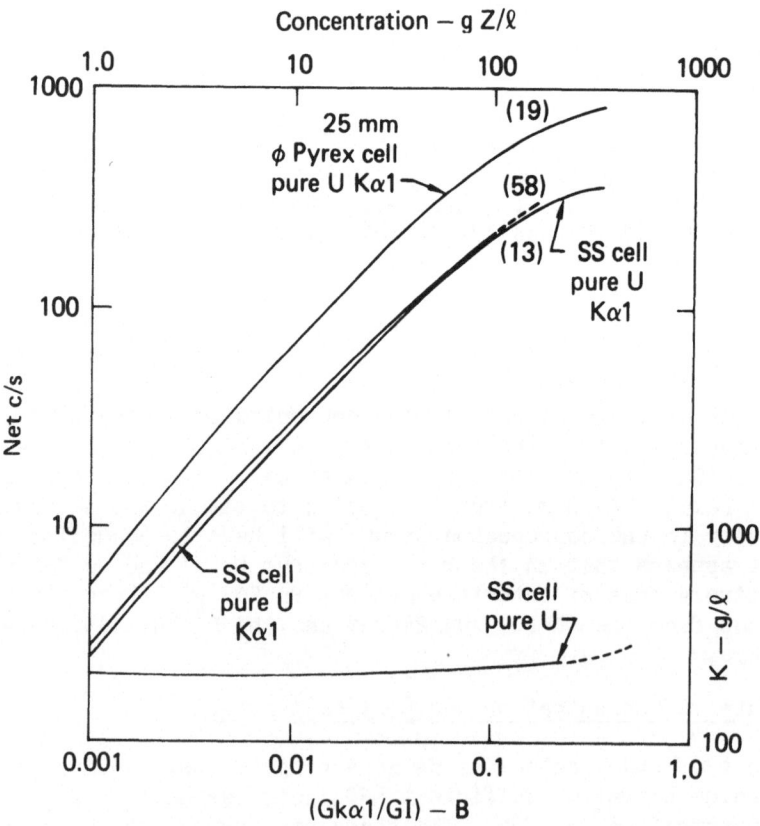

Fig. 5   A log-log plot of the measured net count rate (left)
          versus solution concentration (top) for single ele-
          ment solutions in the Pyrex and stainless steel
          cells. The lower curve shows the behavior of the
          calibration constant K (right) versus the experi-
          mentally determined ratio (bottom).

shaped curves below this give the count rate observed for Pu Kα1 or
U Kα1 for unmixed solutions of plutonium or uranium nitrate respectively, in the SS cell. The small figures in parentheses indicate the analyzer dead times in percent for 100-g/ℓ solution concentrations in each cell type. The Pu solution contained natural radioactivity. The U solution in the SS cell contained less solution
volume than the 25-mm Pyrex cell, hence its lower dead time.

As the solution concentration is increased, there is less than
a linear increase in the count rate. This decline in count rate is
a combination of increasing self-absorption of the Kα1 x ray within
the solution and an effective decrease in the solution volume as the
concentration is increased. Clearly, the net count rate observed
will depend on the ^{57}Co source strength ($T_{1/2}$ = 270 d), the experimental geometry and cell wall thickness, and the HPGe detector efficiency. Such a simple curve showing count rate vs concentration is
not time independent. Furthermore, at high concentrations the rate
of change of count rate with concentration becomes less sensitive
(i.e., a 1% change in count rate corresponds to a 4% change in concentration at 300 g U/ℓ for the 25-mm cell). The observed count
rate is also sensitive to minor changes in geometry and system dead
time, which varies with concentration. Air bubbles in a flowing
stream would also affect the observed count rate. So, it is desirable to define a calibration procedure which is independent of
source half-life and system dead time, insensitive to minor changes
in geometry and stream flow conditions, and more sensitive to concentration changes.

It can be seen from the two spectra in Fig. 3 that as the x-ray
intensity increases, the 140°, incoherently scattered, 122-keV radiation at 86 keV decreases. The ratio of the Kα1 x-ray intensity to
a portion of the spectrum that includes the incoherent peak is almost independent of concentration. The concentration in g/ℓ can be
related to this ratio by the formula

$$C = K \left[\frac{G K\alpha 1}{G I} - B \right]$$

(1)

where GKα1 is the gross count within a window including the Kα1
x-ray peak, GI is the gross count in a window including the incoherent peak, and K is a nonlinear calibration parameter in g/ℓ.
The constant B is the ratio GKα1/GI for pure nitric acid. This
ratio is independent of source exciter half-life, changes in dead
time, and small changes in geometry. Thus, the bracketed quantity
in Eq. (1) is an experimentally measured quantity. If a well-
defined relationship between K and the bracketed quantity can be
established, then their product will yield the concentration.

The lower portion of Fig. 5 shows the behavior of K (right border) vs this ratio (lower border) for the SS cell containing pure uranium nitrate solution. The increase in K at higher concentrations is effectively a result of increasing self-absorption of U Kα1 and decreasing fluoresced volume as the solution concentration increases. A least squares fit to K, expressed as a polynomial function of the natural logarithm of the bracketed quantity, results in the equation shown as an inset at the top of Fig. 6. This figure shows the percentage deviation between the calculated and experimental K as a function of solution concentration in the SS cell. The mean absolute value difference is 0.34% with a root mean square deviation of 0.20%.

Dynamic Concentration Measurements

One of the advantages of the Pyrex cell (coupled to the peristaltic pump via Tygon tubing) is the ability to observe the solution under flow or static conditions while a measurement is in progress. The ND-600 disk-based analyzer system was programmed to carry out an analysis for a preset time Δt s, store the results on disk, clear the memory, and begin a new analysis, again for Δt s. Data transfer and memory clearance required a minimum of 10 s. This cycle could be repeated n times, where n was a preselected integer. Data could be taken in this manner with the solution either static or flowing.

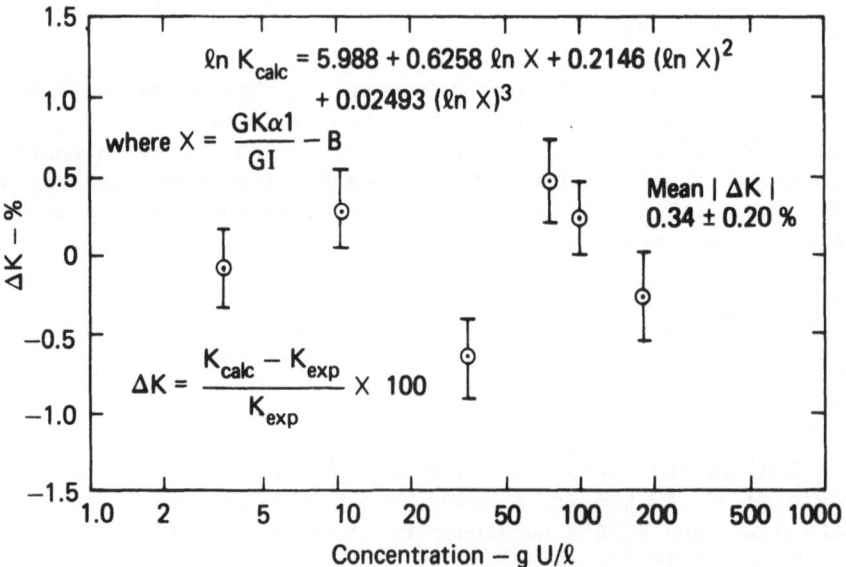

Fig. 6   A semilog plot for the residuals ΔK versus the concentration. The equation used to calculate K is shown. The mean |ΔK| is 0.34 ± 0.20%.

The top section of Fig. 7 shows a set of 24 measurements; however, the first and last six measurements were made with the solution static, while the central 12 measurements were made with the
solution flowing at 80 ℓ/h. The static and flow results overlap
well within the precision of their standard errors, which are
slightly larger (0.70%) than for the 24-cycle static run (0.65%).

Another advantage of the flow system is the ability to demonstrate dynamic concentration measurements. By introducing a known
volume into a dry cell and tubing system, small volumes of pure 3.0M
HNO$_3$ acid can be introduced into the separatory funnel. These volumes are such as to lower the concentration by known small increments. Similarly, by introducing known volumes of a more concentrated solution, the concentration can be increased. In this manner
concentration changes in an actual flowing stream can be simulated.
The lower section of Fig. 7 illustrates such measurements. Initially, 100.0 ml of the 100.0-g-U/ℓ standard solution was introduced

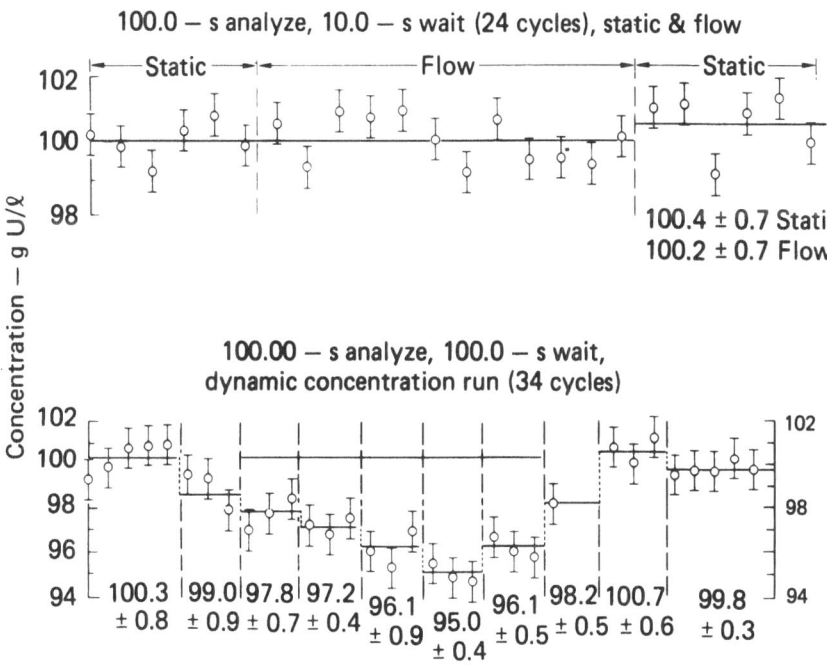

Fig. 7   Two plots of solution concentration versus time.
         Analysis time and pause times vary. The top set
         of data points was taken with the first and last
         six measurements static, and the center 12 measure-
         ments flowing; the lower set is a dynamic concen-
         tration run (see text). The error bars indicate
         1σ statistical counting accuracy only.

into the flow system with the cell and tubing dry. Appropriate volumes of pure $3\underline{M}$ HNO_3 were calculated and measured out so that they could be introduced into the flow system. This caused a concentration reduction of exactly 1.0 g U/ℓ for each of five times. Then, appropriate volumes of more concentrated uranium nitrate solutions were introduced, to increase the entire solution volume 2.0 g U/ℓ for each of three steps. This brought the system to 101 g U/ℓ, whereupon a final calculated volumetric addition of pure $3\underline{M}$ HNO_3 returned the solution to its original 100.0-g-U/ℓ concentration.

The pause time of 100 s was selected on the basis of earlier dynamic concentration runs. It was somewhat longer than the time required for the solution volume to reach equilibrium after introduction of an additional volume. Again each 100-s analysis had approximately the same statistical accuracy of 0.53%; however, the mean values indicated for each successive concentration change are determined from just the measurements made within each interval. Unfortunately for the seventh concentration change (98.2 g U/ℓ), the data storage disk filled up and two cycles were lost. After the next concentration adjustment, data storage was continued on a new disk. Note that the mean of the first and last five dynamic concentrations is 100.05 ± 0.63 g U/ℓ, in excellent agreement with the mean value of 100.13 ± 0.40 g U/ℓ from the first 24 static/flow cycle measurements.

REFERENCES

1.   T. J. Warren, W. E. Prout, M. C. Thompson, M. S. Okamoto, and G. S. Nichols, Technical Data Summary Coprocessing Solvent Extraction Facility, Savannah River Laboratory, Document DPSTD-AFCT-77-9 (December, 1977).
2.   D. C. Camp, An Introduction to Energy Dispersive X-ray Fluorescence Analysis, Lawrence Livermore Laboratory, Livermore, California, UCRL-52489 (June, 1978).

NOTICE

DIRECT DETERMINATION OF NIOBIUM IN URANIUM-NIOBIUM ALLOYS*

Jack L. Long

Rockwell International, Rocky Flats Plant
P.O. Box 464
Golden, Colorado 80401

ABSTRACT

A procedure for determining niobium in uranium-niobium binary alloys was developed. The procedure requires only that a compact be made by pressing turnings, and wet chemical treatment is avoided entirely. By eliminating the wet chemistry, time is reduced by a factor of at least three compared with previous practice.

Standardization was accomplished by separating each batch of machined turnings from specially cast uranium-niobium alloys into individual strands. A composite portion of each strand was analyzed for niobium by wet chemistry. The balance of the turnings from each batch was pressed into a compact. Seven standard compacts resulted. The average intensity ratio of niobium-to-uranium vs. the percent niobium for fourteen determinations on each standard compact was used to calculate a correlation coefficient of 0.999.

The standard deviation of the intensity ratios between duplicate compacts (average of 52 samples) was less than the standard deviation found between aqueous duplicates (average 125 samples) by a factor of 4.

INTRODUCTION

Uranium-niobium alloys are prepared by arc melting. The electrode which is melted is made up of pure uranium, pure niobium and

---

*Work performed under Department of Energy Contract.

scrap alloy. During the process, only a small zone of material is in the molten state at any given time. Thus, conditions favor incomplete mixing which may result in non-homogeneous alloys.

The problem of inhomogeneity is evident in the analysis of niobium. The effect is averaged out by using a large sample, either for wet chemical analysis or x-ray analysis.[1,2] The wet chemical preparation for x-ray fluorescence analysis is shown in Fig. 1. This is a time-consuming procedure, as evidenced by the many steps, and a significant amount of alloy is converted to aqueous waste because of the large sample weight (sixteen grams) because each sample is run in duplicate.

The disadvantages of the aqueous preparation prompted an investigation of a more direct approach. The evaluation criteria were a) precision and accuracy at least equal to the existing method, b) elimination of aqueous wastes, and c) reduced analytical time.

EXPERIMENTAL

The characteristics of the desired procedure are illustrated in Fig. 2. The feasibility of using this approach was demonstrated

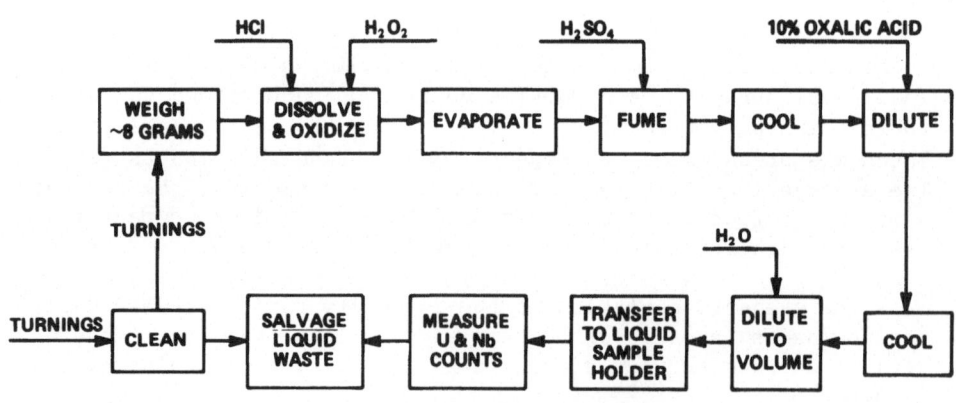

Fig. 1.   Aqueous niobium procedure

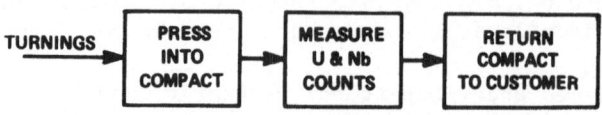

Fig. 2.   Compact niobium procedure

by making intensity measurements on single strands of machined
turnings. The variability of niobium content in the individual
strands was evident. Thus, the major problem to be solved was the
preparation of standards. This was accomplished by having seven
alloys cast, covering the range of 5 weight percent to 6.5 weight
percent niobium. Each alloy was machined into turnings and approximately 30 grams were collected. Each batch was separated into
individual strands which were cut so that two piles resulted; one
containing about one-third of the batch and the other the remaining
two-thirds. The pile containing the smaller amount was submitted for
niobium determination by wet chemical analysis. The larger pile was
pressed into a compact. A pressure of 41,000 pounds per square inch
was used. The seven standard compacts which resulted all had different surface textures due to the variability in machined turnings.

   X-ray intensity data were collected on a GE XRD-6 with modified
electronics for pulse height selection, detection and counting. The
tube was tungsten, operated at 60 kilovolts and 40 milliamperes.
The analyzing crystal was LiF 220. A plot of 2θ vs intensity is
shown in Fig. 3. The 2θ angles are slightly different from the published values.[3] Count data on both the niobium and uranium peaks
were obtained with a gas filled proportional detector operated at
1.800 kilovolts. The baseline for pulse height selection was set at
200 millivolts and the window at 1320 millivolts. A 20-second count
time was used.

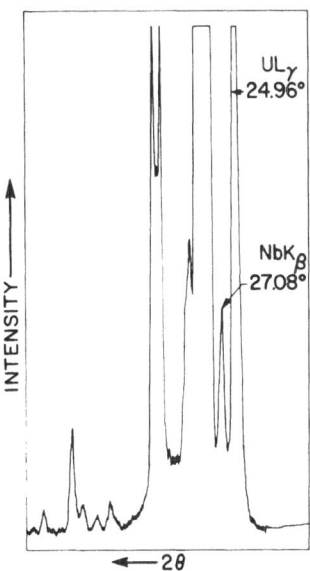

Fig. 3. Wavelength spectra of uranium-niobium alloy

RESULTS AND DISCUSSION

Typical count data as a function of niobium content are presented in Table I. Similar data for the seven standards are
summarized in Table II. Instead of individual counts, a niobium-
to-uranium ratio is given for each niobium concentration. The
standard deviation and percent relative standard deviation were
calculated from 14 measurements taken on each standard. It is
evident from the statistics that compacting the turnings minimizes
the effect of varying surface texture, because the percent standard
deviation is approximately what one would expect from counting error
alone.

TABLE I

TYPICAL COUNT DATA

% Nb	Nb Counts/ 20 Seconds	U Counts/ 20 Seconds
5.07	28660	108425
6.61	30890	99630

TABLE II

U—Nb RATIOS VS. Nb CONCENTRATION

FOR STANDARD COMPACTS

Nb, Wt. %	Nb Intensity/ U Intensity *	S.D.	% RSD
5.07	0.264	0.0028	1.06
5.17	0.269	0.0030	1.12
5.43	0.277	0.0021	0.76
5.56	0.282	0.0018	0.64
5.94	0.299	0.0026	0.87
6.22	0.310	0.0031	1.00
6.61	0.327	0.0032	0.98

*Each value is the average of 14 measurements.

A least squares plot of the above data is given in Figure 4. Each point represents the average of 14 ratios collected over approximately one month and the correlation coefficient is 0.999. For any group of three ratios measured consecutively the correlation coefficients were 0.995 or greater. This kind of correlation indicates that compaction of turnings averages out the inhomogenieties in any given sample so that the x-ray detector "sees" an average composition that is equivalent to the percent niobium obtained by the wet chemical method.

Comparative statistics were obtained for the aqueous x-ray procedure and the compaction procedure. The difference in niobium-to-uranium intensity ratios between duplicate aliquots was determined. For the aqueous procedure, data from 125 samples were used; for compactions, data from 52 samples were used. The results are as shown below:

Procedure	Mean Difference	Standard Deviation
Aqueous	0.0071	0.0068
Compaction	0.00192	0.00156

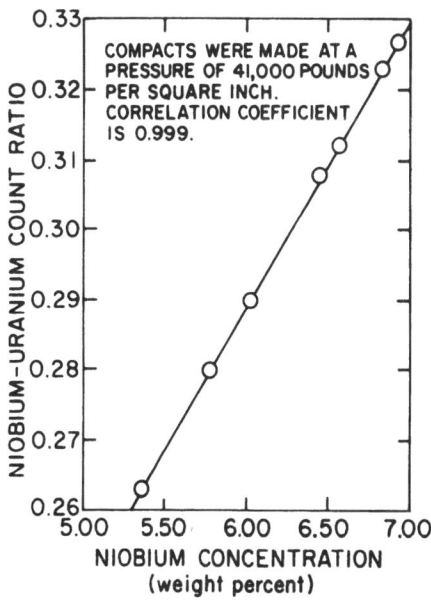

Fig. 4. Least squares plot niobium concentration vs niobium-uranium ratio

These results clearly show a statistical advantage for the compaction procedure. This advantage made it possible to do a niobium distribution study on an individual ingot. When this study was attempted with the aqueous x-ray procedure, the results did not provide useful correlations. The reason is that if the niobium-uranium intensity ratio varies by two standard deviations (2 x 0.0068), then the percent niobium varies by approximately 0.3 and this is much too large because 0.05 percent niobium variations are significant. However, when the study was repeated with the compaction procedure, correlatable results were obtained as shown in Figure 5.

The uniformity in niobium concentration in the center sample as compared to the edge samples is evident. This was expected because the center remains molten longer and thus the niobium has a longer time to melt and/or dissolve and to distribute more uniformly. Precisely why the niobium concentration in the edge samples changes abruptly near the middle of the ingot is not known, but it is consistent with the more rapid freezing occurring near the outside of the ingot.

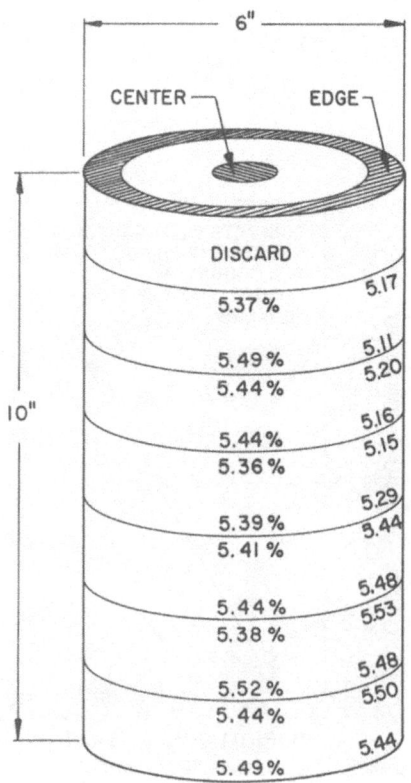

Fig. 5.   Sectioning and analysis of ingot

Use of the compaction procedure reduces both elapsed time and manhours required for niobium analyses. The aqueous procedure required several hours for a single sample but the compaction procedure required less than an hour. The aqueous procedure required 2.5 man-days for 20 samples in duplicate compared to less than 0.7 man-days for 20 samples in duplicate by the compaction procedure. Additional manhours savings are obtained because no time is required to prepare reagents.

SUMMARY

An x-ray fluorescence procedure was devised for analyzing niobium in uranium-niobium alloys, which completely eliminates wet chemical preparation of samples. This was accomplished by pressing turnings into compacts and determining niobium and uranium intensities on the surface of the compact. Standards were prepared by relating x-ray intensity data to precisely selected samples analyzed by a wet chemical procedure.

The statistical variability in niobium results has been reduced to the level where meaningful correlations are obtained when an arc melted ingot is sectioned and analyzed.

In addition to the above benefits, a time savings by a factor of 3 or greater has been realized.

ACKNOWLEDGEMENTS

The author wishes to express his appreciation to Larry Wilson for supplying alloys used for standards, to Amy Miller for the wet chemical analyses and to Clint Heiple for assistance in compaction.

REFERENCES

1. S. Kallman, "Treatise on Analytical Chemistry," Part II, Vol. 6, I. M. Kolthoff, P. J. Elving, and E. B. Sandell, eds., Interscience Publishers, New York (1964) p. 232.

2. Private communication from D. P. Anderson, Rockwell International Golden, Colorado.

3. E. W. White and G. G. Johnson Jr., "X-Ray and Absorption Wave-lengths and Two-Theta Tables," A.S.T.M. Committee E-2, Philadelphia, Pennsylvania (1970).

IN VIVO X-RAY FLUORESCENCE ANALYSIS

FOR MEDICAL DIAGNOSIS

L. Ahlgren, T. Grönberg, and S. Mattsson

Radiation Physics Department
University of Lund
S-221 85 Lund, Sweden

INTRODUCTION

Occupational exposure to lead is common in many industrial applications and hence it is of considerable medical interest to control the body-burden of lead in living man. More than 90 % of the lead in the body is concentrated in bone and hence in vivo measurements of the lead in the skeleton should give the most satisfactory way for estimating the body-burden. The routine method used today for checking on lead contamination is that of measurements on blood samples. However, since the concentration of lead in the blood is a sensitive function of the actual exposure conditions, this method provides only a poor indication of the total body-burden and the integrated lead exposure.

Cadmium is another highly-toxic heavy metal, which has wide spread application, particularly in metallurgy and the plastic industry. The kidneys are the principal site of accumulation in man and hence renal cadmium levels are a good biological indicator of the total body-burden. A simple method for in vivo detection should enable us to assess and control cadmium toxemia. In the present work, we will now discuss the use of in vivo X-ray fluorescence analysis for cadmium determination in the kidneys and lead determinations in the fingers of living man.

After the pioneer work of HOFFER et al[1] on fluorescent scanning of natural iodine in the thyroid several authors have suggested the use of X-ray fluorescence analysis to follow in vivo the distribution and elimination of stable tracers injected prior to the investigation. None of these methods have as yet found any wide spread use as routine methods for man.

Patients with suspected malfunction of the kidneys normally undergo urography, which is an X-ray examination using large amounts of contrast media. For some of these patients a quantitative determination of kidney function must be made and this is nowadays normally carried out by a separate study by measuring the clearance from blood of intravenously injected ^{51}Cr-EDTA or ^{125}I- or ^{131}I-labeled contrast media. If the initial injection of the contrast medium could be followed by in vivo X-ray fluorescence analysis, the injection of radioactive material and the repeated sampling of blood could be eliminated. We will discuss the possibilities of using this method for the clinical evaluation of kidney function.

IRRADIATION AND DETECTION ARRANGEMENTS

a) Lead determinations: The left forefinger was fixed in a perspex holder and about 1 cm^3 of the second phalanx was irradiated 40 minutes with two collimated ^{57}Co sources having a total activity of 0.8 GBq ($\approx$20 mCi). The countrate of characteristic X-rays emitted from lead was measured with a Ge(Li) spectrometer (16 mm diam x 5.2 mm; energy resolution: FWHM= 750 eV at 75 keV). The mean angle between the incident and measured radiation was 90°. The diameter of the bone measured was calculated from two X-ray pictures taken at orthogonal projections. The bone mineral concentration in the finger bone was estimated from the quotient of the coherent and the incoherent scattered primary photons[2] and the lead concentration in the finger bone was then derived from measurements on a finger phantom made of silica paraffin wax and bone ash with known size and bone mineral concentration[2].

b) Cadmium determinations: The right kidney was carefully localized by a B-scanning ultrasonic equipment. The collimated beam from an 11 GBq ^{241}Am-source (300 mCi) and the collimated Ge(Li) detector was then directed towards the centre of the kidney as given by the ultrasonic measurement. The angle between the primary and measured radiation was 110°. The collimators were wide enough to irradiate and measure the whole kidney. To reduce the fluence rate of the low energy photons emitted from the source a 70 μm Mo filter was placed in front of it.

c) In vivo measurements of iodine concentration: The same 11 GBq ^{241}Am source which was used for cadmium measurements under b) was used in this experiment. After injection of the contrast agent, the iodine concentration in the forefinger tip (0.5-1 cm^3) of the patient was studied.

RESULT AND DISCUSSION

a) Lead: The minimum detectable concentration of lead in forefinger bone was 20 μg per g of bone tissue. The mean absorbed dose in the irradiated part of the finger was 2.5 mGy (250 mrad) as

measured with LiF dosemeters and the total energy imparted was estimated to be 0.01 mJ, which is insignificant in comparison to an
ordinary X-ray examination of the hand (compare table 1).

Figure 1 shows the relation between the measured lead concentrations in occupationally exposed persons and our minimum detectable concentration. 22 persons who had earlier worked at a storage
battery plant were studied; 15 of them had measurable concentrations of lead in their forefingers. We have also studied 5 retired
workers from a metal industry, all of them showed significant contamination with Pb.

The correlation between the Pb-concentrations in the blood
and in the skeleton has been shown to be bad for the group of occupationally exposed persons we have studied[4].

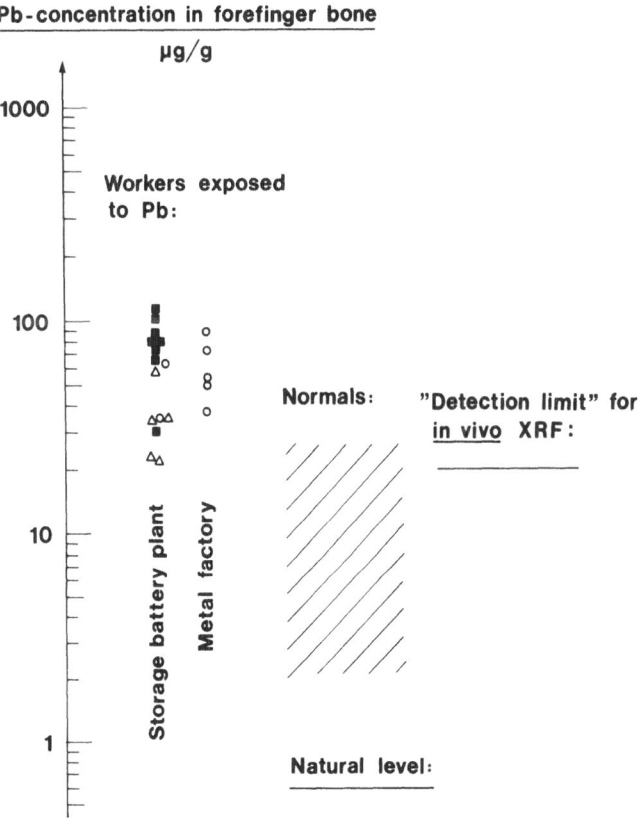

Fig. 1.   Measured lead concentrations in the forefinger bones of
           two groups of occupationally exposed workers with exposu-
           re times 3 - 15 years (Δ), 15 - 30 years (0) and 30 - 45
           years (■). The range of reported lead concentrations in
           the skeletons of normal individuals[3] is also indicated
           in the figure as is the present minimum detectable con-
           centration using in vivo XRF.

Our results show that even with the present X-ray fluorescence technique, it is possible to follow and get quantitative information on the lead concentration in some of the normal individuals. It is, however, very important to lower the detection limits so that new groups can be made accessible for quantitative studies of lead contamination (children living near streets, people working with gasoline etc).

b) <u>Cadmium</u>: The analytical problems in connection with the <u>in vivo</u> measurements of cadmium in the kidney are considerably more difficult than these arising in the determination of lead concentration in fingers. This is mainly due to the difficulties of controlling the irradiation and detection geometry of the kidney. The sensitivity for cadmium detection has been studied in water phantoms and it has been found that the sensitivity and the detection limit are strongly dependent on the depth of the organ. The sensitivity within a kidney of diameter 4 cm will vary by a factor

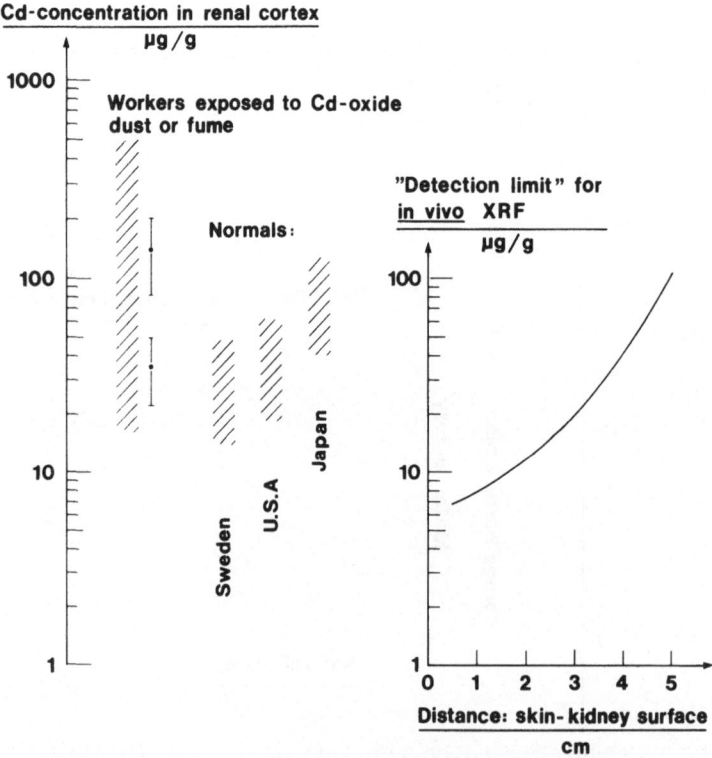

Fig. 2.   A comparison between the reported concentrations of Cd in the renal cortexes of occupationally exposed workers and normals[5] and our minimum detectable Cd concentration in kidney cortex having <u>in vivo</u> X-ray fluorescence analysis. The results of our first two <u>in vivo</u> measurements on persons with known Cd-exposure are also indicated (Ī).

of 30 - 180 depending on the geometrical arrangement. This very
pronounced variation of sensitivity with depth gives us a possibility for a selective measurement of Cd in the kidney cortex. It is,
however, necessary to make a very careful determination of the distance between skin and kidney surface. With our ultrasonic measurement the uncertainty in measured position is ± 3 mm, giving an uncertainty in the absolute concentration of cadmium in the kidney
cortex of about ± 40%. In figure 2 is shown the range of cadmium
concentrations in the renal cortex[5] of workers exposed to cadmium
oxide dust or fume as well as of normals. The two separate values
are the results of our first in vivo X-ray fluorescence measurements on persons with known exposure to cadmium. The figure shows
that the X-ray fluorescence technique is useful for quantitative
measurements of cadmium concentration in the renal cortex of most
of the occupationally exposed persons, if the distance between
skin and kidney is less than 4 cm. Non occupationally exposed persons, preferentially persons with kidneys close to the skin, are
also available for quantitative determination of cadmium concentration in their renal cortexes.

    c) In vivo measurements of contrast agents for the clinical
          evaluation of kidney function: In patients undergoing uro-
graphy it was possible to follow iodine concentrations by in vivo

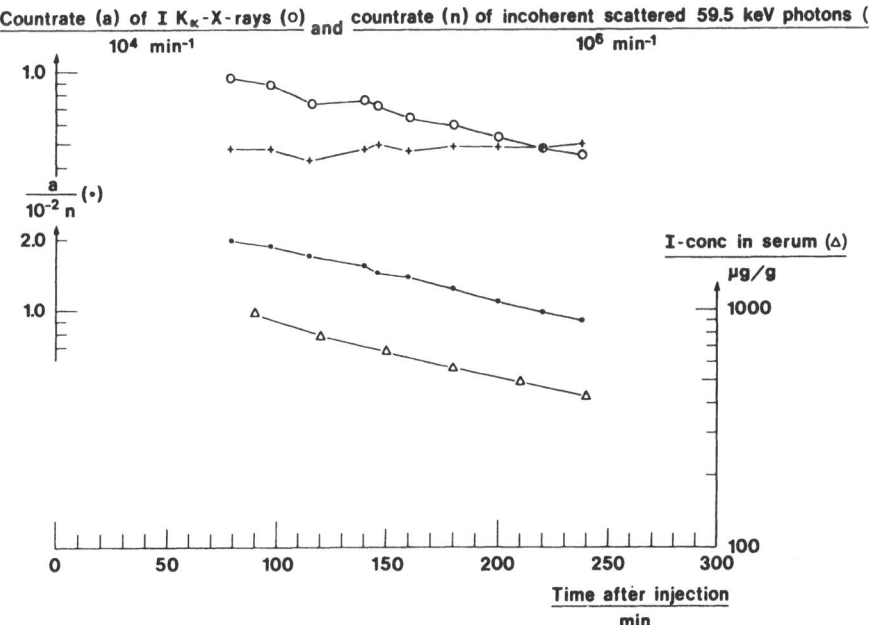

Fig. 3.   Retention of iodine in the finger-tip of a patient who
           was given Isopaque[R] for urography (280 mg I per kg body
           weight). Serum data is given for comparison.

measurements on the finger-tips of patients with normal kidney function up to 6-10 hours after injection. Patients with reduced kidney function can be followed for still longer periods. The important point is that we have found identical retention curves by means of in vivo measurements on the patient´s finger-tip and by means of serum analysis in vitro. In repeated measurements on the finger-tips, it has shown to be very valuable to use the size of the distribution of incoherent scattered photons as an internal standard to correct for varying positions of the finger from measurement to measurement. This is illustrated in figure 3 which shows the uncorrected counting rate in the I K_α-peak (a) and in the distribution of incoherent scattered photons (n) as well as the ratio a/n, which is very insensitive to variations in the measurement geometry.

Table 1.   The mean absorbed dose and the energy imparted in in vivo X-ray fluorescence analysis (XRF). Data for other type of radiological examinations are given as a comparison

Type of investigation	Mean absorbed dose mGy	Energy imparted mJ
XRF of lead in finger bones	2.5	0.01
Ordinary X-ray picture of a hand	0.65	0.10-0.25
XRF of cadmium in a kidney	0.6	0.4
Prompt gamma neutron activation analysis of cadmium in a kidney	0.57^7 (neutrons) 0.29^7 (photons)	$1.1^{x)}$ 0.6
Kidney function studied by XRF of the finger-tip after injection of contrast medium	2.3	0.003
Urography		510
^{51}Cr-EDTA clearence		0.8

x) The biological effects of the energy imparted by neutrons at low absorbed doses may be 10-100 times larger than if the same energy is imparted by photons[8,9].

CONCLUSION   ·

In vivo X-ray fluorescence analysis has important applications
in occupational and environmental medicine for direct measurements
of Pb and Cd contamination. In the future, great efforts will be
directed to lowering the detection limit for these elements so as
to be able to check the contamination of new groups of high risk,
for example children. The use of plane polarized photons for excitation may be valuable in reaching lower detection limits. In studies of organ function there seem to be good possibilities for a
continuous development of X-ray fluorescence techniques in combination with the development of new tracer substances.

REFERENCES

1. P. B. Hoffer, W. B. Jones, R. B. Crawford, R. Beck and A.
    Gottschalk, Fluorescent thyroid scanning: A new method
    of imaging the thyroid, Radiology 90, 342-344 (1968).
2. L. Ahlgren and S. Mattsson, An X-ray fluorescence techni-
    que for in vivo determination of lead concentration in
    a bone matrix, Phys. Med. Biol. 24, 136-145 (1979).
3. P. S. I. Barry, Distribution and storage of lead in human
    tissues, in: The biogeochemistry of lead in the envi-
    ronment, Part B. Biological effects. Ed. by J. O.
    Nriagu, Amsterdam 1978 pp. 97-150.
4. L. Ahlgren, B. Haeger-Aronsen, S. Mattsson and A. Schütz,
    In vivo determination of lead in the skeleton follow-
    ing occupational exposure, Accepted for publ. in
    Br J. Ind. Med.
5. L. Friberg, M. Piscator, G. F. Nordberg and T. Kjellström,
    Cadmium in the environment, Second edition. CRC Press,
    Cleveland, Ohio U.S.A., 1974.
6. T. Grönberg, T. Almén, K. Golman, K. Lidén, S. Mattsson
    and S. Sjöberg, Noninvasive determination of kidney
    function by X-ray fluorescence analysis (in prepara-
    tion).
7. K. J. Ellis, D. Vartsky and S. H. Cohn, A mobile prompt-
    gamma in vivo neutron activation facility, in: Symp.
    Nucl. Activation Techniques in the Life Sciences,
    Vienna, 22-26 May 1978, IAEA, Vienna, 1979, Paper IAEA
    SM-227/78.
8. H. H. Rossi, A proposal for revision of the quality fac-
    tor, Rad. and Environm. Biophys. 14, 275-283 (1977).
9. J. F. Thomson, K. H. Allen and R. J. M. Fry et al, Life
    shortening after exposure to neutrons and gamma rays,
    Annual Report 1977, Division of Biological and Medical
    Research, Report ANL-78-90, Argonne National Laboratory,
    Argonne, Illinois, U.S.A. pp 92-94 (1978).

EFFECT OF CHEMICAL STATE UPON PHOSPHORUS-$L_{2,3}$

FLUORESCENCE SPECTRA

Kazuo Taniguchi

Department of Solid-State-Electronics
Osaka Electro-Communication University
18-8, Hatsumachi, Neyagawa, Osaka 572, Japan

ABSTRACT

The $P-L_{2,3}$ emission spectra of the phosphorus compounds were obtained using a secondary excitation. We found that the spectra depended upon the chemical state, the main peak of $L_{2,3}$ emission spectra shifts to a higher energy, and the peak of intensity of higher energy increases with an increase of oxidation number. It should be noted that the phosphorous spectra of the phosphate anions scarcely reveal influences of the phosphate compounds, except that of H_3PO_4. We concluded that the spectral feature for the phosphorus compounds is influenced by the condition of surrounding atoms but is not influenced by the bond condition.

INTRODUCTION

X-ray spectroscopy has long been a basic method for studying the elemental composition of materials and the electronic structure of atoms. The electronic structure of the initial ionization state of the atom is relatively insensitive to its molecular or condensed matter environment, but that of the final ionized state is strongly affected by this atomic environment. The major features of the associated x-ray emission spectra can yield direct information as to the populations, energy levels, and widths (refer to life time of the excited state) of the outer electronic levels. Transitions between the molecular orbital states and the nearby relatively sharp core level states result in low energy x-ray spectra that typically lie in the 100-200 eV. Such low energy x-ray emission spectra can be directly applied, using a standard sample system, for the identification of unknown valence state, and for the investigations of the electronic structure of chemical compounds. As is well known, $K\beta$ emission spectra result from the transitions from 3p levels to 1s levels. Thus we can investigate only 3p electrons in the valence band for the second-row elements. But as $L_{2,3}$ emission spectra result from the transitions from 3s or 3d levels, we can investigate 3s and 3d character in the valence band by measuring the $L_{2,3}$ emission spectra.[1] Such spectra can sensitively portray the orbital structure of the molecule and the chemical conditions.

As the samples are decomposed under electron bombardment, fluorescence excitation is frequently used in the region of soft x-rays. However, in the region of ultrasoft x-rays, it is difficult to obtain spectra of sufficient intensity with fluorescence excitation. In an earlier work, Henke and Smith[2] were able to demonstrate the feasibility of obtaining the $L_{2,3}$ spectra of chlorine in different chemical states by applying spectrographic techniques of such efficiency as to minimize the effects of radiation decomposition. Recently, $L_{2,3}$ emission spectra of sulfur[1] and chlorine[3] in different chemical states have been more precisely measured and studied by Taniguchi, Henke, Perera and others with fluorescence excitation. In this present work, the phosphorus-$L_{2,3}$ fluorescence spectra for several phorphorus compounds have been obtained using carbon photon excitation and lead lignocerate analyzing crystal.

EXPERIMENTAL

The high efficiency techniques which have been developed by Henke and Tester[4] for secondary excitation of ultrasoft x-ray spectra have made possible the interpretation of light element valence band x-ray spectra.[5] Ultrasoft x-ray excitation leads to reduced radiation damage to samples and provides considerably improved peak to background ratios.

The x-ray spectrometer is a vacuum x-ray spectrometer provided with a soller collimator. The basic spectrographic approach that is used in this laboratory is shown in Fig. 1. After reflection from the plane crystal, the beam from the fluorescent radiator passes the soller collimator before the counter. The most intense and practical line source in the P-$L_{2,3}$ excitations is carbon-$K\alpha$ (44.7Å). It is generated in our demountable sources using a copper anode element which is coated with a colloidal graphite. The source is stable for many hours of operation, typically at 8 kV and 150 mA. The x-ray tube window is constructed with 5µ polycarbonate (Kimfol) with a few hundred angstroms thick aluminum coating which has an effective transmission band for the

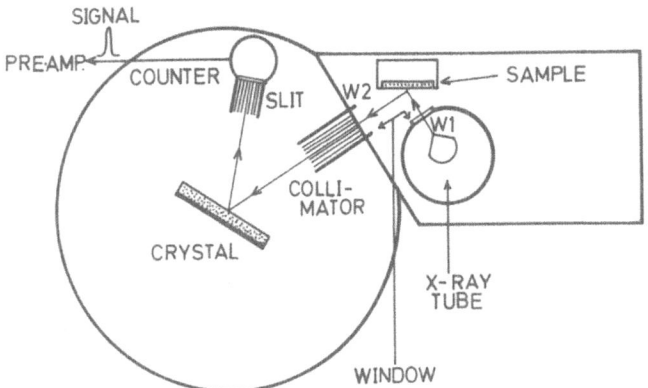

Fig. 1. Schematic diagram of the low energy x-ray spectroscopy
         applying close coupled demountable x-ray source.

carbon $K\alpha$ radiation. The plane crystal is a specially developed
Langmuir-Blodgett type multilayer analyzer of lead lignocerate
which has 2d-spacing of 130Å. The background and impurity lines
of higher order are reduced by using a gas flow proportional
counter with pulse-height discriminator. Propane gas at 50 torr
pressure is used as a counting gas and is sealed in the counter
with polyvinyl-formal (formvar). The solid specimens are pressed
as a fine powder into a smooth aluminum ring, in order to keep the
target sample pure and at constant pressure. The liquid samples
were measured in the frozen condition using a cryostat.

     The spectrum is calibrated by the sharp $M\zeta$ lines of zirconium
of 81.9Å and molybdenum of 64.5Å. A detailed description of these
experimental techniques have been reported elsewhere.[5] The measurement time was distributed among two identical samples in order
to insure that the radiation decomposition effect upon these measurements would be negligible. The total counting times are 50
seconds per point and the counter is step scanned by 1/10 degree.

RESULTS AND DISCUSSION

     The phosphorous-$L_{2,3}$ fluorescence spectra of H_3PO_4, Li_3PO_4,
KH_2PO_4, Na_3PO_4, and K_3PO_4 are shown in Fig. 2. It should be noted
that the spectra of the central phosphorus atom of phosphate anions

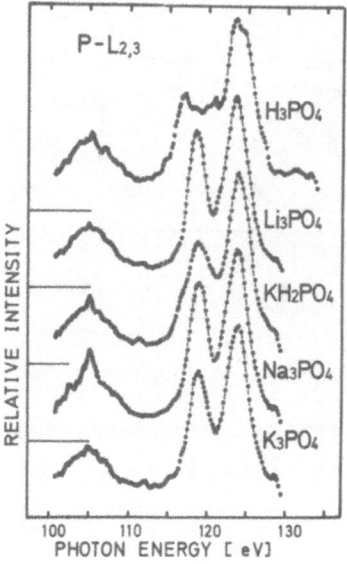

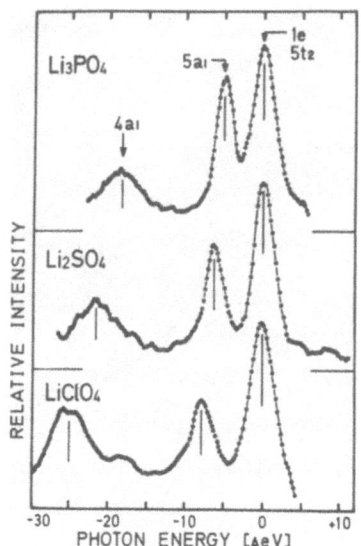

Fig. 2. Comparison of the feature of the phosphorus-$L_{2,3}$
emission spectra for the several
phosphate compounds.

Fig. 3. Comparison of the $L_{2,3}$
emission spectra of the central atom of Td symmetry for
Li_3PO_4, Li_2SO_4, and $LiClO_4$.

scarcely reveal any influence of the cations of these phosphate
compounds except for H_3PO_4. This suggests that the interaction
between the phosphate anion and the cation is relatively small,
and that the crystal effect be neglected. In the case of H_3PO_4,
however, the spectral feature differs from other phosphate compounds, but the peak positions at the lower energy side and higher energy side are much the same. The phosphate anion for strong
localization has a tetrahedral symmetry (Td point group). But
the phosphate anion of H_3PO_4 does not have a tetrahedral symmetry
because it is not strong localization between PO_4^{---} ion and three
H^+ ions. It then seems reasonable that H_3PO_4 has C_{3v} molecular
symmetry in due consideration of three hydrogen.

In general, there are similar features of the $L_{2,3}$ emission
spectra between molecules having Td symmetry of the different
central atom, although the energy range of the emission spectra
differs. The comparison of the $L_{2,3}$ emission spectra of the central atom of Td symmetry for Li_3PO_4, Li_2SO_4, and $LiClO_4$ is shown
in Fig. 3. There are seven occupied orbitals in the valence band
for the Td molecular symmetry, so that transitions into the $L_{2,3}$-
level are allowed only from a_1, t_2 and e_1 orbitals with the
dipole selection rules. The main peak at higher energy has to be

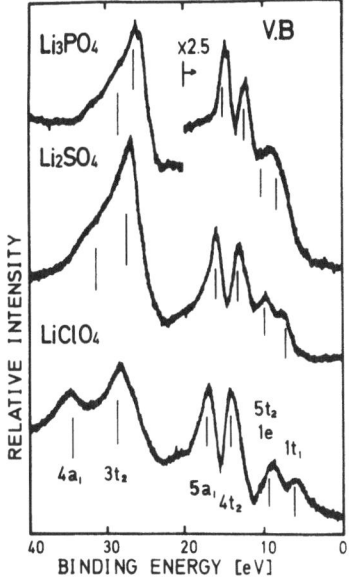

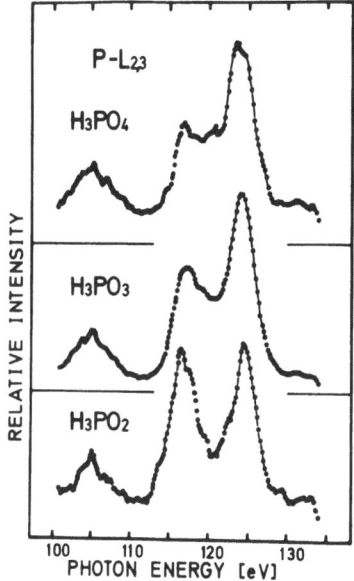

Fig. 4. Comparison of the x-ray photoelectron spectra for the valence band region from Li_3PO_4, Li_2SO_4, and $LiClO_4$.

Fig. 5. Comparison of the feature of the phosphorus-$L_{2,3}$ emission spectra for H_3PO_4, H_3PO_3, and H_3PO_2.

assigned $1e$ and $5t_2$ orbital with its comparatively high 3d population. The peak at lower energy and middle energy should be assigned $4a_1$ and $5a_1$, respectively, of which the molecular orbital consists mostly of the 3s population. The x-ray photoelectron spectra for the final level (valence band) for Li_3PO_4, Li_2SO_4, and $LiClO_4$ are shown in Fig. 4. We found that the molecules having Td symmetry of different central atom shows the similar feature in XPS. For this reason, there is a similar feature of the $L_{2,3}$ emission spectra between the molecules having Td symmetry of the different central atom.

The comparison of the phosphorus-$L_{2,3}$ emission spectra for phosphoric acid (H_3PO_4), Phosphorous acid (H_3PO_3) and hydrophosphorous acid (H_3PO_2) is shown in Fig. 5. Each spectrum consists of three major peaks at about 105 eV, 117 eV and 124 eV. We found that the measured spectra changed depending upon the chemical state. In proportion to the increase in the number of oxygens, the peak at about 117 eV decreases, and the peak at about 124 eV increases and shifts to low energy side just a little.

The compositional formulae for these compunds are shown in Fig. 6. In the constitutional formula of H_3PO_4, the central P

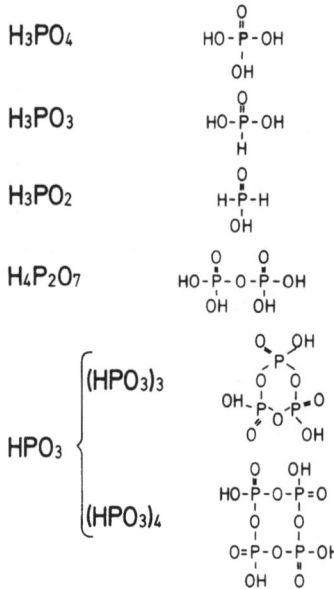

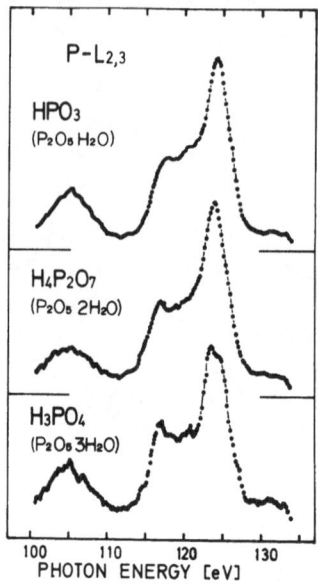

Fig. 6. The chemical constitutional formulas for H_3PO_4,
H_3PO_3, H_3PO_2, $H_4P_2O_7$, and HPO_3.

Fig. 7. Comparison of the feature of the phosphorus-$L_{2,3}$
emission spectra for HPO_3,
$H_4P_2O_7$, and H_3PO_4.

atom combined with O atoms only. In the H_3PO_3, the central P atom
combines with three O atoms and one H atom. But in the H_3PO_2, the
central P atom combines with two O atoms and two H atoms. I believe that the peak intensity is changed with the number of oxygen
atoms. However, in the constitutional formulas for metaphosphoric
acid (HPO_3), pyrophosphoric acid ($H_4P_2O_7$) and phosphoric acid
(H_3PO_4), the central P atom combines with oxygen atoms only.
Fig. 7 shows the comparison of the P-$L_{2,3}$ emission spectra for
HPO_3, $H_4P_2O_7$ and H_3PO_4 which the central P atom combines with O
atoms only. In this case, the spectral features of these compounds have something in common with each other. It then seems
reasonable to consider that the spectral feature for these compounds is influenced due to the condition of surrounding atoms and
is not influenced due to the bond condition.

We should like to consider the relation of the number of
oxygen atoms. We know that the peak of higher energy side of $L_{2,3}$
spectra grows in proportion to the increase in the number of oxygen atoms for the phosphoric acid, phosphorous acid and hypophos-
phorous acid. We can see the same phenomenon for chlorine and
sulfur compounds. Fig. 8 shows the comparison of sulfur-$L_{2,3}$

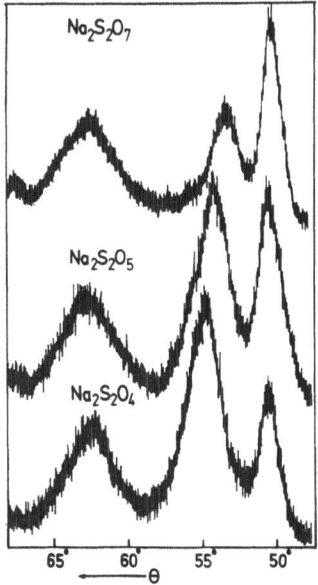

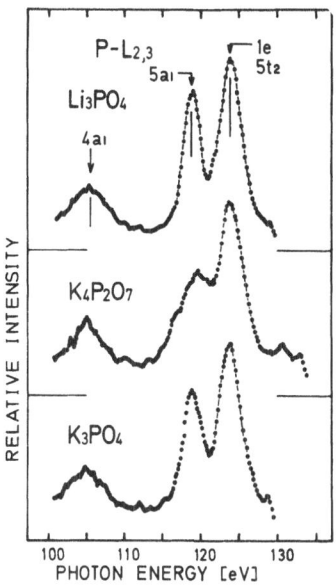

Fig. 8. Comparison of the feature of the sulfer-$L_{2,3}$ emission spectra for $Na_2S_2O_7$, $Na_2S_2O_5$, and $Na_2S_2O_4$.

Fig. 9. Comparison of spectral feature of phosphoric acid in which hydrogen atom is change to potassium atom.

emission spectra for $Na_2S_2O_7$, $Na_2S_2O_5$, and $Na_2S_2O_4$. In proportion to the increase in the number of oxygens, the peak of higher energy increases and the middle peak shifts to higher energy and decreases. In this work, we cannot interpret transitions from the molecular orbital with the help of a molecular orbital calculation. However, it seems reasonable to consider that the peak of higher energy side is related to the number of oxygen atoms.

   Next, we require further examination of the spectral feature of phosphoric acid. In Fig. 2, the spectrum of H_3PO_4 differs from other phosphates, especially at the middle energy range. If you look carefully at Fig. 2, you can see that the spectral feature of KH_2PO_4 differs just a little from the other phosphates and that there is a hump at the low energy side of the middle peak. It then seems reasonable that this peak is related to the hydrogen atom. Fig. 9 shows the comparison of spectral features of potassium prophosphate and potassium phosphate in which the hydrogen atoms of pyrophosphoric acid and phosphoric acid are changed to potassium atoms. The middle peak of these compounds is shifted to the high energy side compared with phosphoric acid, and the spectral feature of potassium phosphate is similar to Li_3PO_4 which has the Td symmetry.

CONCLUSION

In this work we have shown that a spectral feature is closely connected with the chemical state, and that $L_{2,3}$ emission spectra are very useful for the chemical state analysis. We have found that the main peak of $L_{2,3}$ emission spectra shifts to higher energy side and the peak of higher energy side increases according to an increase of oxidation number. It is convenient to think that the oxygen 2p electrons diffuse to the 3d atomic orbital of the central atom in the valence molecular orbital when the central atom combines with oxygen. We have found these techniques to be useful in the study of the electronic structure of the molecular orbital by combining the results of the $L_{2,3}$ emission spectra. In this work, we can interpret as transitions from the molecular orbitals with the help of a molecular orbital calculation. We would like to apply the intensity approximation using molecular orbital theory for $L_{2,3}$ emission spectra. These are subjects for future study in our work.

ACKNOWLEDGMENT

The main program of this work had been begun by the author during his stay at the University of Hawaii. He expresses his sincere thanks for the kind guidance of Professor B. L. Henke. We gratefully acknowledge some very useful discussions with Dr. Ikeda and Dr. Adachi of Osaka University, and the undergraduate students in their laboratory.

REFERENCES

1. K. Taniguchi and B. L. Henke, "Sulfur $L_{2,3}$ Emission Spectra and Molecular Oribtal Studies of Sulfur Compounds," J. Chem. Phys. 64:3021-3035 (1976).

2. B. L. Henke and E. N. Smith, "Valence Electron Band Analysis by Ultrasoft X-Ray Fluorescence Spectroscopy," J. Appl. Phys. 37:922-923 (1966).

3. B. L. Henke, R. C. C. Perera, and D. S. Urch, "Cl-$L_{2,3}$ Fluorescent X-Ray Spectra Measurement and Analysis for the Molecular Orbital Structure of ClO_4^-, ClO_3^-, and ClO_2^-," J. Chem. Phys. 68:3692-3704 (1978).
K. Taniguchi *et al.*, "Molecular Orbital Studies of Chlorine Compounds by Cl-$L_{2,3}$ Emission Spectra and Photoelectron Spectra," Report of Osaka Electro-Communication University 14:165-171 (1978).

4. B. L. Henke and M. A. Tester, "Techniques of Low Energy X-Ray Spectroscopy (0.1 to 2 keV Region)," *in* W. L. Pickles, Editor, Advances in X-Ray Analysis, Vol. 18, p. 76-106, Plenum Press (1976).

5.   B. L. Henke and K. Taniguchi, "Quantitative Low-Energy X-Ray
     Spectroscopy (50-100 Å Region)," J. Appl. Phys. 47:1027-1037
     (1976).

X-RAY STUDY OF THE BAND STRUCTURE IN STANNIC OXIDE

A. A. Bahgat* and K. Das Gupta

Department of Physics
Texas Tech University
Lubbock, Texas 79409

INTRODUCTION

In this article we present the experimental data on L_3-absorption spectra of metallic tin and tin in SnO_2. We have used (i) a bent crystal spectrometer[1] and (ii) a high resolution flat crystal spectrometer[2] with a microfocus target as the source for the continuum radiation. We have observed a "white line" on the low energy side of the primary L_3-absorption edge of Sn in SnO_2. Similar "white lines" in absorption spectra were reported earlier by M. Siegbahn[3] in rare earth elements with f-vancies, who designated such lines appearing immediately on the low energy side of the primary absorption edge as "white lines" in absorption spectra. It is interesting to note that metallic tin does not reveal the "white line", while Sn in SnO_2 clearly revealed a sharp white line on the low energy side of the primary absorption edge.

EXPERIMENTAL

The present study was performed with two different soft x-ray vacuum spectrographs. The first was a bent crystal spectrograph of radius 508 mm designed to work at large Bragg angles and was described earlier.[2] The second was a flat silicon crystal spectrometer with a microfocus x-ray source, the geometrical setup was described recently[1] and is shown in Fig. 1. The absorption spectra were taken in the second order of quartz to obtain a higher resolution using the bent crystal spectrograph and in the first order of silicon using the flat crystal and the continuum from the microfocus target.

*Present address: Dept. of Physics, Al-Azhar Univ., Cairo, Egypt.

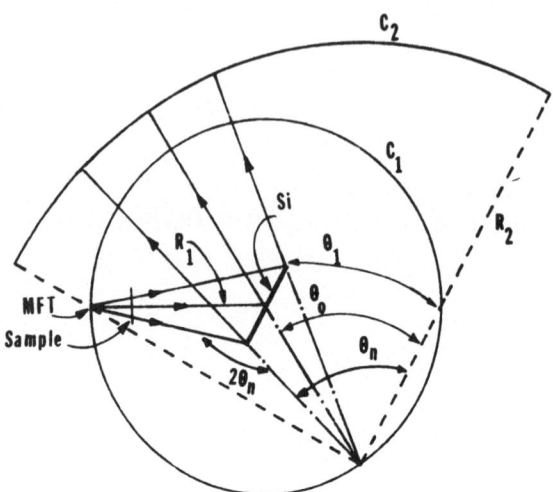

Fig. 1.   A schematic representation of the flat crystal spectrometer.
          R_1=36cm: is the radius of the circle C_1 centering the
                  analyzing crystal of silicon.
          MFT: is the position of the microfocus target 10μm x 10μm
                  as the source of the continuum radiation.
          R_2=67cm: is the radius of the circle C_2 centering the
                  vertual image of the MFT on the circle C_1.
          2L=7cm: is the horizontal width of the exposed part of the
                  analyzer crystal of silicon, θ_0=30°.

     The x-ray tube of the bent crystal spectrometer was operated
at 6 kV, 30 mA.   The sample was kept at a distance of 5 cm from the
analyzing crystal.   The vacuum was maintained at 10^{-6} torr.   The
purity of the sample was 99.99 per cent.   The sample thickness for
different sets of experiments varied between 1-2 micron. $K\alpha_1$, $K\alpha_2$
lines of copper were taken as reference lines.   The microfocus tube
was operated at 15 kV, 0.4 mA when the flat Si(111) crystal was
used.

     The dispersion is 7 eV mm^{-1} and 10 eV mm^{-1} for quartz and for
silicon crystals respectively.   The line width of SnLb$_2$ is 7 eV at
half maximum.   The instrumental broadening is small compared with
the accepted data.

RESULTS AND DISCUSSION

     Figure 2 represents the L_3-absorption spectra of Sn in metal
and in SnO$_2$ as taken with the quartz crystal in the second order.
Figure 3 represents the absorption of SnO$_2$ as taken by the flat
crystal spectrometer.   The zero of the energy scale is taken at

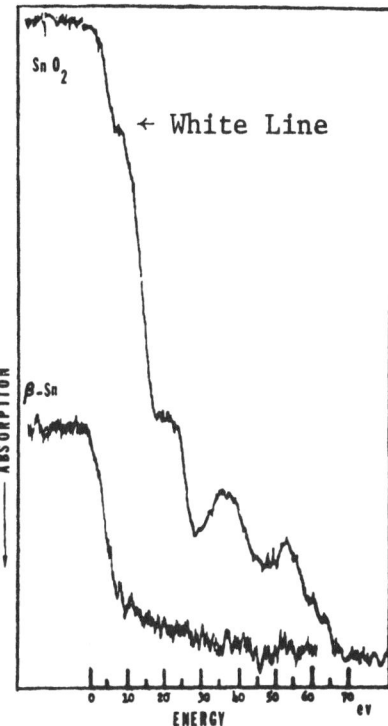

Fig. 2. L_3-absorption spectra of Sn in metal and oxide, using a
           bent crystal spectrometer.

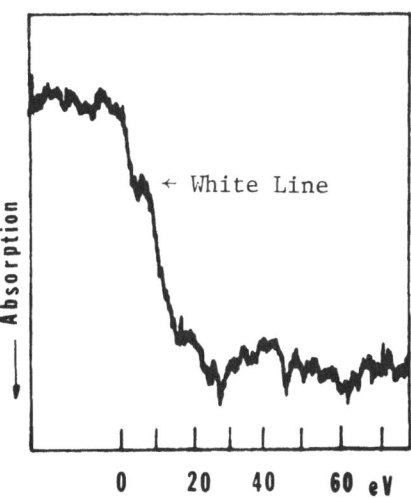

Fig. 3. L_3-absorption spectrum of Sn in oxide using the flat
           crystal spectrometer.

the first indication of absorption, and it is assumed that it corresponds to the bottom of the conduction band of Sn. The L_3-absorption edge of metallic tin as shown in Fig. 2 does not reveal the white line observed in the absorption spectrum of SnO_2, although there is a slight indication of a hump in L_3-absorption spectra of tin which in Fig. 2 is almost unnoticable. The position of the white line has been marked in Fig. 2 which in the orginal negative appears as a distinct white line. The L_3-absorption spectrum of tin in SnO_2 shows a frequency shift by 2.5 eV relative to that of tin metal. The white line appears at 7 eV and extended fine strutures are prominent in the absorption spectra of SnO_2 and is absent in metallic tin in the same region. The appearance of a prominent white line in the vicinity of the primary absorption edge indicates a high transition probability of 2p electrons of Sn to the vacant levels of the valence band of SnO_2. It is believed that valence electrons from Sn are transferred to oxygen centers in SnO_2 and the resultant valence band should be considered a characteristic feature of the valence band of SnO_2. In L_3-absorption spectra of tin in tin oxide the initial state is a vacancy in the 2p level which increases the effective z value of tin and the unoccupied levels immediately after the filled band is high in SnO_2. If we consider this absorption process in the final state is a transfer of 2p electron to an unoccupied level centering Sn nucleus then the wave-function of the unoccupied levels of high density that would give rise to the white line should have 5s or 5d character.

This indicates a high density of vacant nS-levels within the conduction band. Such a result can be correlated and confirmed to that obtained through the internal conversion and Mössbauer effec-isomer shift[4] measurements which lead to the conclusion that 5s (valence-shell) electron density near the nucleus is about 30% smaller in SnO_2 than in tin.

ACKNOWLEDGEMENT

This research supported by the Robert A. Welch Foundation.

REFERENCES

1. S.S. Wald and S.K. Cheng, Frequency Shift of L-Valence Band Spectra of Niobium and Tin in Nb_3Sn Compound, in "Advances in X-Ray Analysis", C.S. Barrett, D.E. Leyden, J.B. Newkirk, and C.O. Rudd eds., Plenum Press, New York (1978) Vol. 21, p. 241.
2. A.A. Bahgat and K. Das Gupta, A New Type of X-Ray Absorption Spectrometer, Rev. Sci. Instrum. 50(8):1020 (1979).
3. B.K. Agarwal, "X-Ray Spectroscopy. Springer Series in Optical Sciences, Vol. 15," David L. MacAdam, ed., Springer-Verlag, Berlin Heidelberg (1979) p. 274.

Oxide Crystal, Brit. J. Appl. Phys. 16:195 (1965).

2. F.J. Arlinghaus, Energy Bands in Stannic Oxide, J. Phys. Chem. Solids 35:931 (1974).

3. J.L. Jacquemin and G. Bordure, Band Structure and Optical Properties of Intrinsic Tetragonal Dioxides of Group-IV Elements, J. Phys. Chem. Solids 36:108 (1975).

4. S.S. Wald and S.K. Cheng, Frequency Shift of L-Valence Band Spectra of Niobium and Tin in Nb_3Sn Compound, in "Advances in X-Ray Analysis", C.S. Barrett, D.E. Leyden, J.B. Newkirk, and C.O. Rudd eds., Plenum Press, New York (1978) Vol. 21, p. 241.

5. A.A. Bahgat and K. Das Gupta, A New Type of X-Ray Absorption Spectrometer, Rev. Sci. Inst., [August] (1979).

6. B. Nordfors and E. Noreland, The X-Ray L-Absorption Spectra of 48 Cd-52 Te, Arkiv För Phy. 20:1 (1961).

7. L.V. Azaroff and D.M. Pease, X-Ray Absorption Spectra, in "X-Ray Spectroscopy", L.V. Azaroff, ed., McGraw-Hill, New York (1959).

8. L.G. Parratt, Electronic Band Structure of Solids by X-Ray Spectroscopy, Rev. Mod. Phys. 31:616 (1959).

9. J.-P. Bocquet, Y.Y. Chu, O.C. Kistner, and M.L. Perlman, Chemical Effect on Outer-Shell Internal Conversion in Sn^{119}; Interpretation of the Mössbauer Isomer Shift in Tin, Phy. Rev. Letters 17:809 (1966).

ENERGY DISPERSIVE XRF COMPOSITION PROFILING USING CRYSTAL COLLIMATED INCIDENT RADIATION

W. J. Boettinger, H. E. Burdette, and M. Kuriyama

National Bureau of Standards
Center for Materials Science
Washington, DC 20234

ABSTRACT

In order to measure changes in composition as a function of distance (macrosegregation) in directionally solidified two phase samples, a well collimated incident x-ray beam is required for XRF analysis. This is accomplished using Bragg diffraction of AgKα radiation from a highly perfect Si crystal. Because the incident beam is also monochromatic, additional advantages are realized: a) the backgrounds caused by Compton and thermal diffuse scattering (TDS) of the incident beam are well localized in the energy spectrum and do not interfere with the fluorescent peaks, b) the TDS can be used as a monitor of the incident photon flux and hence eliminates often substantial errors caused by incident beam intensity fluctuations.

Using several prepared standards, the ratio of PbL counts to TDS counts was found to be a function of the total Pb content of the two phase microstructure, with a reproducibility determined only by counting statistics. Furthermore, the function was found to be nearly linear over a wide range of compositions. Standard methods of absorption or enhancement correction can be employed using this ratio. The spatial resolution, determined by profiling a sharp discontinuity between two metals, was 0.5 mm.

Macrosegregation data is presented for Pb-Sn two phase alloys whose compositions range from 35 wt % Pb to 70 wt % Pb. Comparison of compositions with those determined by a titration method agrees to within 2 wt % for most of the metallurgical structures present in the work. Somewhat larger deviations were found for samples with high Pb contents with extremely coarse two phase microstructures.

INTRODUCTION

In the application of x-ray fluorescence chemical analysis to the measurement of macrosegregation in cast alloys several unique requirements occur. First, composition information must be obtained from materials which are frequently multiphase. Second, the composition should be averaged over the multiphase structure yet with good spatial resolution ($\sim$ 1mm). Third, accuracies of $\sim$ 0.5% or better are required. Attempts at composition profiling using the electron microprobe have proven unsatisfactory[1] presumably due to the presence of multiphase structures and shallow penetration of the electron beam.

The present approach to the measurement of macrosegregation employs the use of a monochromatic incident beam prepared by diffraction of characteristic, tube-excited x-rays by a highly perfect Si crystal. This method gives a reduction of the background as well as angular collimation of the incident beam for good spatial resolution. Results using this method for one-dimensional composition profiling on directionally solidified Pb-Sn two phase alloys are presented.

EXPERIMENTAL

The general x-ray optical alignment employed is shown in Fig. 1. Radiation from a small focus conventional (1kW) tube with a Ag target passes through a vertical slit (0.3 mm) and a horizontal slit (1 mm) 20 cm from the source. These slits are used to reduce the size of the beam. The beam strikes a perfect silicon crystal which is set for 220 diffraction of AgKα in the surface reflection geometry. This diffraction makes the beam quite narrow in the

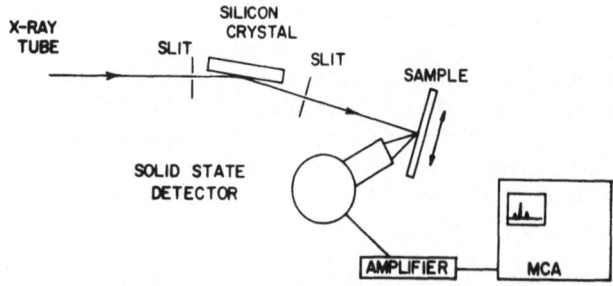

Fig. 1. X-ray optical alignment for Energy Dispersive XRF Chemical Profiling Technique. Characteristic radiation is diffracted from a Si crystal to produce a monochromatic and well collimated beam for incidence on the sample.

horizontal direction. The diffracted beam then passes through a
relatively large vertical slit (1 mm) and horizontal slit (3 mm) to
form the incident beam for XRF analysis. These slits do not reduce
the size of the diffracted Kα beam in the horizontal direction but
greatly reduce the amount of scattering from the primary slits.
Perfect silicon crystal boules for this purpose can be obtained
commercially and are routinely prepared by slicing with a diamond
saw and careful metallographic mechanical polishing[2].

As shown in Fig. 1 the sample is placed perpendicular to the
incident beam and is translated in a horizontal direction to perform composition profiling. Fluorescence from the sample is detected by a Si (Li) detector with a 10 mm diameter window placed
1.5 cm from the sample. The tube was run at 40 keV, 20 mA although
operation above 45 keV can be used to excite (440) diffraction from
the Si crystal to create a bichromatic incident beam if desired.

In the present study of macrosegregation[3] samples consisted of
3 mm diameter cylinders of Pb-Sn alloy which had been directionally
solidified under various conditions of fluid flow. These samples
were previously mounted and sectioned longitudinal to the cylinder
axis (growth direction) for microstructural determination. These
samples were placed at the sample position with the axis horizontal
for composition profiling in the solidification direction.

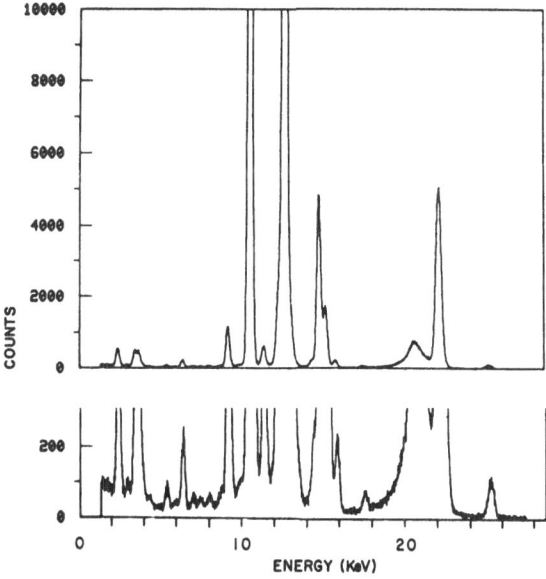

Fig. 2. Typical spectrum obtained for a Pb-Sn sample with low
background because of the use of a monochromatic (22.1 keV)
incident beam. PbL lines, 8.5 - 16 keV; TDS, 22.1 keV;
Compton 20.7 keV; PbM, 2.1 keV; SnL, 5.1 keV.

RESULTS

    An example of a spectrum from a sample is shown in Fig. 2. In
addition to fluorescence lines, two peaks are seen at 20.7 and 22.1
keV; viz., Compton scattering and thermal diffuse scattering (TDS)
of the incident 22.1 keV x-ray beam respectively. These two
scattering processes are the major source of background in normal
white beam XRF analysis. Because of the use of monochromatic
incident x-rays this scattering is well localized in the energy
spectrum and separate from the fluorescent peaks. This greatly
improves the signal to background ratio. An indication of this
improvement is the observation at 9.18 keV and 11.35 keV of the
weak PbL_l and PbL_η lines respectively shown in Fig. 2.

    Five standard samples were prepared for calibration. Care was
taken to ensure a fine microstructure by rapid cooling, and chemical
analysis was performed using a titration method[4]. Replicate XRF
runs were performed on these standard samples to determine the
reproducibility of the measurements. It was found that due to
variations in incident beam flux, time was not a useable normalizing
factor for the fluorescent yield. As a more workable alternative,

Table 1.   Check on Reproducibility of X-Ray Chemical Analysis
           Technique

Analysis No.	Composition w/o Pb	Integrated Intensity (counts) PbL	TDS	Ratio PBL/TDS	
105	37.8	206,747	24,580	8.41	
106	37.8	223,252	26,105	8.55	8.46±.06
107	37.8	207,524	24,526	8.46	
108	37.8	192,960	22,923	8.42	
109	40.7	212,595	22,722	9.36	
110	40.7	213,104	22,580	9.44	9.43±.06
111	40.7	213,708	22,717	9.41	
112	40.7	213,863	22,496	9.51	
113	43.7	258,690	25,571	10.12	
114	43.7	256,742	25,762	9.97	
115	43.7	263,423	25,948	10.15	
116	43.7	245,714	24,410	10.07	
117	43.7	232,170	22,874	10.15	10.11±.08
118	43.7	225,439	22,135	10.19	
119	43.7	222,242	22,094	10.06	
120	43.7	229,763	22,897	10.04	
121	43.7	224,851	24,006	10.20	

A 0.4 w/o Pb change corresponds roughly to a $\frac{PBL}{TDS}$ ratio change of .08.

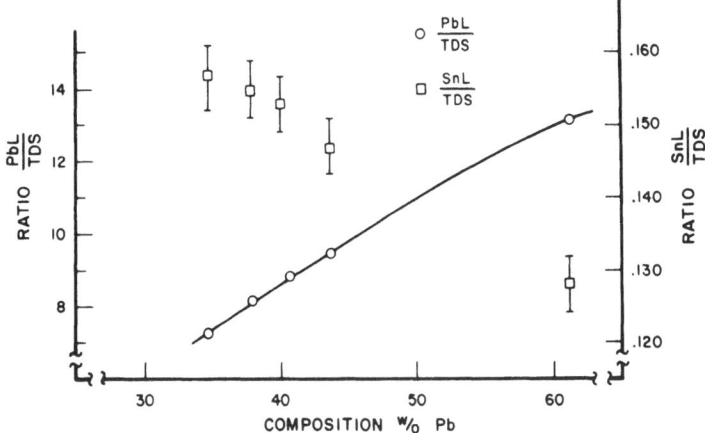

Fig. 3.   Normalized fluorescent yield for PbL and SnL as a function
          of composition.

the ratio of the fluorescent yield to the integrated TDS yield permitted the reproducible measurement of a parameter nearly linear in
composition. In principle the TDS yield is sensitive to composition
also, however, its dependence on composition is different than
fluorescent yield. Additionally, the TDS yield integrated over the
entire scattering sphere is relatively insensitive to composition
and its workability as a good beam monitor is perhaps related to the
relatively large solid angle received by the detector. Compton
and/or TDS have previously been employed to normalize fluorescent
yield by others[5]. In the present work Compton scattering was not
used because of the relatively low intensity compared with the TDS.

     Results for some of the calibration runs are shown in Table 1.
The ratio of the standard deviation of the normalized fluorescent
yield to the mean value is 8×10^{-3}. This is almost identical with
expected standard deviation of a ratio expected from the counting
statistics (9×10^{-3}). For a fixed sample, it is our assertion that
the precision of the method is only limited by counting statistics.

     Fig. 3 shows a plot of the normalized PbL and SnL (α, β and γ)
yields versus composition over the range 35-60 w/o for calibration
runs made after some realignment of the optics. The SnL yield was
unusable because of the small slope compared to the errors due to
counting statistics. For the level of counts shown in Table 1
(corresponding to a counting time of about 16 hours) and the slope
of the curve in Fig. 3, the one-sigma error in Pb content is about
0.4 w/o for alloys of composition near 40 w/o.

The spatial resolution of the presently employed method was measured by profiling a sharp discontinuity between a Pb and Au sample. The normalized Lα yield for Au and Pb are shown in Fig. 4. The spatial resolution is about 0.5 mm.

Examples of macrosegregation data obtained with this method are shown in Figs. 5 and 6 with a comparison to results obtained by the titration method. Samples for wet chemistry were approximately 100 mg and occupied 2 mm of length in the profiling direction with an expected accuracy of 0.25 wt %. The results are plotted against fraction solidified, which is position normalized to the total length of the sample ($\sim$ 10 cm). This is a parameter important in theories of macrosegregation. The data shown in Fig. 5 are in better agreement with wet chemical data than are Fig. 6 . Previously, it was mentioned that the expected one-σ error in composition based on counting statistics alone would be 0.4 wt %. Most of the data in Fig. 5 agree to with $\pm 2\sigma$ of the x-ray data. The results of Fig. 6 however fall outside of this expected experimental scatter.

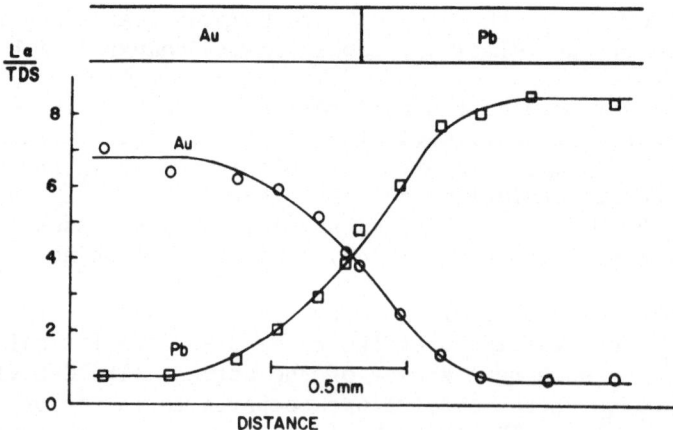

Fig. 4.   Normalized fluorescent yield for AuLα and PbLα versus distance across a planar boundary separating Au from Pb. The spatial resolution of the method is about 0.5 mm.

Examination of the microstructure of the four samples provides some insight into this discrepancy. Figs. 7a and 7b show microstructures typical of samples in Figs. 5 and 6, respectively. The former is a structure of fine (4μm) plates of Pb- and Sn-rich phase while the latter additionally contains large (50-100 μm) dendrites of Pb-rich phase. Effects due to absorption as well as averaging over the structure with a 0.5 mm incident beam are different for the two microstructures. Measurements with a wider beam as well as attempts at particle size correction may be useful in improving results for coarse microstructures.

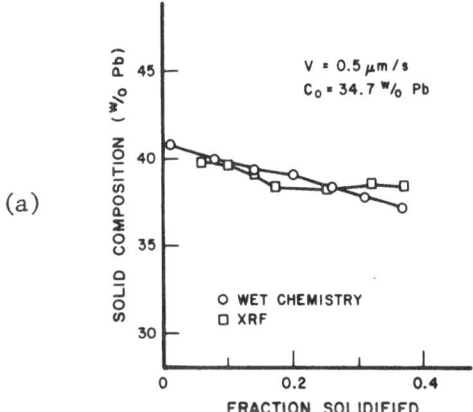

(a)

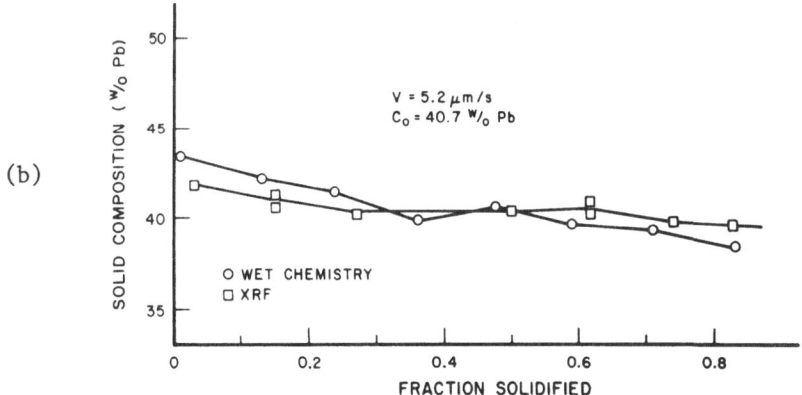

(b)

Fig. 5.   Solid composition versus normalized distance for two samples solidified under conditions producing little macrosegregation. Agreement of XRF measurements to wet chemistry is approximately within $\pm 2\sigma$ expected from counting statistics.

(a)

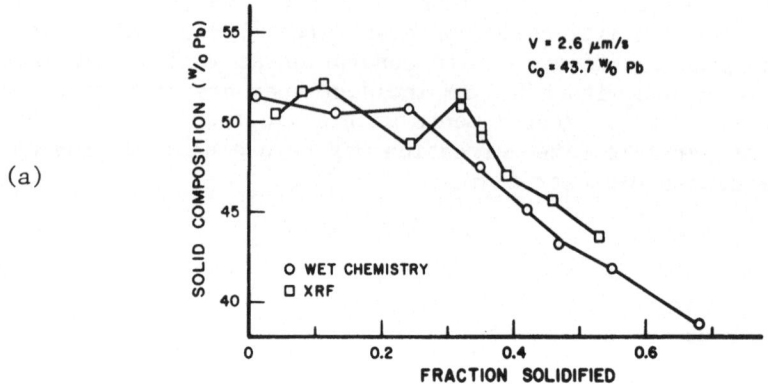

(b)

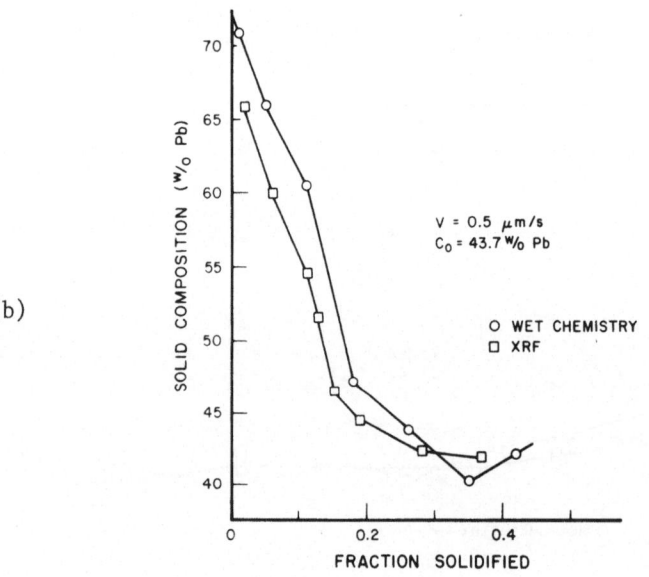

Fig. 6.   Solid composition versus normalized distance for two
samples solidified under conditions producing significant
macrosegregation. Disagreement may be caused by extremely
coarse microstructure of these samples.

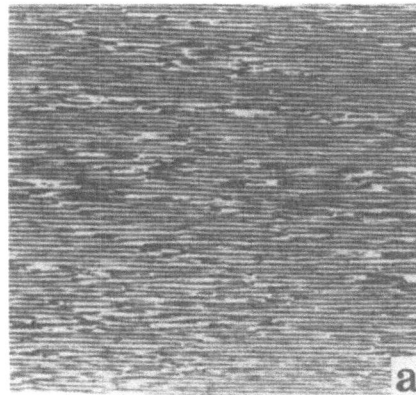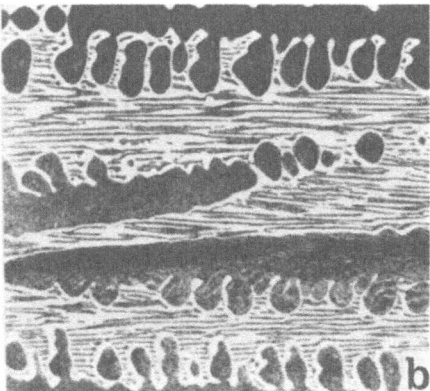

Fig. 7. Metallographic sections of samples typical of those analyzed
by XRF. Dark etched phase is Pb solid solution. Light
etched phase is Sn solid solution.
Pictures are from areas of sample 1/3 mm x 1/2 mm.
a) Fine microstructure typical of samples shown in Fig. 5.
b) Coarse dendritic microstructure typical of samples
shown in Fig. 6.

ACKNOWLEDGEMENT

This work was partially sponsored by the Materials Processing
in Space Program, National Aeronautics and Space Administration.

REFERENCES

1. R. M. Sharp and A. Hellawell, The Microscopy and Composition
    of Quenched Solid-Liquid Interfaces, J. Cryst. Growth
    5:155 (1969).
2. W. J. Boettinger, H. E. Burdette, M. Kuriyama and R. E. Green,
    Jr., Asymmetric Crystal Topographic Camera, Rev. Sci.
    Instrum. 47:906 (1976).
3. W. J. Boettinger, S. R. Coriell, F. S. Biancaniello and
    M. R. Cordes, Solutal Convection and Liquid Diffusion
    Coefficients in: "NBS: Properties of Electronic Materials,"
    J. R. Manning (ed) NBSIR 78-1483 and NBSIR 79-1767.
4. M. J. Tschetter and R. Z. Bachman, Rapid EDTA Determination
    of Pb, Talanta 21:106 (1974).
5. See for example C. G. Clayton and T. W. Packer, Some
    Applications of Energy Dispersive X-ray Fluorescence
    Analysis in Minerals, Exploration and Process Control, in
    "Advances in X-ray Analysis," Vol. 23 (1980).

ENERGY DISPERSIVE X-RAY FLUORESCENCE

ANALYSIS OF INKS ON PAPER

J.C. Russ

EDAX International, Inc.

P.O. Box 135, Prairie View, IL 60069

INTRODUCTION

Interest in the analysis of inks on paper has come primarily, although not exclusively, from forensic scientists. Two classes of problems are encountered: comparison of inks in different regions of a single document, to detect additions (such as adding a "ty" to your handwritten check for "nine" dollars) made with a visually similar but hopefully compositionally distinct ink; and verification of authenticity in documents (including stamps, stock certificates, currency, etc.) by the presence of specific tag elements in certain of the inks present.

For both applications, the principle analytical difficulties arise from the fact that the ink is present as a small sample, both in lateral dimension and thickness. This means that the X-ray spectrum is dominated by the substrate, typically paper containing clays, titanium oxide brighteners, or other materials with a variety of elements. The X-ray fluorescence analysis can be confined to a small dimension laterally by the use of a small collimator. In the MAX system, a 0.5 mm modified pinhole collimator produces a 0.5 x 0.7 mm elliptical excited spot on the sample, which can be positioned on a selected inked line using a medium-power binocular microscope. The high efficiency of the energy-dispersive detector makes it possible to obtain adequate count rates from such small areas, even using a low power air-cooled tube (operated at 15 watts in these examples).

DETECTION LIMITS

To ascertain practical detection limits for tagging elements in inks, a series of ink samples containing 0.5% to 10% of several elements as oxides were prepared, and deposited on papers in thicknesses under 2μm. Table 1 summarizes the compositions of the inks. Each specimen was analyzed for

100 seconds in a standard MAX system using continuum excitation from a
tungsten target X-ray tube operated at 30 kV, 0.5μA. Other tube targets
and perhaps the use of primary filters could improve the results for specific
elements. Larger collimators could give increased count rates in proportion
to their area, but in most cases would be incompatible with the size of inked
markings.

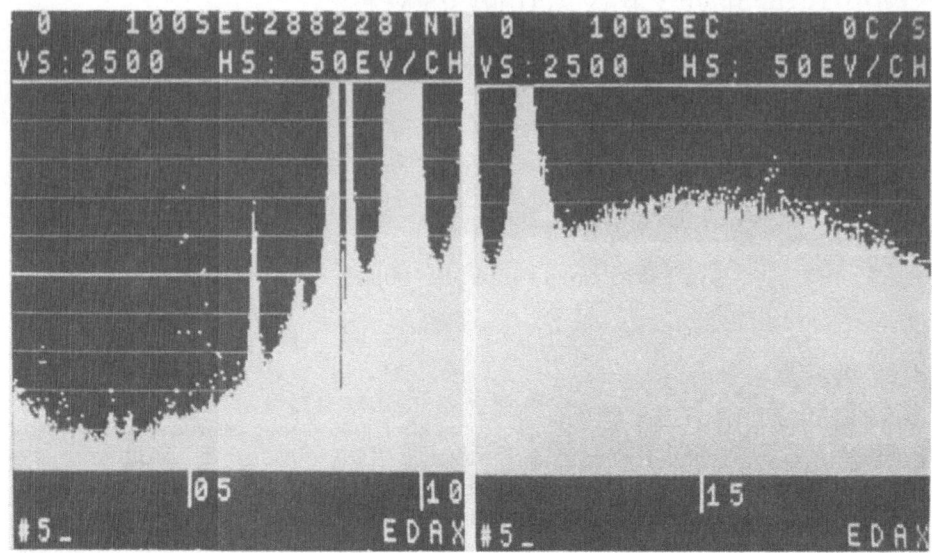

Figures 1 and 2: Comparison of ink (dots) and paper (bars) spectra on sample #5.

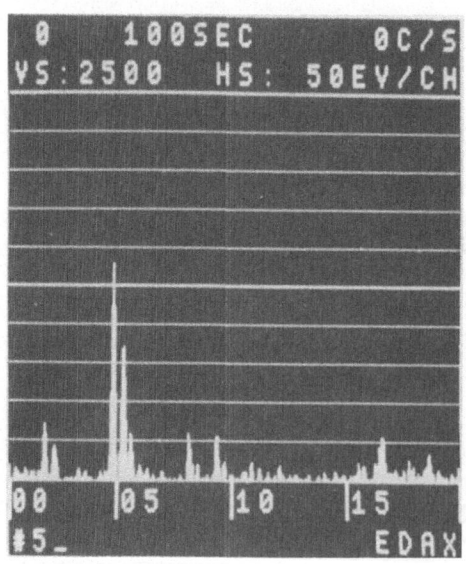

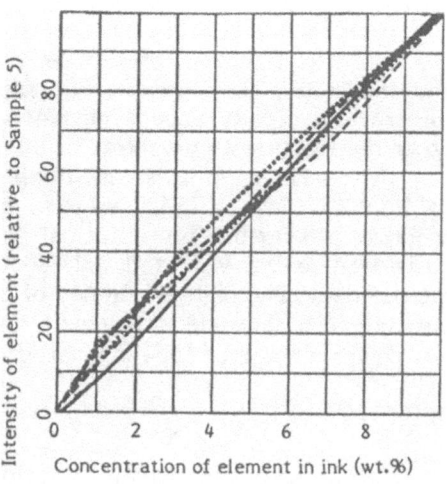

Figure 3: Net ink spectrum from
sample # 5.

Figure 4:   Intensity/concentration
relationship for additive elements.

Table 1 Elements and Concentration (wt%) in each sample.

Sample Nr.	1	2	3	4	5
Additive					
Al_2O_3	0.5	1.0	3.0	7.0	10.0
Nb_2O_5	0.5	1.0	3.0	7.0	10.0
$Ce(OH)_4$	0.5	1.0	3.0	7.0	10.0
Ta_2O_5	0.5	1.0	3.0	7.0	10.0
Varnish	97.5	95.5	87.5	78.5	59.5
Codrier	0.5	0.5	0.5	0.5	0.5
Total:	100.0	100.0	100.0	100.0	100.0
Specific gravity of spread ink	1.016	1.046	1.118	1.288	1.476
Thickness of ink film (um)	2.00	1.90	1.84	1.94	1.85

By comparing spectra from the inked paper to a 'blank" spectrum measured on the adjacent paper alone, the elemental peaks were readily seen. Figures 1 and 2 show the Ce and Nb peaks, respectively. Subtracting the "blank" spectrum by normalizing on the scattered tungsten tube line leaves behind the net elemental peaks, as shown in Figure 3. Table 2 lists the net intensities obtained by this method, and Figure 4 shows the linear intensity/concentration relationships. Detection is possible at the lowest concentration, which is below that actually used for tag elements in inks.

PRACTICAL APPLICATIONS

Applying the same analytical procedure to a real document, the spectra in Figures 5 and 6 were obtained. The "blank"spectrum was measured adjacent to the printed line and then subtracted from the spectrum measured on the line to give Figure 7. The net spectrum clearly shows the presence of Si, S, Cl and Fe in addition to the tag element Ba in the ink.

Another example is shown in Figure 8. The Australian dollar has fine transverse lines printed to form the background of the illustration. A genuine bill (dot spectrum) exhibits iron and calcium, and less titanium, than does a counterfeit (bar spectrum). By positioning the point of analysis between the lines, to analyze the paper only, it was easy to show that the difference in iron was related to the inks used, while the calcium and titanium differences reflected the use of different papers.

CONCLUSION

Practical analysis of inks on paper can be readily performed using a small excitation spot. It is generally necessary to measure and subtract the blank paper spectrum to isolate the elements in the ink. These can be detected in the concentrations usefully present and used to confirm authenticity and identify forgeries and counterfeits.

Table 2 Net element intensities

Sample	Al	Ce	Ta	Nb(L)	Nb(K)
5(10%)	181	9937	5174	896	2061
4(7%)	121	7453	3695	652	1561
3(3%)	67	3121	1649	254	784
2(1%)	28	1126	812	61	229
1(.5%)	15	304	322	32	161

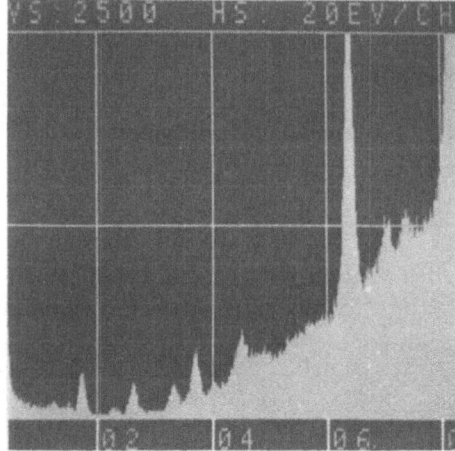

Figure 5: "Blank" paper spectrum measured between lines.

Figure 6: Spectrum from red ink line on paper.

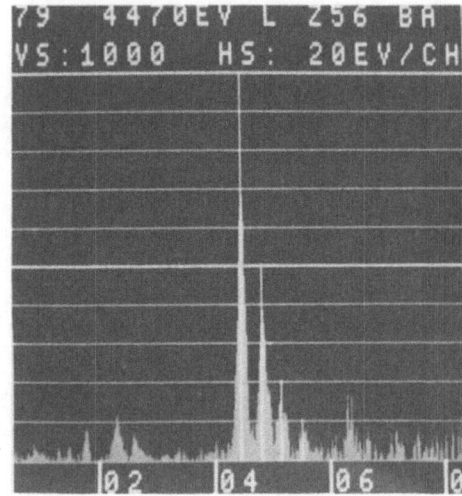

Figure 7: Net spectrum from red ink.

Figure 8: Comparison of spectra from genuine (dots) and counterfeit (bars) bills.

DETERMINATION OF THE THICKNESS OF SiO_2-LAYERS ON Si BY

X-RAY ANALYSIS AND BY X-RAY PHOTOELECTRON SPECTROSCOPY

Maria F. Ebel, H. Ebel and J. Wernisch

Institut für Technische Physik

Technische Universität Wien, Vienna, Austria

It is feasible to investigate the thickness of oxide layers on silicon wafers by X-radiation in the 0.1-10 nm thickness range. For example, X-ray photoelectron spectroscopy (XPS) is a well-applicable technique, with information depth of a few nm. Fig. 1 presents the principle of this method. An impinging characteristic X-radiation $h\nu$ (*e.g.* Al Kα) count rate ejects Si 2p photoelectrons from the Si-substrate (d), with count rate n_1, which, on their way to the electron spectrometer, have to pass through the SiO_x-interface (c), the SiO_2-layer (b) and the contamination overlayer (a), whereas Si 2p photoelectrons ejected from the SiO_2-layer, with count rate n_2, have just to penetrate the contamination overlayer. The Si 2p electrons originating from the SiO_x-interface, for the situation shown in Fig. 1, can be added to the substrate count rate. (The line position and peak height make it difficult to separate the two signals). As they are caused by different chemical binding energies, the energies of the Si 2p signals from the substrate and from the oxide layer show a difference of 4.3 eV and consequently the photoelectron lines can be easily separated. The following equation describes the relation between the ratio of the photoelectron count rates r ($r = n_2/n_1$) and the takeoff angle β. ·

$$r = R \cdot (\exp D/\lambda_{ox} \cdot \cos \beta - 1)$$

where D is the SiO_2-layer thickness. Obviously this ratio is free from the influence of the contamination overlayer. The meaning of R is given by

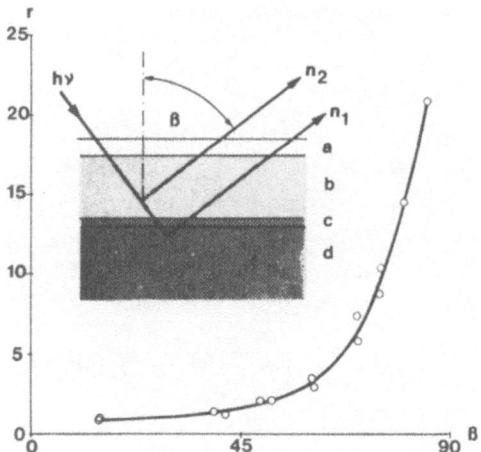

Fig. 1.   Experimental set-up for determination of the thickness of
an oxide layer by XPS and experimental results (count
rate ratio r depending on take-off angle β)

$$R = \frac{A_{Si}}{A_{Si} + 2 \cdot A_0} \cdot \frac{\rho_{ox}}{\rho_{Si}} \cdot \frac{\lambda_{ox}}{\lambda_{Si}} \cdot$$

A_{Si} and A_0 are the atomic weights of silicon and oxygen, ρ_{Si} and
ρ_{ox} are the densities of silicon and silicon oxide, and λ_{Si} and
λ_{ox} are the mean free paths of the Si 2p photoelectrons in silicon
and oxide respectively. Experimental results[1] on the dependence of
the count rate ratio r on the take-off angle β are shown in Fig. 1,
for D=3.8 nm (measured by an ellipsometric technique). Together
with a value of R=0.66 and the definition of the error v_i

$$v_i = r_i - 0.66 \cdot (\exp D/\lambda_{ox} \cdot \cos \beta - 1),$$

where r_i stands for the measured count rate ratio of take-off angle
β_i and

$$d(\Sigma \ v_i^2)/d(D/\lambda_{ox}) = 0,$$

the reduced thickness has been evaluated for six nearby r_i-values.
Fig. 2 shows the dependence of D/λ_{ox} on $\bar{\beta}$, where $\bar{\beta}$ is the mean
value of the take-off angles β_i. A wrong value of R (R=0.66)
could be responsible for this angular dependence of D/λ_{ox}. To
clarify this assumption the error v_i is expressed by

$$v_i = r_i - R \cdot (\exp D/\lambda_{ox} \cdot \cos \beta - 1)$$

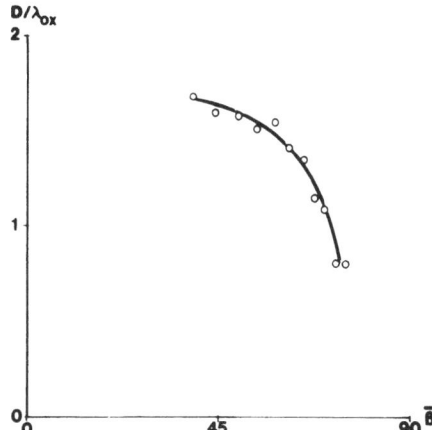

Fig. 2.   Dependence of the reduced thickness of the oxide layer
          (calculated from the experimental results of Fig. 1) on
          the take-off angle

Hence, D/λ_{ox} and R are determined by

$$\partial(\Sigma \; v_i{}^2)/\partial R = 0 \qquad \text{and} \qquad \partial(\Sigma \; v_i{}^2)/\partial(D/\lambda_{ox}) = 0$$

Fig. 3 depicts the relationship $D/\lambda_{ox} = f(\bar{\beta})$ and $R = g(\bar{\beta})$.   Here,
a comparable dependence of D/λ_{ox} on $\bar{\beta}$ is obtained with small D/λ_{ox}-
values corresponding to large R-values.   This correlation also
holds for negative values, which are the result of the mathematical procedure, without any physical significance.   Thus, the actual
thickness D (D/λ_{ox} respectively) of the oxide layer is related to
a measured thickness.   Surface roughness causes this angular dependence of measured D/λ_{ox}-values and, as the mean free path of photoelectrons in matter is extremely small, profile depths in the range
of a few nm have a marked influence.

This statement,[2] based on model calculations, can be verified
by X-ray fluorescence analytical experiments.   In Fig. 4 the principle is outlined.   A silver layer of approximately 0.3 μm is
evaporated onto a silicon wafer.   The angle between the incident
monochromatic Mo Kα-radiation and the take-off angle of the fluorescence radiation is 90°.   The attenuation of the Mo Kα-radiation
by Ag and Si is two orders of magnitude less than the attenuation
of the Ag- and Si-fluorescence radiation.   For this experimental
situation (Fig. 4) $1/\mu \cdot \rho$ is equivalent to the mean free path λ
and its range corresponds to the thickness of the silver layer.
To show the influence of surface roughness, silver has been evaporated under identical conditions onto two silicon wafers, one substrate with a smooth surface and one with a rough surface.   A

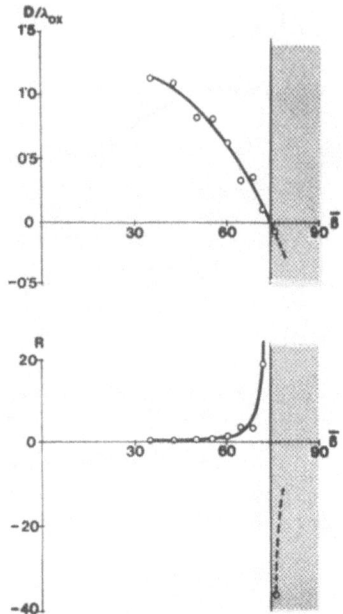

Fig. 3.   Dependence of the reduced thickness of the oxide layer
          (calculated from the experimental results of Fig. 1) and
          the factor R on the take-off angle

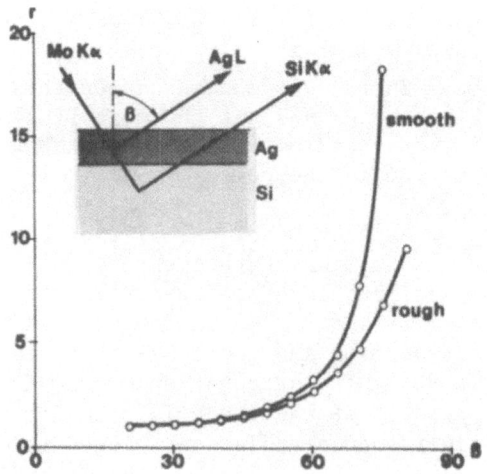

Fig. 4.   Experimental set-up for X-ray fluorescence analytical
          experiments and experimental results for samples with
          smooth and rough surfaces

profile depth of the rough surface of approximately 1 μm is in
accordance with the range of the layer thickness and the mean free
path of the fluorescence radiation. These special experimental
details guarantee the transferability of fluorescence results to
XPS. Here, count rate ratio is given by $r = n_{Ag}/n_{Si}$. The $r(\beta)$-
curves are depicted in Fig. 4. For large take-off angles β the
experimental results allow a clear distinction between smooth and
rough sample surfaces. When calculating the layer thickness D and
its dependence on $\bar{\beta}$, with two unknown values R and D, comparable
to Fig. 3, the resulting relation can be seen from Fig. 5. For
the smooth surface, D is constant (regarding a statistic significance), for the rough surface a decrease, comparable to Fig. 3,
is observed. The product R·D in dependence on R can be approximated
by a quadratic curve, as shown in Fig. 6. For a known D--it can be
measured by atomic absorption spectrometry (AAS)--the intersection
between the curve $y = R \cdot D_{XRFA} = A + B \cdot R + C \cdot R^2$ (from X-ray fluorescence
analytical measurements XRFA) and the straight line $y = R \cdot D_{AAS}$ gives
the R-value, R=4.39, in Fig. 6. After calibration (either for a
smooth or a rough surface) any thickness D of a Ag-layer on silicon can be evaluated, when for R=4.39 (on the abscissa) the corresponding value $R \cdot D_{XRFA}$ of the respective curve is determined and
subsequently is divided by 4.39.

    Finally, experience gained by X-ray fluorescence analytical
measurements can be applied to XPS investigations. Here, the layer
thickness can be determined by ellipsometric technique (D_e),
including the interface and the contamination overlayer (see Fig.
1) and thus a difference ΔD between D_e and the thickness of the
oxide layer, determined by XPS, exists. The curves given by

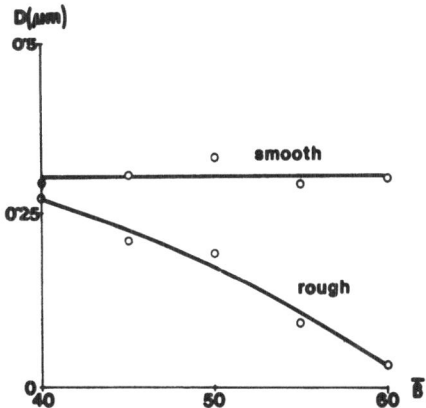

Fig. 5. Dependence of layer thickness (calculated from the experi-
         mental results of Fig. 4) on the take-off angle for a
         sample with smooth and a sample with rough surface

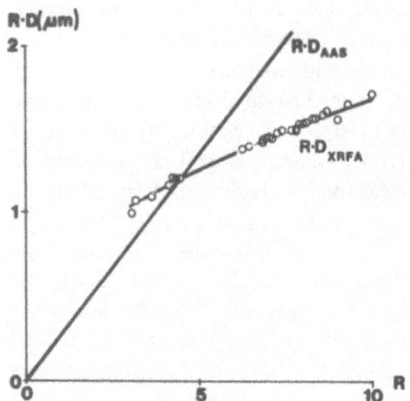

Fig. 6.   Different presentation of the results of Fig. 5.   R·D
depends on R with D=D$_{XRFA}$

$$(R \cdot D / \lambda_{ox})_i = a_i + b_i \cdot R + c_i \cdot R^2$$

describe oxide layers of different thicknesses (labeled by the
index i).   A modification of this equation together with the above
given meaning of R leads to the following expression for the error
v_i:

$$v_i = D_{e,i} - \Delta D - [A_i + B_i \cdot \frac{\lambda_{ox}}{\lambda_{Si}} + C_i \cdot (\frac{\lambda_{ox}}{\lambda_{Si}})^2] \cdot \lambda_{Si}$$

$$A_i = \frac{a_i}{\dfrac{A_{Si}}{A_{Si} + 2 \cdot A_0}} \cdot \frac{\rho_{ox}}{\rho_{Si}}$$

$$B_i = b_i$$

$$C_i = c_i \cdot \frac{A_{Si}}{A_{Si} + 2 \cdot A_0} \cdot \frac{\rho_{ox}}{\rho_{Si}}$$

Together with

$$\partial(\Sigma v_i^2)/\partial \Delta = 0, \qquad \partial(\Sigma v_i^2)/\partial \lambda_{Si} = 0, \qquad \partial(\Sigma v_i^2)/\partial (\frac{\lambda_{ox}}{\lambda_{Si}}) = 0$$

the unknown values ΔD, λ_{Si} and $\lambda_{ox}/\lambda_{Si}$ can be evaluated.   Three
curves allow the determination of three unknowns.   Fig. 7 gives

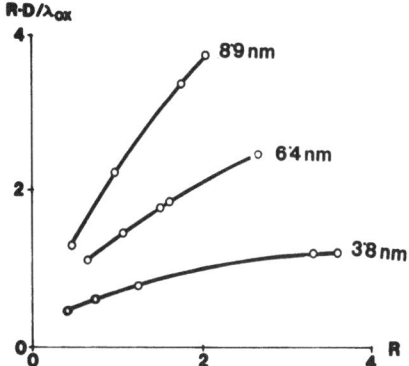

Fig. 7. $R \cdot D/\lambda_{ox}$ depicted as a function of R, for three SiO_2 layers, for Al Kα-radiation. The parameters of the curves are the thicknesses D_e, determined by ellipsometric technique

the results of measurements performed on three SiO_2 layers. The evaluation method described in this paper leads to results, not influenced by surface roughness:[3]

$$\Delta D = 0.85 \text{ nm}, \quad \lambda_{Si} = 2.59 \text{ nm}, \quad \lambda_{ox}/\lambda_{Si} = 1.22.$$

According to Penn[4] $\lambda_{Si} = 2.54$ nm.

A comparison of the two methods illustrates the problems caused by surface roughness (in the nm range) in applying XPS to thickness determination. Furthermore, it is shown that the influence of surface roughness (in the μm range) on X-ray fluorescence analytical measurements of layer thicknesses must not be neglected.

ACKNOWLEDGEMENT

The authors are glad to acknowledge the financial support of this work by the Fonds zur Förderung der wissenschaftlichen Forschung in Österreich (Projekt Nr. 1567, Projekt Nr. 3226).

REFERENCES

1. C. S. Fadley, "Solid State- and Surface-Analysis by Means of Angular-Dependent X-Ray Photoelectron Spectroscopy," Progress in Solid State Chemistry 11:265–343 (1976).
2. M. F. Ebel, "Zur Bestimmung der reduzierten Dicke D/λ dünner Schichten mittels XPS," J. Electron Spectrosc. Relat. Phenom. 14:287–322 (1978).

3. M. F. Ebel and W. Liebl, "Evaluation of XPS-Data of Oxide
    Layers," J. Electron Spectrosc. Relat. Phenom. 16:463-470
    (1979).
4. D. R. Penn, "Quantitative Chemical Analysis by ESCA," J. Elec-
    tron Spectrosc. Relat. Phenom. 9:29-40 (1976).

THE EFFECTIVE USE OF FILTERS WITH DIRECT EXCITATION OF EDXRF

Ronald A. Vane and William D. Stewart

United Scientific Corporation

1400D Stierlin Road, Mountain View, CA 94043

ABSTRACT

Primary beam transmission filters in energy dispersive X-ray fluorescence (EDXRF) analysis are used to shape the spectral output of the X-ray tube. The effective use of these filters allows the optimization of excitation conditions for each different analysis. Filters are used in two basic ways in EDXRF; either as edge filters or as white filters. The proper choice of filter and excitation conditions optimizes the analysis of a particular element or spectral region by shaping the primary radiation to reduce background and to maximize excitation.

INTRODUCTION

Optimum analysis conditions in energy dispersive X-ray fluorescence (EDXRF) analysis are obtained with minimum background under the peaks of interest and with exciting radiation located as closely as possible above the absorption edges of the analyte elements. The optimum excitation conditions in EDXRF differ greatly from wavelength dispersive systems since energy dispersive systems operate with comparatively low excitation fluxes due both to lower dispersion losses and amplifier count rate limitations. The reduction of scattered primary radiation in the spectrum lowers the background to improve the peak to background ratio and also reduces the total count rate load on the linear amplifier. Not processing unwanted counts reduces pulse pileup and deadtime, allowing operation at longer time constants with improved resolution or at higher net peak count rates with improved statistical precision. The most

efficient way to achieve these results is by the use of filters in
the primary beam from the X-ray tube.

Filters selectively remove various X-ray energies from the
primary X-ray beam to prevent their scattering from the sample.
The scattered bremsstrahlung, characteristic lines, and diffraction
peaks can thus be removed from or minimized in the resulting spectrum. Filters can be used in two ways; either as edge filters
where the location of the absorption edge is used to selectively
filter out part of the tube radiation, or as "white" filters in
which the thickness and mass of the filter is used to stop radiation below a certain energy.

METHODOLOGY

X-ray filters are used by placing a thickness of filter material in the X-ray beam. The absorption in' the filters alters the
X-ray energy distribution of the beam as a function of energy. In
general, the absorption coefficient decreases with increasing energy
except at absorption edges where a sharp increase occurs. A typical set of absorption coefficient curves are shown in Fig. 1. Since
X-ray absorption follows Beer's Law ($I = I_o \exp - \rho\mu X$, where ρ =
density, μ = mass absorption and X = thickness), the same amount of
total absorption can be achieved with different filter materials by
varying the thickness of the filters.

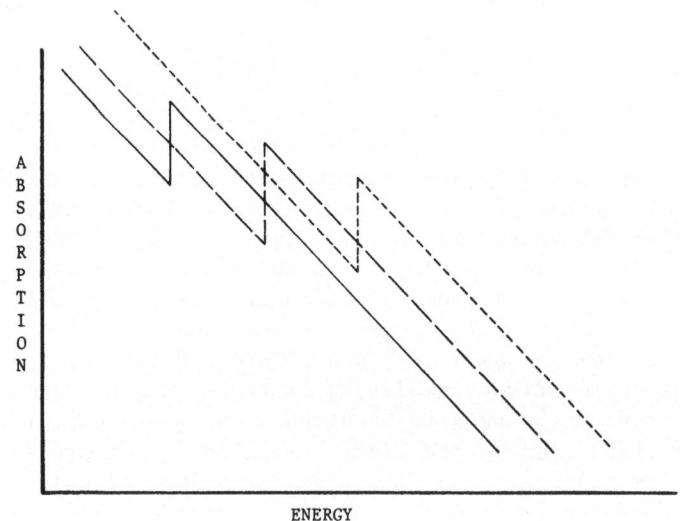

Figure 1. Typical Absorption Coefficient Curves with K Absorption
          Edges

A typical unfiltered output spectrum of an X-ray tube consists of the characteristic lines of the X-ray tube plus bremsstrahlung. The filter will remove most of the X-rays up to a "cutoff"
energy above which higher and higher proportions of primary X-rays
pass through the filter. At the absorption edge, these higher
energy X-rays are strongly absorbed, but with increasing energy,
a second high energy cutoff-point is again reached and they begin
to pass.

The low energy cutoff point of the primary bremsstrahlung continuum can be controlled by the thickness, density and choice of
material. Changing the filter moves the low energy cutoff for the
scattered background to any point in the spectrum. The upper end
of the bremsstrahlung output of the tube is controlled by the setting of the X-ray tube high voltage. Thus the upper and lower
edges of this bremsstrahlung hump can be moved at will, which allows creation of a quasi-monochromatic hump of primary radiation
for excitation at any location in the spectrum. This hump of radiation can be moved close to the absorption edge of any element of
interest and still remove most of the scattered background under
the peak.

Filters used in this manner can be called "white" filters,
since they act in a smooth manner on all energies of X-rays. The
filtering properties of one white filter can be duplicated by
another filter of different material and thickness if there are
no absorption edges of the filter materials in the spectral region.

Edge filters involve the use of the filter material's absorption edges. The location of absorption edges is unique for each
element and at the edge the absorption coefficient jumps, thus absorbing X-rays of higher energy more strongly than those with energy
just below the edge. The absorptivity function of edge filters
away from the edge is identical to a white filter. Thus an edge
filter can act like a crude band pass filter: X-rays below the absorption edge and above the low energy cutoff are passed.

The most common use of an edge filter is in conjunction with
the characteristic lines of an X-ray tube. A filter of the same
material as the tube anode passes its own characteristic lines quite
well but strongly absorbs the bremsstrahlung above the characteristic
lines. This allows the production of primary radiation containing
just the characteristic lines of the tube which results in nearly
monochromatic excitation. The tube should be operated approximately
10 KV above the absorption edge of the anode for best results in
this mode.

EXPERIMENTAL

The use of white filters with changing excitation potential for a Mo side window X-ray tube to create a quasi-monochromatic primary radiation is illustrated in Figs. 2-7. Fig. 2 shows unfiltered X-ray tube excitation at 49 KV of a sample of an oil standard containing 100 ppm of several metals. The broad bremmstrahlung band excites the full range of elements present, but the exciting radiation is strongly scattered by the sample creating a high background level throughout the spectrum. The scattered Mo characteristic lines from the tube prevent the analysis of Mo in the sample and the high background would prevent the analysis of small traces.

To remove the scattered Mo characteristic lines from the spectrum, a thick (.38mm) Cu filter is used as a white filter to remove all primary radiation up to an energy of about 25 KeV including the Mo K X-rays. Fig. 3 shows the spectrum of a USGS rock standard which demonstrates the removal of the scattered bremsstrahlung background in the region from 8 KeV to 25 KeV. The sensitive analytical band under these conditions extends from about 14 KeV to 25 KeV. The Sn peak in this sample is generated from 3.2 ppm Sn. Note that there are no scattered Mo peaks in the spectrum, which allows for the analysis of Mo under these conditions.

In Figs. 4-7, the sample is a piece of polyethylene which contains traces of Zn, Ti and Cl. Fig. 4 shows the use of a .127 mm Al filter with an excitation voltage of 19 KV. The bremsstrahlung hump of exciting radiation runs from 8 KeV to 19 KeV. The Zn peak is excited very efficiently and trace levels of Fe, Co, and Ni can be determined due to the low background and high fluorescent yield. In Fig. 5, a thick cellulose filter (7 layers of Whatman-41 filter paper) and 14 KV on the tube has moved the bremsstrahlung hump down in energy to give a low cutoff at 5 KeV for the analysis of K, Ca, Ti, V, and Cr. Fig. 6 shows the bremsstrahlung hump moved to between 3 KeV and 8 KeV by use of a thin cellulose filter (2 layers Whatman-41 filter paper). This is the best condition for Si, P, S, and Cl analysis. Fig. 7 shows operation at 5 KV with no filter. Note the strong Mo L lines which are available to excite Na, Mg, Al very efficiently.

Fig. 8 shows the use of Mo edge filters in conjunction with the characteristic lines of the X-ray tube. The thinner .05 mm filter shows a wider band pass of bremsstrahlung with the low energy cutoff at about 11 KeV than does the thick .127 mm filter with the cutoff at 13 KeV. The Mo absorption edge cuts off the bremsstrahlung sharply above 20 KeV for both filters. A higher X-ray flux on the samples is obtained with the thinner 0.05 mm Mo filter, at the expense of a higher background.

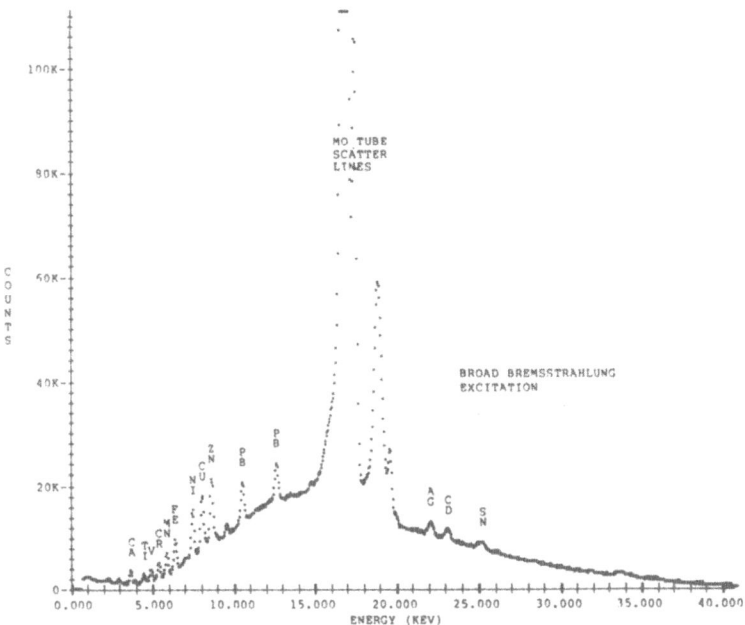

Figure 2. Broad band bremsstrahlung excitation with Mo tube at
49 KV and no filter of an oil sample with several metals at 100ppm.

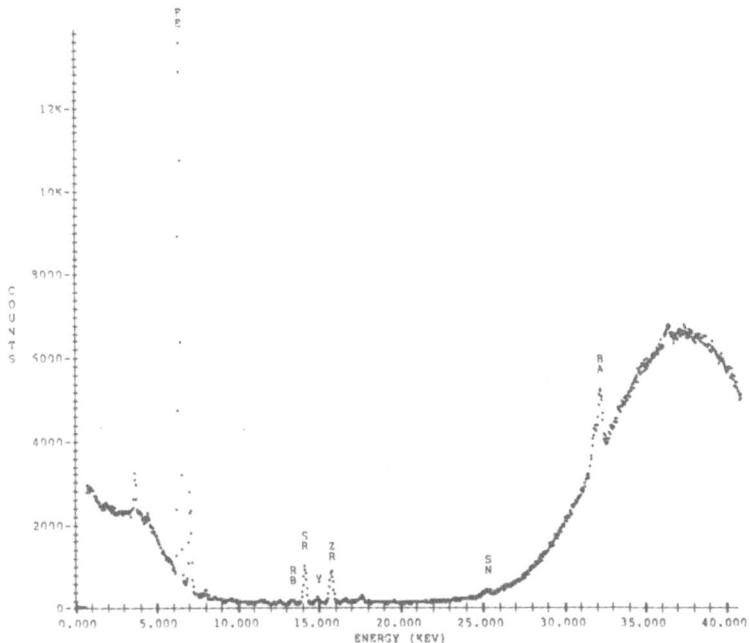

Figure 3. Excitation at 49 KV with a Mo tube and .38 mm Cu filter
of USGS rock standard W-1. Mo characteristic lines are removed.

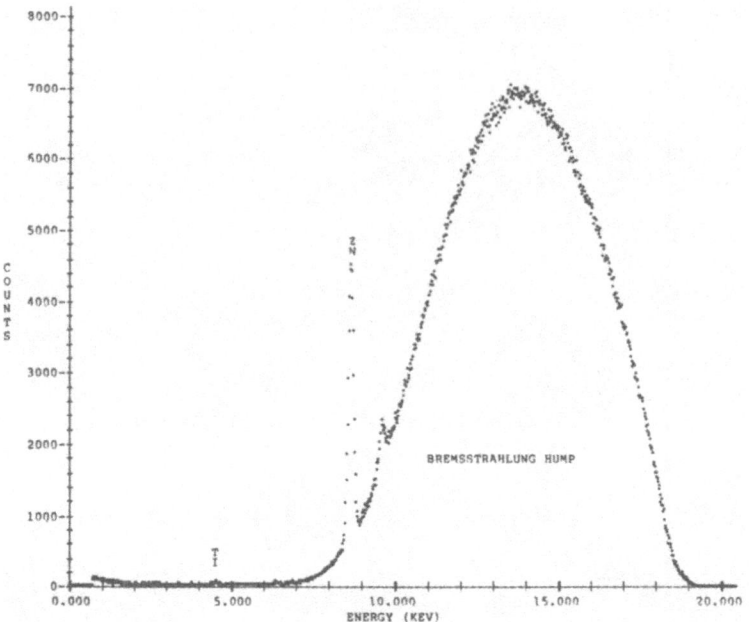

Figure 4.   Excitation at 19 KV with Mo tube and .127 mm Al filter
of polypropylene sample.   Bremmstrahlung cutoff is at about 8 KeV.

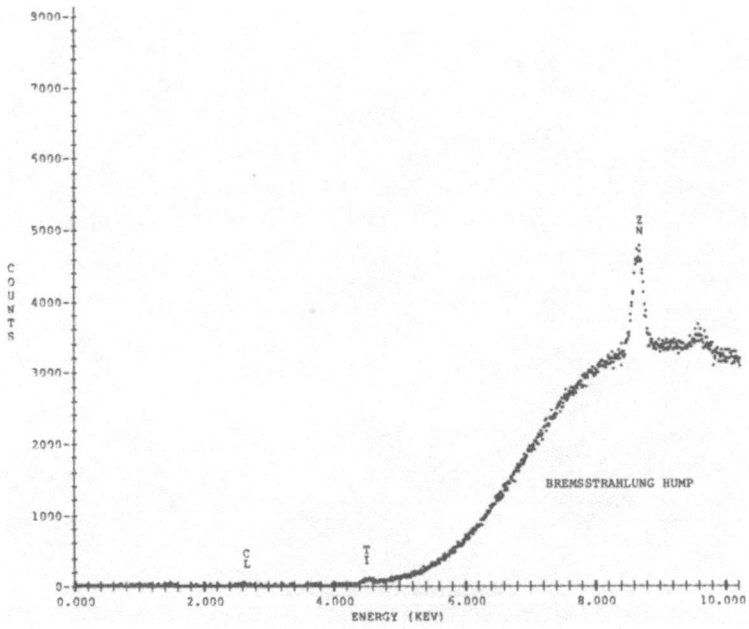

Figure 5.   Excitation at 14 KV with Mo tube and thick cellulose filter of polypropylene sample.   Bremsstrahlung cutoff is at about 5KeV.

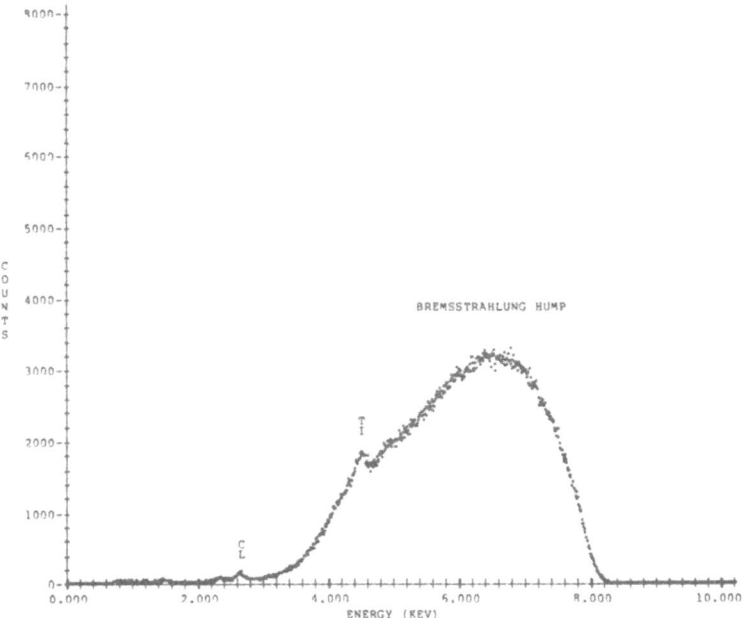

Figure 6.   Excitation at 8 KV with Mo tube and thin cellulose filter of polypropylene sample.   Bremsstrahlung cutoff is at about 3KeV.

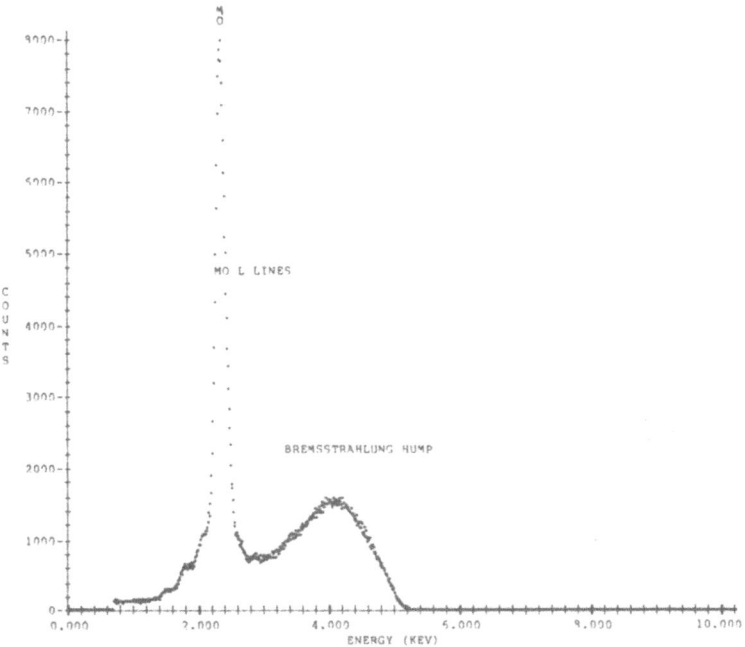

Figure 7.   Excitation at 5 KV with Mo tube and no filter of polypropylene sample.   Note the strong Mo L lines from the tube.

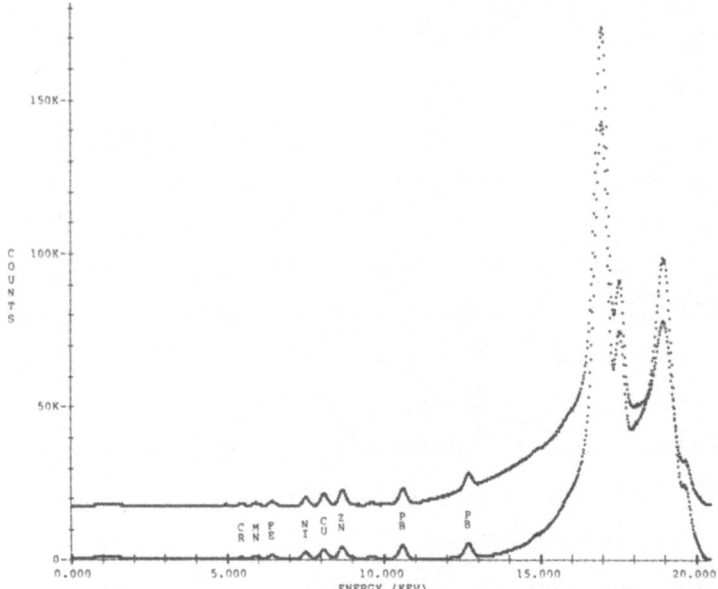

Figure 8.   Excitation at 30 KV with Mo tube of oil sample with
100 ppm of several metals.   Top spectrum done with .05 mm Mo filter and lower spectrum with .127 mm Mo filter. Note that the
bremsstrahlung cutoff changes with filter thickness.

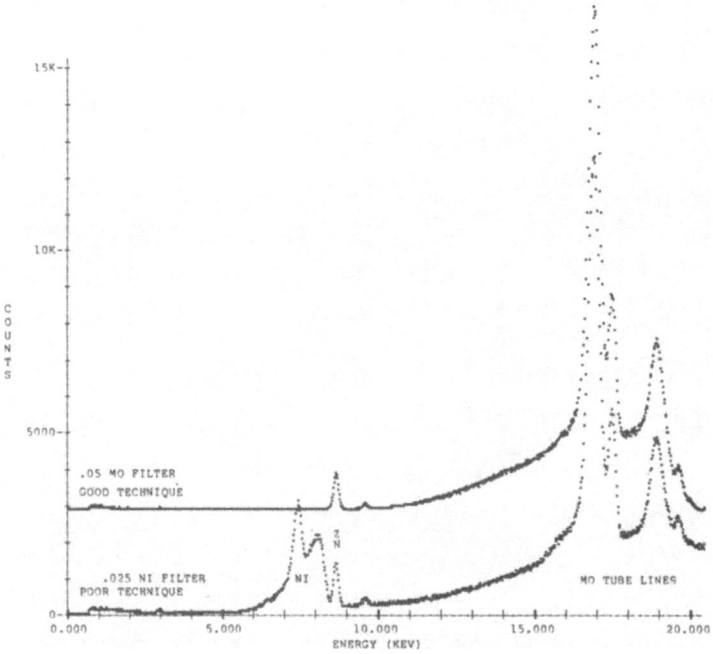

Figure 9.   Excitation of Zn in polypropylene using two types of edge
filters; .05 mm Mo filter and .025 mm Ni.   Ni filter adds structure
to spectrum.

Edge filters can create two excitation regions in a spectrum. This is useful for specific problems where two elements need to be analyzed which have greatly differing energies. Careful selection of an edge filter and excitation potential can result in primary radiation passing through the filter in two regions; one region just below the edge and the second at higher energy. An example is the analysis of Ag Br with a Mo tube at 49 KV and a Mo filter. The Mo characteristic lines excite Br and the high energy bremsstrahlung excites Ag with a low background under each peak.

Care should be taken when using edge filters away from characteristic lines of the X-ray tube. Edge filters used in this mode will create a second band of primary radiation below the edge which will include characteristic lines from the filter. This can add unnecessary background to the spectrum if this second band of radiation is not being used for excitation of lower energy lines. Fig. 9 shows an example of trace Zn analysis with two types of edge filters. Both filters remove background under the Zn peak, but the Ni filter adds background to the spectrum below Zn.

CONCLUSION

Filters are used to modify the primary spectrum to remove the scattered background under peaks of interest and to provide excitation radiation closely above the absorption edges of the elements of interest. Filters should be chosen first by the radiation they let through and then by the radiation removed. The correct choice of filters and excitation potential moves the quasi-monochromatic radiation to almost any point in the spectrum to achieve the optimum excitation conditions for any given element.

A PLASMA CONTROLLED X-RAY TUBE*

James F. McGee and Timo Saha**

Saint Louis University

St. Louis, Missouri 63103

ABSTRACT

Many x-ray tubes, used by crystallographers and others, operate
with the aid of a tungsten filament in the region of 2500°K. The
high operating temperature results in evaporation of the filament
material with two serious consequences. The first is a finite but
relatively short lifetime. The second is contamination of the target and windows with tungsten. In addition, if the tube is of the
demountable type, connected to an oil-diffusion pump and a mechanical fore-pump, carbonaceous deposits can be a problem. In a typical tube, the filament is mounted within a centimeter or two of the
target. The resulting radiant heating of the target presents additional cooling problems especially with low melting-point targets.
Many if not all of the above objectional features are circumvented
by a plasma controlled x-ray tube using a low pressure atmosphere
of helium and a cage-like cathode fabricated from nickel wire-mesh.
An experimental model has been operated for several hours at 15 kv
and 10 ma on an aluminum target. Scaling up of the apparatus will
permit power dissipations in the kilowatt range limited mainly by
the available power source or vaporization of the target material.

INTRODUCTION

The plasma controlled x-ray tube is basically a gas-type
tube[1,2] unlike the more conventional Coolidge or high-vacuum sealed-
off tubes used in x-ray crystallography and elsewhere. Histori-

---

*Work supported by AFOSR Grant No. 78-3480
**On leave from University of Turku, Turku, Finland

cally, gas tubes were used successfully by Seeman, Siegbahn,
Wyckoff and other well known x-ray crystallographers and x-ray
spectroscopists. While high vacuum x-ray tubes enjoy the advantage
of independence of the applied voltage and current settings, they
nevertheless have some well known drawbacks for general use. Their
filament lifetime is finite and often very short. The tungsten
evaporated from the filament will coat the target thereby changing
the spectral output. Coating of the window of the x-ray tube results in reduced output due to absorption. The latter can have
serious consequences in the case of quantitative experiments unless
the output is separately monitored. For a target with fixed cooling capacity the radiant heating of the target by the filament
limits the x-ray power which can be extracted from the target.

THE ANODE STRUCTURE

The plasma controlled x-ray tube like most x-ray tubes of un-
complicated design is basically a diode with an anode and a cathode
which are contained in a glass and metal envelope. The latter may
be highly evacuated as in the Coolidge type of tube or may contain
gas at a pressure determined by the overall operating conditions.
The anode of the experimental tube is made from copper with the
usual provision for circulating a coolant, ethylene glycol, through
several feet of Tygon tubing from a refrigerating system. This arrangement permits the target to be operated at high voltages without significant loss of current through the high electrical impedance of the cooling lines. The anode was cut so as to provide a
flat surface oriented at 45° to the electron beam with provisions
for a window placed so as to receive the x-rays emerging in the
vicinity of 45° to the face of the anode. For the production of
the aluminum Kα lines, a thin sheet of aluminum was soldered to the
copper anode. Longer or shorter K lines are available by attaching
with good thermal contact other materials of appropriate Z number.

THE CATHODE STRUCTURE

In general the cathodes were formed from 30 to 50 mesh nickel
screen. Molybdenum screening if available would have been more
suitable. Cathodes have been made in a variety of shapes mainly
hollow cylinders, spheres and cones. These cathodes are to be dis-
tinguished from the hollow cathodes made of solid material which
are used as high intensity radiation sources by spectroscopists
(non-x-ray). Such cathodes operate at less than the kilovolt range
and at high currents so are basically a low impedance device. The
structure shown at B in Fig. 1 represents a cylindrical cathode.
The end facing the target T has a hole approximately 1/4 inch in
diameter symmetrically located on axis. The other end of the wire
basket is connected to a metal rod which passes through a high
voltage insulator in the metal end plate which is suitably attached

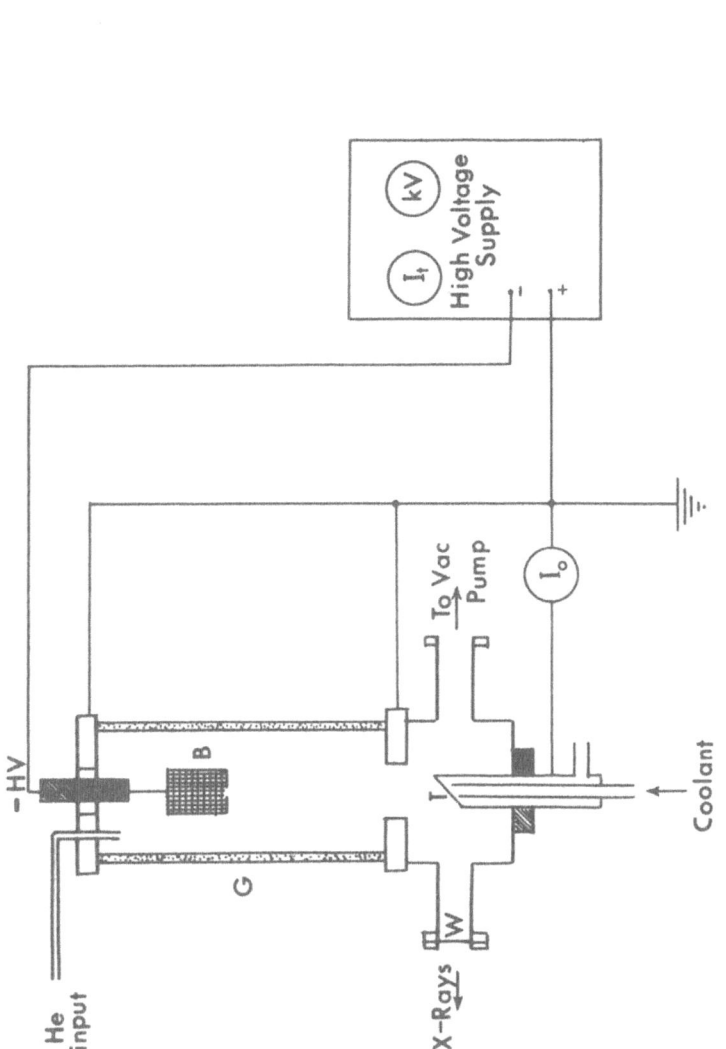

Fig. 1. The glass housing G of the x-ray tube is a 4 x 12 inch Pyrex tube (Corning). The smaller flanges are standard Ultex high-vacuum components. The target support-structure is a modified high-current water-cooled lead-through also by Ultex. Since the lead-through is insulated from the standard flange the actual current to the target I_o is readily measured. The hollow wire-mesh basket at B is in the form of a cylinder 2 inches long and 2 inches in diameter. It is made from #40 mesh nickel. The remaining components were fabricated in our shop.

to the glass envelope G. The latter is a Corning process glass-
pipe measuring 12 inches in length and 4 inches in diameter. The
dimensions of a basket currently in use are 2 inches in length and
2 inches in diameter with a 1/4 inch aperture in the bottom.

PRINCIPLE OF OPERATION

As indicated in Fig. 1, provision is made in the top of the
tube for the entry of helium from a throttling valve. A vacuum
pump continually removes the gas. By appropriate adjustment of the
throttling valve, preferably by a simple servo system receiving its
signal from the emission current, the pressure can be maintained at
some nominal operating value, e.g., 30 μm of Hg. The application of
a potential difference of a few thousand volts between the cathode
B and anode T causes the few positive ions already present to be
accelerated toward the cathode B. The bombardment of the cathode
surfaces produces secondary electrons within the basket structure.
Because of the hole in the bottom face of the basket the electric
field penetrates the aperture to remove and help focus the electrons into a beam which is accelerated toward the target T of Fig.
1. Electrons outside the basket would be accelerated away from the
cathode and subsequently may collide with neutral gas molecules to
create more positive ions. Others may strike the glass walls and
cause it to fluoresce and heat up. Experimentally one observes a
greenish-blue column about 1/4 to 1/2 inch wide originating in the
basket and terminating at the surface of the target. As evidenced
by the fact that x-rays are emitted and that the greenish-blue
beam can be deflected by a magnet it obviously contains electrons.
The fact that the beam does not significantly diverge over a distance of eight or more inches suggests that some constraining force
is at work.

ELEMENTS OF PLASMA PHYSICS

Imagine a cylindrical column of plasma in existence between
plane end electrodes raised to a potential difference V. The established electric field E causes the current I to flow parallel to
the axis of the cylinder. In virtue of Ampere's circuital law a
tangential magnetic field of induction B encircles the current. If
v is the particle velocity in the same direction as the current, the
Lorentz force v x B will be in a radial direction helping to contain
the charged particles within the column. This is the well known
pinch effect of plasma physics. Should some small variation in
cross section of the beam occur, then the beam will experience a
stronger inward push since B varies inversely with radius. The
result is that the perturbation is enhanced and instability sets
in. Much has already appeared in the literature on the subject of
plasma physics including the subject of stability. The reader may
find the three volumes Reviews of Plasma Physics, edited by M. A.
Leontovich and published by Consultant's Bureau, N. Y. 1965, an

interesting introduction to the subject.[3]

OPERATION MODES

     If the pressure of helium in the x-ray tube is maintained in
the vicinity of 100 to 1000 μm of Hg, then the discharge is of the
ordinary glow type described in many text books on atomic physics.
Anode voltages of less than one thousand volts easily produce currents exceeding the capability of ordinary high-voltage power supplies. At lower pressures of the order of 50 microns or less at
least two different modes of operation are possible. In either mode
the path of the electron beam is made visible by the luminescence of
the excited helium atoms along its path. Typically the beam cross-
section is approximately the size of the aperture in the cathode.
However pressure and the applied potential difference are also a
factor in determining the beam diameter. The graphs of Fig. 2 give
typical operating voltages and target currents for operation in
each of two modes A, B at 12 and 21 microns of pressure respectively.

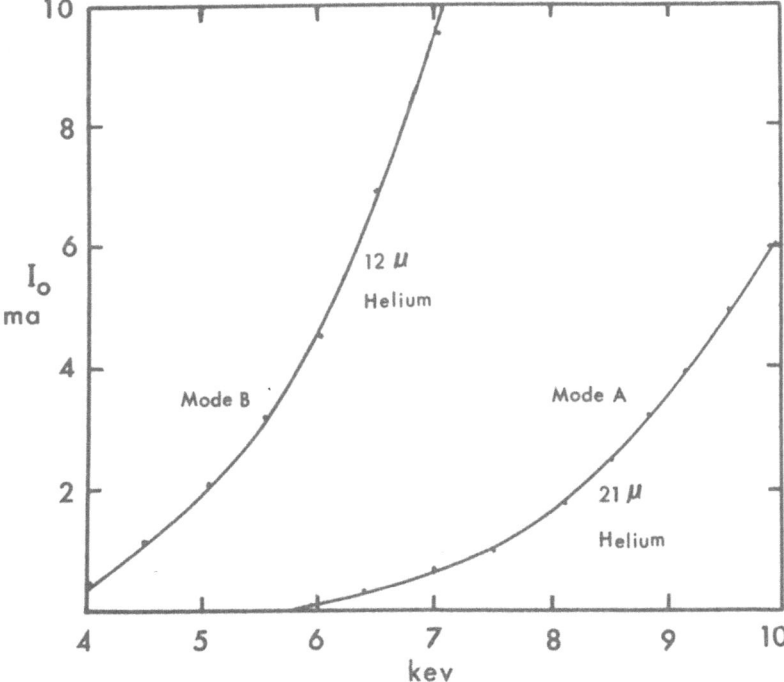

Fig. 2.   Characteristic curves of target current I_o versus target
           potential at constant gas pressure. Two distinct modes
           are evident.

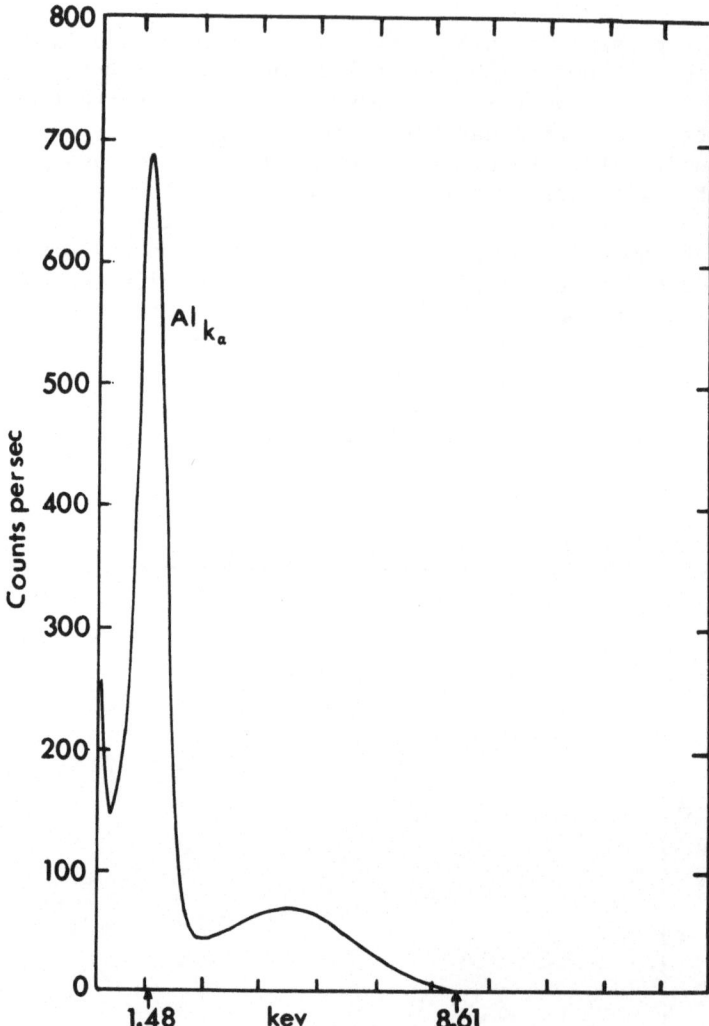

Fig. 3.   Energy spectrum recorded with tube operating at a poten-
            tial difference of 8.61 kev and a target current of 5 ma.
            The spectrum was recorded using a proportional counter
            with P-10 gas and a North American Philips' multi-channel
            analyser (IC 2000 Series). So as not to exceed the per-
            missible counting rate of the proportional counter, a
            pin-hole was used.

ENERGY SPECTRUM

The window of the x-ray tube was made from aluminum foil
0.00045 inch thick. It covers an aperture of 0.25 inch diameter.
An energy spectrum derived from a proportional counter filled with
P-10 gas is shown in Fig. 3. The actual spectrum was made with a
North American Philips' multichannel analyser (IC 2000 Series).
Within the limits of resolution of the recording apparatus the
spectrum appears free of any contaminating radiations, showing only
the unresolved $K\alpha$ lines of aluminum and the Bremsstrahlung. As
indicated the applied potential was 8.6 kev with a tube current of
8 ma. The resulting high intensity of x-rays was cut down by in-
serting a small pin-hole at the window of the x-ray tube so as not
to exceed the counting rate of the proportional counter.

CONCLUSIONS

A gas discharge tube whose electron beam is controlled by
plasma dynamics has been shown to be operable at power inputs
comparable with many sealed off Coolidge type x-ray tubes with low
atomic number targets such as vanadium and chrome. Unlike the
tungsten filament type, the wire mesh cathode is almost indestruc-
tible under ordinary use. Contamination of the target is greatly
reduced. Physical scaling of the cathode structure makes it possible to operate at much higher power inputs limited of course by
the possible vaporization of a particular target if the power
density to it becomes excessive.

REFERENCES

1. Solomon, J. S. and Baun, W. L., Rev. Sci. Instrum. 40,
    1458-60 (1969).
2. Vanhatalo, J., Kaihola, L., and Suoninen, E., Journal of
    Physics E: Scientific Instruments 9, 1156-57 (1976).
3. Leontovich, M. A., editor, Reviews of Plasma Physics,
    Consultant's Bureau, N. Y. 1965.

X-RAY FLUORESCENCE ANALYSIS AT ROOM TEMPERATURE WITH AN ENERGY DISPERSIVE MERCURIC IODIDE SPECTROMETER (1)

M. Singh, A.J. Dabrowski (2), G.C.Huth, J.S.Iwanczyk(2),
B.C. Clark*,and A.K. Baird**
University of Southern Calif.,Medical Imaging Science
Group, 4676 Admiralty Way,Marina Del Rey, CA 90291

* Martin-Marietta Corporation,Denver,Colorado 80201
**Department of Geology,Pomona College,Claremont,California 91711

INTRODUCTION

We have previously reported on the uniqueness and potential of room-temperature spectrometry of low-energy x-rays with a mercuric iodide (HgI_2) detector (1,2,3). In this paper we emphasize the use of HgI_2 detectors for x-ray fluorescence (XRF) analysis.

Because no vacuum plumbing or cryogenic cooling is required, the design of a mercuric iodide room-temperature x-ray spectrometer is extremely simple. Our present design consists of coupling a detector directly to the first-stage FET in a modified Tennelec 161 D preamplifier and making the configuration "light-tight".Aside from providing a suitable entrance window, there are no other requirements for routine spectroscopy. This preamplifier can be miniaturized with present microelectronics technology, and we anticipate development in the near future of a detector-preamplifier package with a total volume of only a few cubic centimeters (a truly portable spectrometer!)

ENERGY RESOLUTION

A detailed analysis of energy resolution relevant to mercuric iodide spectroscopy has been performed by Dabrowski and Huth (4). Presently, as shown in Fig. 1, we obtain a value of 380 eV (FWHM) for the 5.9 keV Fe-55 peak with an all room temperature HgI_2 spectrometer and 300 eV for an electronic pulser.

1.Research supported by Department of Energy funding under Contract
  DE-AS03-76-SF00113.
2.On leave from Institute of Nuclear Research, Swierk, Poland.

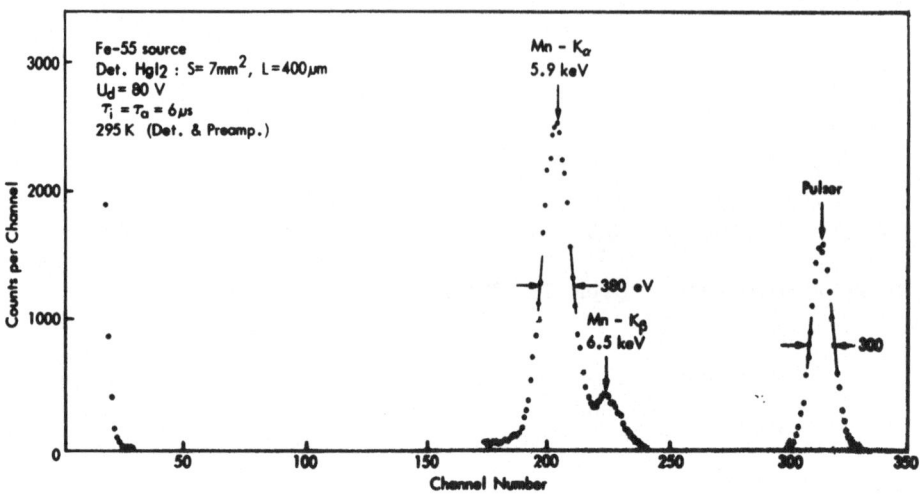

<u>Fig.1</u>   Fe-55 spectrum obtained with HgI_2 at room temperature in air

The electronic noise of this system without detector was 240 eV
(which is equivalent to 205 eV for a silicon detector). An improvement in energy resolution of about 50–100 eV should become possible
upon thermoelectric (Peltier) cooling of the input FET chip to about
$-50^0C(2)$. The ultimate resolution attained with a room-temperature
mercuric iodide detector is therefore expected to be quite close to
that obtained with cryogenic cooled silicon and germanium detectors.

XRF   ANALYSIS – GENERAL CONSIDERATIONS

One of the major applications of energy-dispersive mercuric
iodide spectrometry lies in fabrication of a truly portable and
versatile XRF analyzer. The high stopping power of HgI_2 provides a
detection efficiency suitable for observing K and L x-rays over the
entire elemental range (2).
In addition to the characteristic x-ray peaks, a typical XRF
spectrum shows "escape peaks" and "scatter peaks", and background
in the form of a general "continuum". The escape peaks result from
production and subsequent escape of characteristic x-rays of the detector material. The scatter peaks are produced by a coherent or
incoherent scattering event between the exciting radiation and the
sample matrix. The continuum background is produced via several
modes such as escape of Compton scattered photons, escape of photoelectrons, incomplete charge collection, etc. It should be noted
that similar background and escape peak effects are also observed
with other semiconducting detectors (5).

ESCAPE PEAKS

In a mercuric iodide detector, escape peaks result from production and subsequent escape of K and L x-rays of iodine, and K, L and

M x-rays of mercury. We have developed a general method for computing escape peak probabilities in any given detector as a function of
incident photon energy. The model used for this calculation is shown
in Fig.2. It is assumed that the photon is incident normally onto
the center of a cylindrical detector with diameter "d" and length"l".
The following expression has been derived for the escape peak probability P_{es} (E) as a function of incident energy E.

$$P_{es}(E) = \int_0^\ell \int_0^\pi \mu_x(E) \exp\{-\mu_\tau(E) \cdot x\} dx \cdot d\theta / \pi \cdot \exp\{-\mu_\tau(E_x) \cdot s(x,\theta)\} \quad \cdots \cdot (1)$$

In this expression:
$\mu_x(E)$ = probability (cross-section) of producing a characteristic
              x-ray of the detector material within the detector.
$\mu_\tau(E)$ = total absorption coefficient for the incident photons.
    x = depth at which an interaction occurs.
$E_x, \mu_\tau (E_x)$ = energy and absorption coefficient respectively of the
                 characteristic x-ray produced within the detector.
    θ = angle at which this x-ray is emitted.
$s(x,\theta)$ = distance traversed by this x-ray before escaping from the
              detector.
       The results of this computation are presented in Figs. 3-5. Also
shown are escape peak probabilities for typical germanium and silicon
detectors computed from eq (1). Note that escape peaks in HgI_2 are
comparable or even lower than germanium or silicon detectors in some
energy regions.
       Experimental measurements of escape peak intensities in HgI_2
have been made with several excitation sources. A few examples are
shown in Fig.6. The spectrum obtained from a filtered Am-241 source
is shown in Figs.6a and 6b. The 59.5 keV peak and iodide $K\alpha$ and $K\beta$
escape peaks obtained therefrom at 31.0 and 27.2 keV are seen. The
probabilities of $K\alpha$ and $K\beta$ escape obtained from this spectrum are ~3%
and ~1% respectively, which are in very good agreement with theoretical results of Fig.3. The peaks obtained from a Cd-109 source are
shown in Figs.6c and 6d. The escape peaks arise from the escape of
mercury $L\alpha$ and $L\beta$ x-rays, and are estimated to be ~1% and 1.5% respectively, which agree quite well with the graph of Fig.4. The Fe-55 spectrum is shown in Figs.6e and 6f. Escape peaks are now obtained from
iodine L x-rays and mercury M x-rays. The probabilities as estimated
from Fig.6f are ~0.5% for iodine L escape and ~0.1% for mercury M escape. Again, these values are consistent with theoretical values ob-
ained from Fig.5.

CONTINUUM BACKGROUND

       The following processes are mainly responsible for producing a
continuum background:
1) Compton scatter of incident photons within the detector and subsequent escape of the scattered photons. The probability of such events,

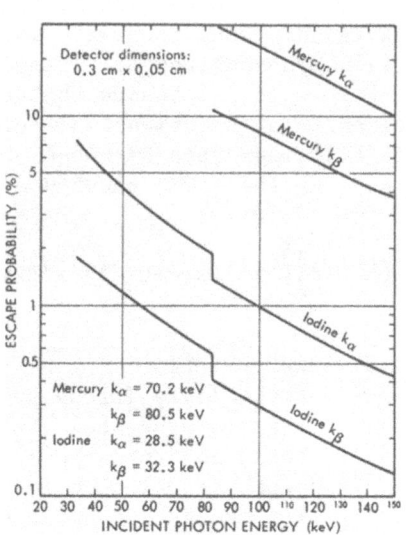

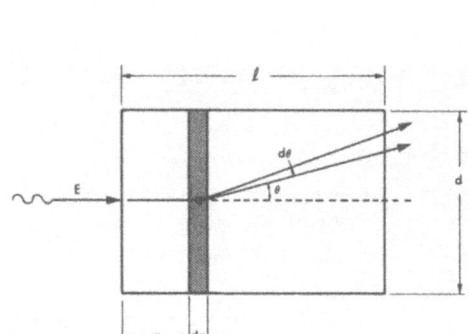

Fig.2 Model for computing interaction probabilities within the
detector.

Fig.3 Theoretical probabilities
for escape of mercury and iodine
K x-rays from HgI₂.

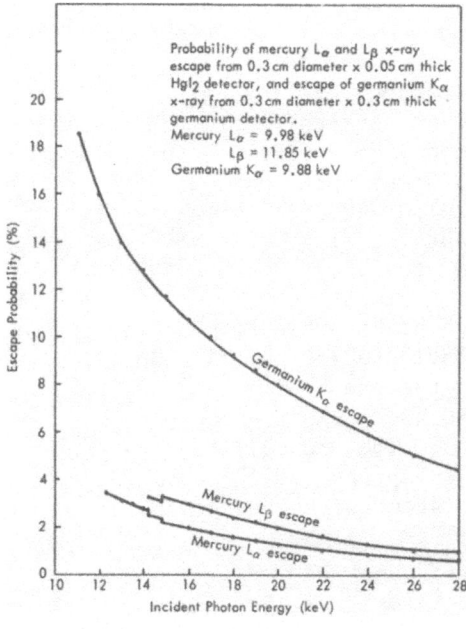

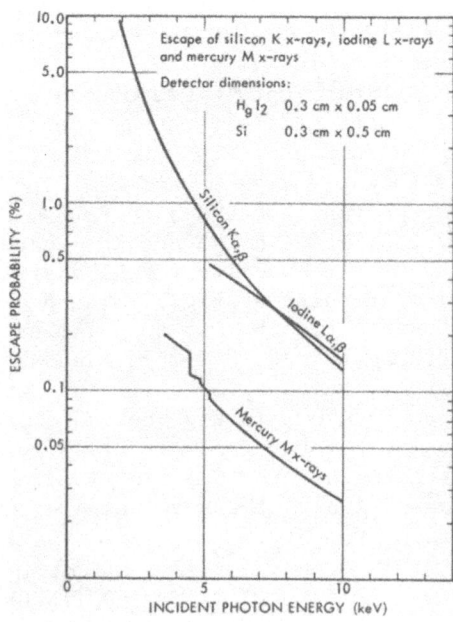

Fig.4 Theoretical probabilities
for mercury L_α and L_β escape
from HgI₂, and K_α escape from
a germanium detector.

Fig.5 Theoretical probabilities
for iodine L escape and mercury
M escape from HgI₂, and silicon
K escape from a silicon detector.

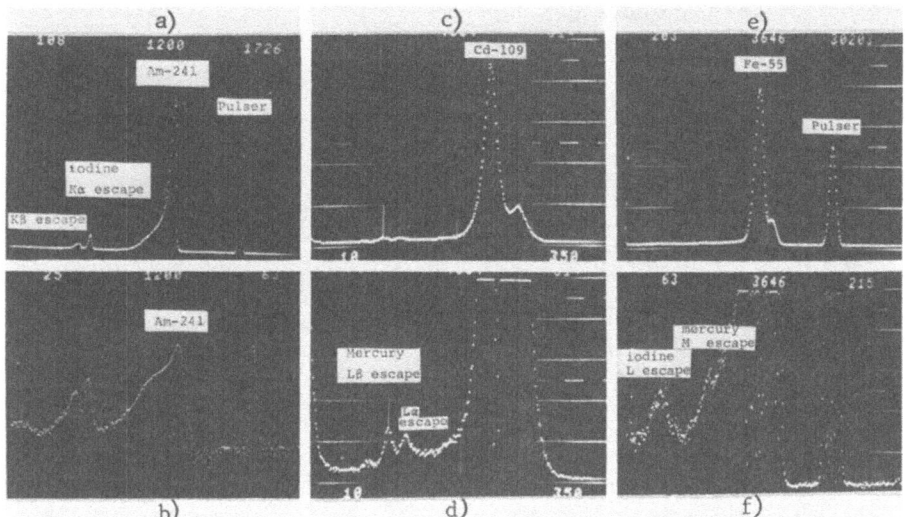

<u>Fig.6</u> Experimental spectra from various excitation sources showing
escape peaks in HgI$_2$.a)Filtered Am-241 spectrum, b)Am-241 on
log scale, c)Cd-109, d)Cd-109 on magnified scale, e)Fe-55,
f)Fe-55 spectrum on magnified scale.

denoted P$_C$, has been computed using the same model as shown in Fig.2
and is given by the following expression as a function of incident
energy E.

$$P_C(E) = \int_0^\ell\int_0^\pi \mu_C(E)\exp\{-\mu_\tau(E)\cdot x\}dx \cdot P_1(\theta)d\theta \cdot \exp\{-\mu_{PE}(E')\cdot s(x,\theta)\} \quad \cdots (2)$$

The symbols which have not been defined previously in eq.(1) are:
$\mu_C(E)$ = Compton absorption coefficient within detector
$P_1(\theta)$ = probability that the photon will be scattered between angles
θ and $\theta + d\theta$. This is obtained from the Klein-Nishina formula properly normalized to unity when integration over all angles is
carried out.
E', $\mu_{PE}(E')$ = energy and photoelectric absorption coefficient respectively for the scattered photon.
    Equation (2) yields the probability of background produced from
upto two Compton interactions. Higher order interactions are considered negligibly small.
    The results of this calculation for incident energies up to 1 MeV
are shown in Fig.7 for typical mercuric iodide, germanium and silicon detectors. It is noteworthy that among these three detectors
the background from this effect is lowest for mercuric iodide.
    2) Escape of a photoelectron with partial deposition of energy
in the detector. The photoelectron is ejected in random directions

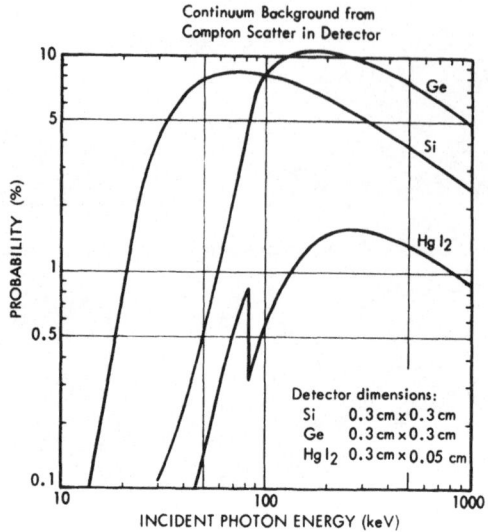

Fig.7 Theoretical estimate of continuum background from Compton scatter of incident radiation within the detector.

and loses energy along its track. Because of the high stopping power of HgI$_2$, only those electrons which are ejected in a backward direction can produce background (neglecting electron channeling effects). Using an approximate expression for the range of an electron (5), it is estimated that the maximum probability for producing background from this process is ~1% for 20 keV photoelectrons in HgI$_2$.

3) Incomplete charge collection is another, and in our opinion the dominant mechanism that produces a continuum background. Non-uniformities in the detection region, distortions in the electric field, particularly near the surface of the detector, and various defects that lead to charge-trapping are some of the main contributing factors. At this point in time there are several unknowns in the HgI$_2$ crystal-growth and detector-fabrication technology that make it difficult to accurately calculate background from these effects. Defects in charge-collection become increasingly important at higher energies where hole transport plays a significant role. Much progress is being made in crystal and detector technology, and we expect that many problems associated with defects in charge-collection and detector non-uniformity will be solved in the near future. Charge collection problems associated with electric field distortion can be minimized by either masking the detector, which we currently do, or by using a guard-ring (6), which is under development.

Experimentally it is observed that the continuum background in a typical XRF spectrum obtained with a HgI$_2$ detector and Cd-109 excitation, is at an approximately 1% level of the scattered peak height.

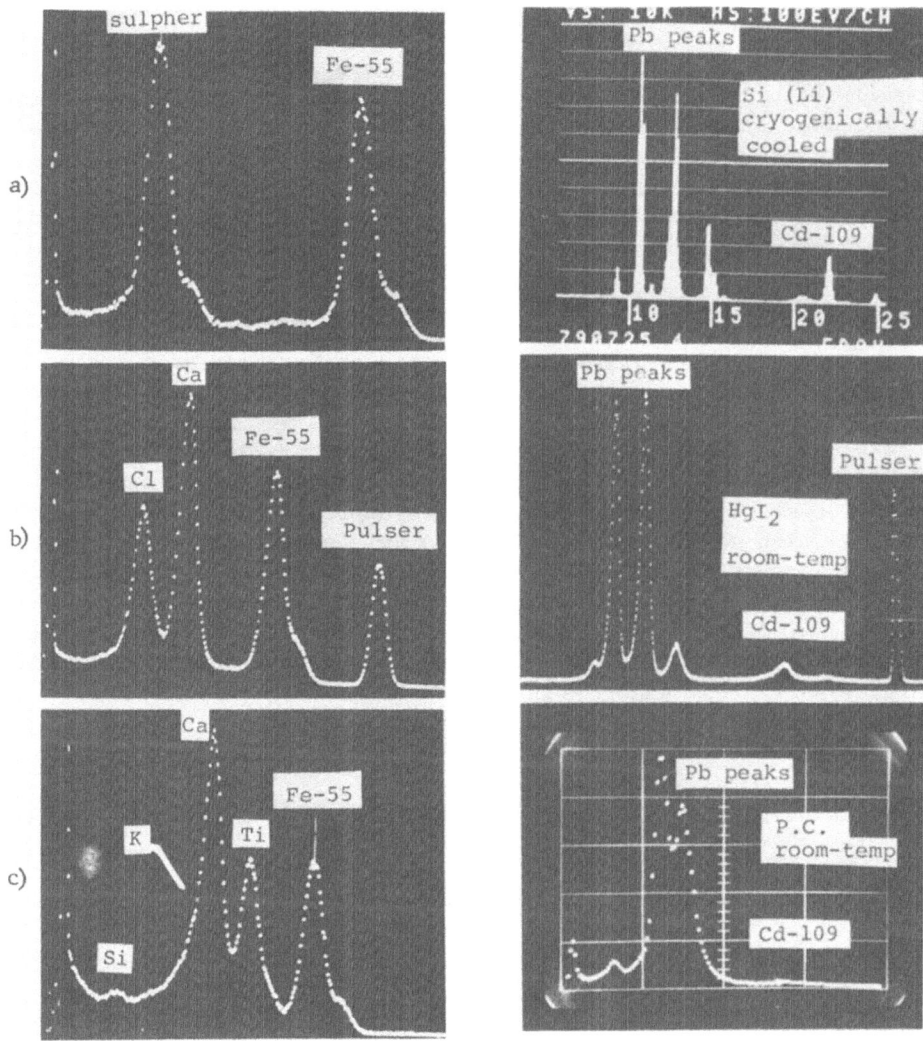

Fig.8 XRF spectra from a)sul-
pher, b)chlorine and calcium,
c)BCR Basalt samples using
Fe-55 source and HgI2 detector.

Fig.9 Comparison of cryogenically cooled Si(Li), room temp.
HgI2 and proportional counter
for detecting lead spectrum
using Cd-109 excitation.

XRF MEASUREMENTS - EXPERIMENTAL

XRF spectra from several samples have been obtained with a room-
temperature mercuric iodide detector (7mm^{2}x400 μm) using Fe-55 and
Cd-109 as the excitation sources. In addition to single-element
samples such as magnesium, silicon, sulfur, iron, nickel, copper,

barium etc. several multi-element samples and rock and ore samples, including uranium, have also been investigated. For comparison,XRF spectra were also obtained for some samples with a cryogenically cooled Si (Li) detector, and a standard room-temperature proportional counter system (energy resolution ~1.1 keV at 5.9 keV). Some of these results are shown in Figs.8 and 9. The significant improvements over a proportional counter are obvious. The performance of the room-temperature HgI_2 rivals that of the cryogenically cooled Si (Li) detector for practical x-ray fluorescence spectroscopy.

CONCLUSIONS

A study of XRF analysis with room-temperature mercuric iodide detectors has been conducted. Factors which contribute to background in XRF spectra have been investigated in detail. One finds that the background produced in mercuric iodide detectors is comparable, and in some instances, even lower than that produced in germanium or silicon detectors. An energy resolution of 380 eV at 5.9 keV has been obtained with an all-room-temperature HgI_2 spectrometer, and further improvements of 50-100 eV are expected. A miniature detector-preamplifier package is under development. It is envisioned that this detector will have unique advantages in a portable elemental analyzer and could simplify instrumentation for space-borne experiments involving soft x-ray spectroscopy.

ACKNOWLEDGMENT

We thank the group at EG & G Santa Barbara for detector fabrication.

REFERENCES

1. A.J. Dabrowski, G.C. Huth, M. Singh et al: Appl. Phys. Lett. 33 (2), 211-213 (1978)
2. G.C. Huth, A.J. Dabrowski, M. Singh et al: Adv. in X-ray Analysis, 22, 461-472 (1978)
3. A.J. Dabrowski, M. Singh, G.C. Huth, J.S. Iwanczyk: Invited Paper, NBS Workshop on Energy Dispersive X-ray Spectrometry, Gaithersburg, Maryland, Aug. 23-25, 1979 (In Press)
4. A.J. Dabrowski, G.C. Huth: IEEE Trans. Nucl. Sci. NS-25 (1), 205-211 (1978)
5. F.S. Goulding, J.M. Jaklevic: LBL-5367 (1976)
6. F.S. Goulding, J.M. Jaklevic: UCRL-20625 (1971)

NEW MOLD DESIGN FOR CASTING FUSED SAMPLES

Joseph E. Taggart, Jr., and James S. Wahlberg

U.S. Geological Survey

Box 25046, MS 928, Denver, Colorado 80225

Geological samples prepared for analysis by X-ray fluorescence (XRF) are particularly prone to intrasample and intersample phase variations creating significant inhomogeneity and variations in matrix effects. These features may best be eliminated or diminished by fusion of the sample with a suitable flux (1,2). Presentation to the X-ray spectrometer may be as a briquet of the comminuted fusion bead (1,3,4,5) or as a molded glass disc or wafer (1,2,5,6). Because of the stability of glass discs and the necessity to eliminate possible contamination during grinding and making briquets, molded glass discs appear to be the best and easiest means of presentation.

Commercial molds used for casting fusion discs are made from platinum-gold (95:5) sheets one-half millimeter thick. During use, the molds become deformed; this deformation produces discs whose analytical surfaces may be concave or convex. Many authors have observed the effect of uneven sample surfaces on analytical precision (1). Carpenter (7) showed that sample positioning is the single largest source of system error. Flattening and refinishing the casting molds are difficult and are only partially successful.

A common method for correcting the uneven and variable analytical surface is the resurfacing of each glass disc before its use. Fabbi and Elsheimer (6) polished discs successively on 600- and 1200-grit diamond-embedded grinding wheels; the U.S. National Bureau of Standards polishes discs on 600 grit (8). This method has been shown (6) to be highly successful for improving analytical precision. However, if polishing is done manually, this method is very time consuming and tedious. The U.S. Geological Survey lab in Menlo Park, CA., is currently using an automatic lapping machine.

Even this adds considerably to the preparation time for each sample.

A mold has been designed that creates discs having a smooth flat surface and eliminates the need to resurface each glass disc (Fig. 1). The two-piece mold consists of a ring and disc and has been fabricated and tested. The mold can be taken apart, and the part of the Pt disc that creates the analytical surface can be ground flat and polished smooth. The ring retains the molten flux mixture until it cools and hardens to a glass disc. In addition, the mold has been made considerably thicker than the commercial varieties. The 1/8" thickness of the disc part keeps the mold rigid but still allows a considerable thickness to be removed during periodic grinding and polishing. The mold is made of a 95% Pt-5% Au alloy (Englehard 7070*) because of its superior heat resistance and non-wetting properties. Owing to the considerable initial investment, the thickness of the Pt disc can be adjusted to achieve a trade-off between rigidity and cost. The amount of Pt lost with each resurfacing is only about 1/3 gram (about $4.00 at current prices).

To the unaided eye, the surface of the sample discs produced with the new mold are comparable to those that are individually ground and polished. To test the precision of the samples from the new mold, we mixed 40 grams of basalt spiked with Na and P (to ensure sufficient count rates) with 320 grams of flux (90% lithium tetraborate, 10% lithium carbonate) and fused the mixture in a large Pt dish at 1100°C. The contents of the dish were periodically swirled to homogenize the sample. After quenching, the fusion mixture was ground to -300 mesh and rolled to further ensure homogeneity. Fourteen seven-gram aliquots were used to produce fourteen sample discs from used commercial molds. These discs were then counted for a period of 500 seconds on a Phillips PW1600* simultaneous wavelength dispersive XRF unit employing an end window rhodium tube at 50 Kv 50 ma. All the net counts were sufficiently high to ensure good counting statistics (Table 1). Coefficients of variation were determined for each of eight elements. The discs were then ground flat and polished with successively finer grits down to 0.05 micron alumina. These samples were then counted under the same conditions as those listed above. To test the precision of the instrument, we counted one of the 0.05 micron polished discs 14 times.

New aliquots of the basalt/flux mixture were used to prepare 14 sample discs from the new mold; the discs were then counted.

---

*Any trade names used in this paper are for descriptive purposes only and do not constitute endorsement by the U.S. Geological Survey.

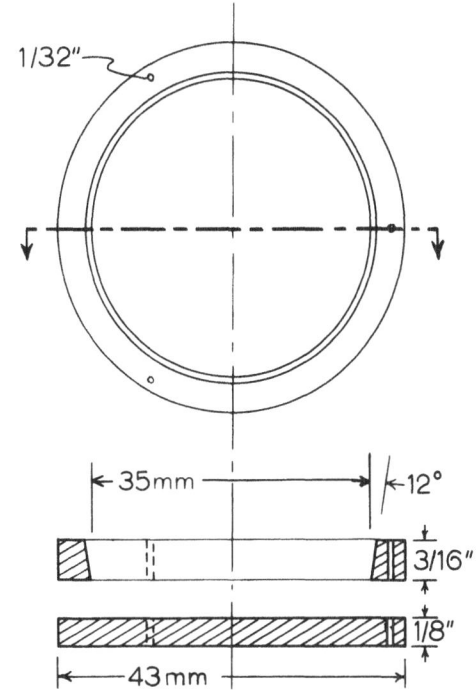

Fig. 1. Mechanical drawing of mold

The coefficients of variation for all the determinations are shown
in Table 2. Clearly, the discs from the new mold are superior to
those from used commercial molds and are comparable to those samples
that were meticulously polished.

Table 1. Instrumental conditions Phillips PW1600*

LINE	CPS/%	NET COUNTS 500 SECONDS	CRYSTAL
Na Kα	50	0.2×10^6	Semi Focus – TAP
Mg Kα	170	0.2×10^6	Semi Focus – TAP
Al Kα	1100	4.0×10^6	Semi Focus – PET
Si Kα	800	8.5×10^6	Semi Focus – PET
P Kα	2500	1.4×10^6	Semi Focus – GE
K Kα	2000	3.1×10^6	Semi Focus – PET
Ca Kα	1800	3.8×10^6	Semi Focus – LIF200
Fe Kα	2700	8.4×10^6	Semi Focus – LIF200

Table 2.   Comparison of coefficients of variation for three types
           of discs.

	Na	Mg	Al	Si	P	K	Ca	Fe
Discs from commercial mold -	-3.4	1.5	2.2	2.1	2.3	1.1	1.0	0.8
Same discs after polishing - - - - -	-0.9	0.3	0.2	0.2	0.3	0.1	0.2	0.4
Discs from new mold- - - - - -	-1.1	0.5	0.5	0.4	0.5	0.3	0.3	0.3
Replicate runs on one sample- - - - -	-0.6	0.2	0.1	0.1	0.2	0.2	0.1	0.1

In order to determine the relative importance of flatness vs.
polish and the frequency that the mold will have to be repolished,
we ground the polished glass discs on increasingly coarser silicon
carbide paper and then counted them. The results (Table 3) indicate that flatness is more important than polish as long as the
surface is ground to 400 grit or finer. Therefore, the Pt disc
needs to be ground to only about 600 grit and does not have to be
polished. The insignificance of the effect of scratches on the
surface of the mold indicates that the mold needs refinishing very
infrequently.

Table 3.   Discs from commercial mold ground to various finishes;
           comparison of coefficients of variation.

	Na	Mg	Al	Si	P	K	Ca	Fe
240 grit ----------	0.8	0.5	0.4	0.6	0.6	0.4	0.5	0.5
400 grit ----------	0.7	0.5	0.3	0.5	0.4	0.3	0.3	0.5
600 grit ----------	0.8	0.4	0.3	0.3	0.3	0.1	0.2	0.5
Polished ----------	0.9	0.3	0.2	0.2	0.3	0.1	0.2	0.4
Replicate runs on one sample---------	0.6	0.2	0.1	0.1	0.2	0.2	0.1	0.1

An added benefit in using the new mold was a much lower incident rate of disc failure due to cracking. Observation of the disc
under crossed poloroids showed that the discs made in commercial
molds are under stress; those from the new molds did not appear to
have any internal stress. Furthermore, the discs from the new mold
could be labeled with a diamond scriber if desired, but the discs
from the commercial mold shattered.

Because the surface of the individually ground discs and the discs from the new mold are flat, either type may be ground on a piece of silicon carbide paper to remove a thin layer of radiation damage or hydration with no noticeable change in the count rate. This feature is especially useful to extend the life of a monitor disc. Discs with a non-flat surface cannot, of course, be ground flat without changing their count rate.

REFERENCES

1.   E. P. Bertin, Principles and Practice of X-ray Spectrometric Analysis, Plenum Press, New York, 2nd ed., p. 1079 (1975).

2.   K. Norrish and J. T. Hutton, "An Accurate X-ray Spectrographic Method for the Analysis of a Wide Range of Geological Samples," Geochim. et Cosmochim. Acta. 33, 431-453 (1969).

3.   B. P. Fabbi, "A Refined Fusion X-ray Fluorescence Technique and Determination of Major and Minor Elements in Silicate Standards," Am. Min., 57, 237-245 (1972).

4.   B. P. Fabbi, "X-ray Fluorescence Determination of Sodium in Silicate Standards Using Direct Dilution and Dilution Fusion Preparation Techniques," X-ray Spectrometry, 2, 15-17 (1973).

5.   R. J. Parker, "Factors Affecting the Quality of Major Element Rock Analysis by X-ray Fluorescence Combined with Flux-Fusion Sample Preparation," London University, Imperial College, Tech Rept. XRF-2B, p. 35 (1977).

6.   B. P. Fabbi and H. N. Elsheimer, "Evaluation and Application of an Automatic Fusion Technique to Major-Element XRF Analysis of Silicate Rocks." Paper #267, Abstracts 3rd Annual Federation of Anal. Chemistry and Spectroscopy Socs. Meeting Nov. 15-19, 1976.

7.   D. Carpenter, "Sample Positioning Effects in X-ray Spectrometry," Unpublished Data, Rockwell International, Golden, Colo. (1977).

8.   P. A. Pella, Oral Communication, U.S. National Bureau of Standards, Washington, D.C., (1978).

X-RAY IMAGING

N. Gurker

Institut für Technische Physik
Technische Universität Wien (Vienna), Austria

INTRODUCTION

Element mapping has been so far an analytical field covered by those investigation methods using a scanned particle beam on the excitation side of the specimen. These methods include localized interaction processes, as in the electron microprobe, secondary ion mass spectrometer and Auger electron spectrometer. On the other hand, methods which use a stationary radiation beam on the input side (non-localized interaction) usually give summarized information about the sample area (X-ray fluorescence spectrometry, photoelectron spectrometry, ...). This paper deals with an approach to extending the application field of X-ray fluorescence analysis (XRFA) to element mapping. Fig. 1 shows the main components of the X-ray imaging system. Since modern XRF systems often work under computer control, the only additional hardware components of the system are a scanned sample holder and an X-ray source producing a line-shaped X-ray beam. The computer is usually not used for ease, speed or convenience of operation, but is used to generate the element image and process all the collected fluorescence data.

An intensity profile showing a line resolution of the selected element is produced by generating a line-shaped X-ray beam and translating the sample stepwise (by means of a computer-controlled stepping motor) across the fixed X-ray beam while measuring the fluorescence intensity in a specified energy window at each translation step.

Fig. 2 shows schematically the sample holder performing

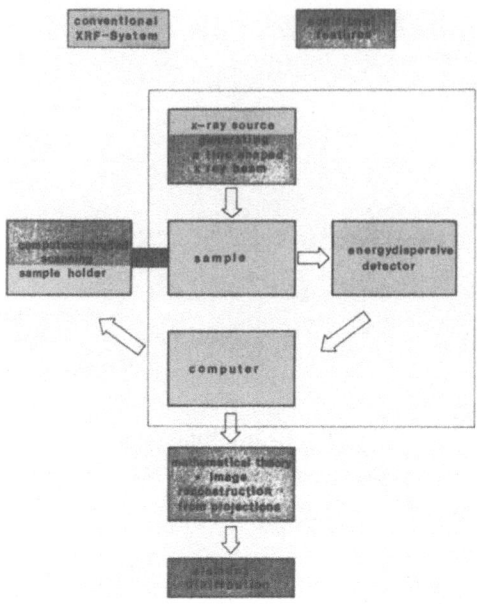

Fig. 1. Components of the X-ray imaging system indicating the
         conventional (unshaded) and newly added (shaded) features

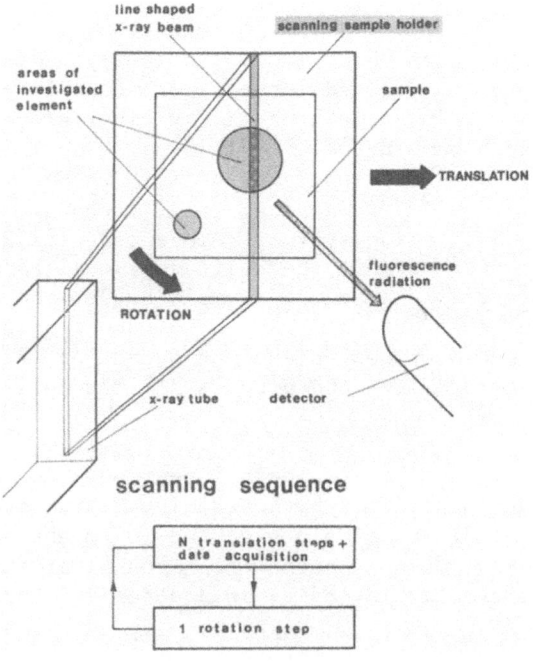

Fig. 2. Experimental arrangement: sample holder performing scan-
         ning movements in a fixed X-ray tube/detector assembly

scanning movements (translations, rotations) in a fixed X-ray
tube/detector assembly. Fig. 3 gives a view of the experimental
arrangement and of the scanning unit. Acquiring a translation profile of an element distribution is equivalent to taking its
'projection' in a given angular orientation. Data acquisition will
be determined if many such projections, equally spaced in an angular range of 180°, have been performed (Fig. 4).

The computer converts the line resolutions of the sampled
translation profiles into a point resolution (= element map) making use of a mathematical theory known as 'image reconstruction
from projections' or 'representation of a function by its line
integrals'.[1,2] The best known application of this theory at present is its use in medical computed tomography (CT, Head- and
Whole Body-Scanners) revealing the internal structure of the human
body in terms of attenuation coefficients in a quantitative form.[3]

IMAGE RECONSTRUCTION FROM PROJECTIONS

Introducing the idea of 'back projection' of projection pro-
filed over an image matrix, one will get a blurred image of the
scanned element distribution when simply using measured profiles
(Figs. 5 and 6: computer-simulation, 100 translation profiles).
There are reconstruction errors simulating element contributions
outside the scanned element areas. Starting from a mathematical
formulation of the translation profile, sampling Fourier trans-
formations lead to a modified concept of back-projection
('filtered' back projection, FBP) giving reproductions without
reconstruction errors,[4] (Figs. 7 and 8).

The 'filtered' profiles p* result from a convolution integral
of measured profiles p with a correction function q.

$$p* \ (x, \theta) = \int_{-\infty}^{+\infty} p \ (x', \theta) \ q \ (x-x') \ dx'$$

Reconstruction formulas for filtered back projection are derived
from continuous type image plane and translation profiles, and
then have to be modified for discrete experimental realization;[4]
iterative reconstruction techniques start from the discrete character of data sampling and reconstruction matrix (Fig. 9). The
iterative solution is based on correction factors c resulting from
differences between measured (p^m) and computed (p^q) profile data.[5]

Iterative reconstruction algorithms, although much more time

Fig. 3. View of the experimental arrangement and of the scanning
         unit

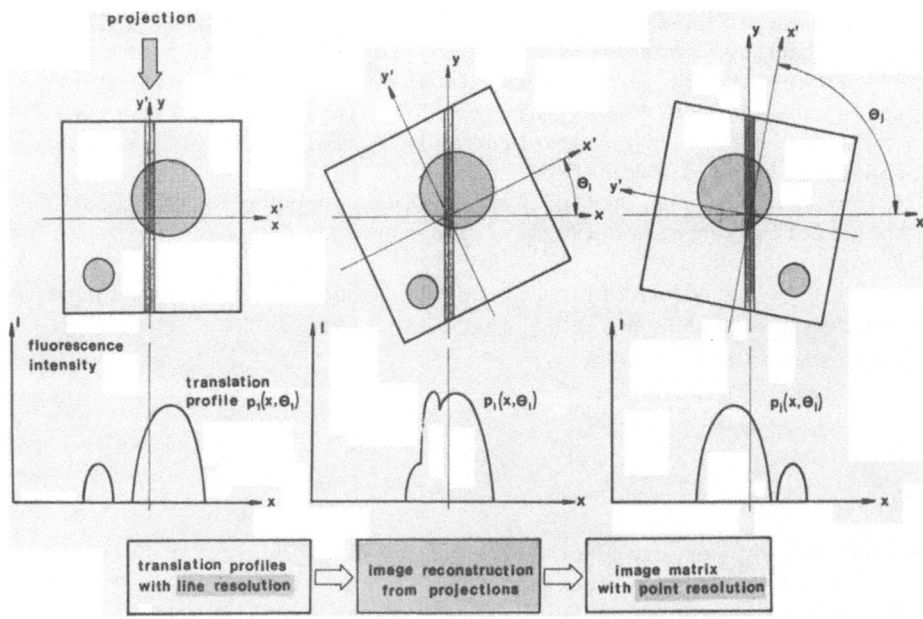

Fig. 4. Acquisition of translation profiles (projections) at
         different angular orientations of the sample

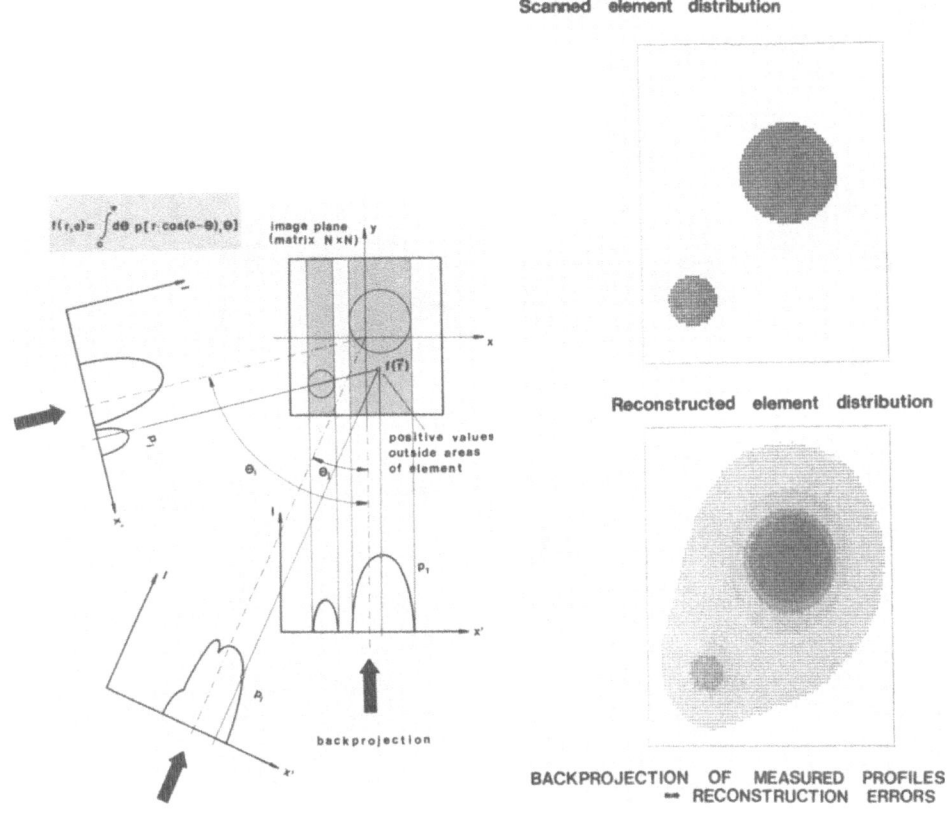

Scanned element distribution

$f(r,\theta) = \int_0^{\pi} d\theta \; p[r \cdot \cos(\theta-\Theta), \theta]$

image plane
(matrix N×N)

positive values
outside areas
of element

backprojection

Reconstructed element distribution

BACKPROJECTION OF MEASURED PROFILES
→ RECONSTRUCTION ERRORS

Fig. 5   Fig. 6

Back projection of measured profiles

consuming than Fourier-like techniques ($\sim$ 10x), will be superior
in case of 'incomplete data' (*i.e.*, where there is a small number
of projections or poor statistics for translation profiles).

The image quality (spatial resolution) is given by:

1) x-ray beam width
2) translation step size
3) rotation step size
4) data acquisition time/translation step (statistics)

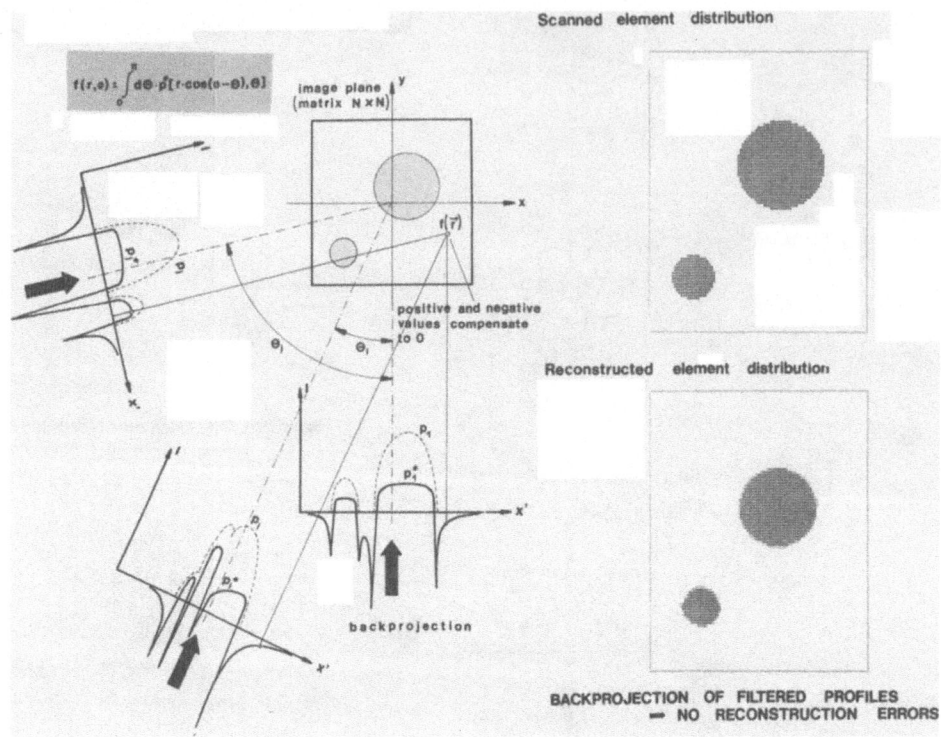

Fig. 7   Fig. 8

Back projection of filtered profiles

EXPERIMENT

Imaging capabilities of a scanning X-ray system based on the method of 'reconstruction from projections' have been verified by using test structures formed by a 30 μm Cu-film on an epoxied substrate (Fig. 10). The Mo X-ray tube operating parameters have been 30 kV, 5 mA; the X-ray beam width and translation step size have been varied from 0.1 mm to 0.25 mm, and the rotation step size has been varied from 1.8° to 14.4°. Fluorescence data (CuKα-radiation) have been collected by a Si(Li) detector, data reduction has been done on a Tracor Northern TN-11 system. Profile data have been slightly smoothed before being processed by back projection (FBP) or iterative techniques (ART). Reconstructed Cu element maps for varied scanning parameters are given in Figs. 11-14.

Any attempt to improve image quality is followed by an increase in data collection time. From experimental and simulated scans[6] the following minimum scanning parameters for a spatial resolution of Δx may be derived:

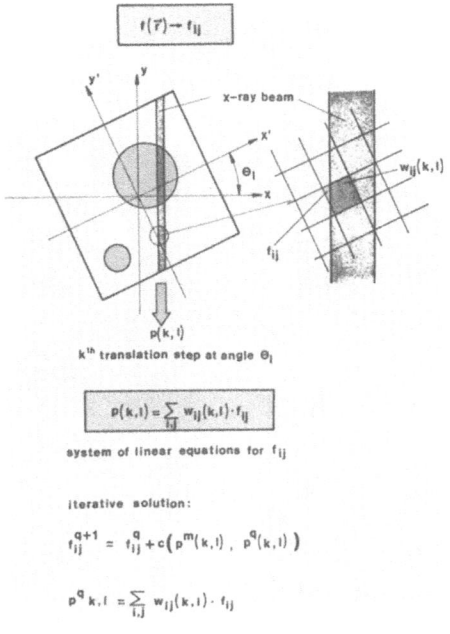

Fig. 9. Iterative (Algebraic) reconstruction technique (ART)

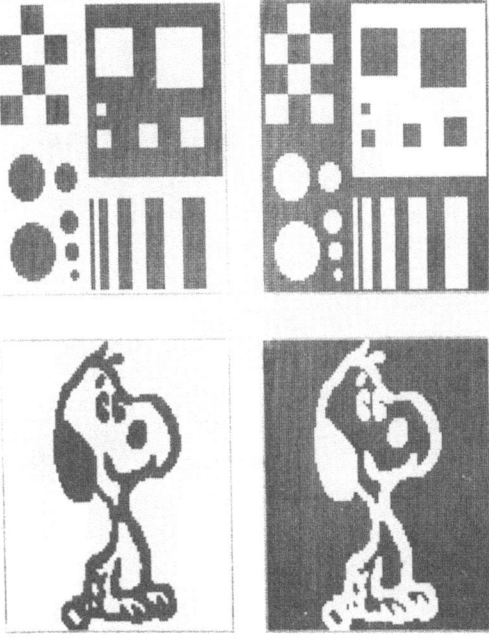

Fig. 10. Test images formed by a 30 μm Cu-film on an epoxied
substrate (sample size 10 x 10 mm)

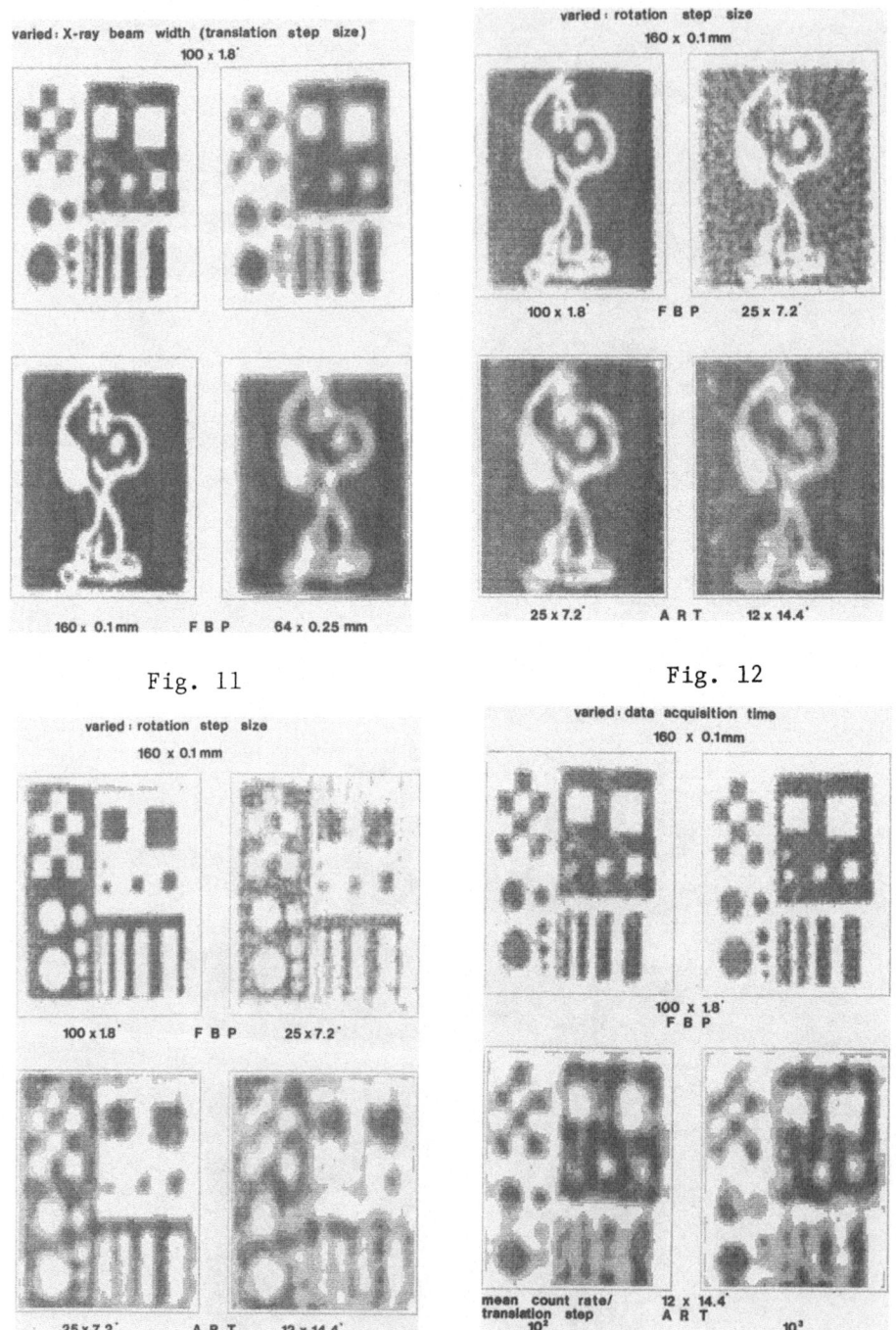

Fig. 11

Fig. 12

Fig. 13

Fig. 14

Reconstructed Cu element maps for varied scanning parameters

1) X-ray beam width, translation step size $\lesssim \Delta x$
2) $\gtrsim$ 25 translation profiles
3) mean count rate/translation step $\gtrsim 10^2$.

CONCLUSION

   Why do we scan the sample with a line-shaped X-ray beam and perform image reconstruction from projections and not simply scan the sample across a point-shaped X-ray beam?

   Because the minimum scanning parameters show that scanning time will be less ($\sim$ 1/2) when using a line-shaped beam instead of a point-shaped beam, just as the intensity resolution of the image matrix will be better in case of a line-shaped X-ray beam due to the fact that one image element is measured not once but a number of times (equal to the number of translation profiles), thus reducing statistical errors.

   Element mapping methods applying scanned particle beams usually work on a micron scale concerning both spatial resolution ($\gtrsim$ 1 μm) and investigated sample areas ($\sim$ 100 x 100 μm). X-ray scanning, however, works on a macro scale (spatial resolution $\gtrsim$ 100 μm, investigated sample areas = several cm^2). Thus X-ray scanning (X-Ray Macroscopy = XMAC) may expand element mapping facilities to application fields not yet covered by existing element mapping methods.

ACKNOWLEDGEMENTS

   Dipl. Ing. K. Zeiner and Dipl. Ing. H. Schandl are gratefully acknowledged for experimental and construction assistance. Research work has been financially supported by the Austrian "Fonds zur Förderung der wissenschaftlichen Forschung" (Project No. 3746).

REFERENCES

1. J. Radon, "Berichte über die Verhandlungen der Königlich Sächsischen Gesellschaft der Wissenschaften zu Leipzig," Mathem. Phys. Klasse, 69:262-266 (1917).
2. A. M. Cormack, "Representation of a function by its line integrals with some radiological applications," I, J. Appl. Phys. 34:2722-2727 (1963); II, J. Appl. Phys. 35:2908-2913 (1964).
3. G. N. Hounsfield, "Computerized transverse axial scanning (tomography) Part I: Description of the system," Br. J. Radiol. 46:1016-1022 (1973).

4. G. N. Ramachandran, A. V. Lakshminarayanan, "Three-dimensional
    Reconstruction from Radiographs and Electron Micrographs:
    Applications of Convolutions instead of Fourier Transforms,"
    Proc. Natl. Acad. USA 68, 9:2236-2240 (1971).
5. R. Gordon, "A Tutorial on ART," IEEE Trans. Nucl. Sci. 21:78-
    93 (1974).
6. N. Gurker, "Element Mapping by a Scanning X-Ray System," X-Ray
    Spectrom., to be published.

THE SEARCH-MATCH PROBLEM

Monte C. Nichols
Sandia Laboratories
Livermore, California   94550

Quintin Johnson
Lawrence Livermore Laboratory
Livermore, California   94550

INTRODUCTION

Identification or recognition of an unknown is a problem encountered
in areas as diversé as X-ray diffraction (1-4), signature and voice
verification (5), fingerprint identification (6), and spectroscopy
(mass (7), infared (IR) (8), ultraviolat (UV), Auger). Features
uniquely characteristic of the unknown, which are used in the
identification process, are called the "pattern" of that unknown.
In a typical identification, the one or more sub-patterns (components) making up the unknown pattern are identified by a search
through a set of known (standard) patterns to yield potential
matches. This searching and matching can be done sucessfully by
hand, especially if the number of patterns in the file of standards
is not too great (9,10). When the number of phases which consititute
the unknown pattern becomes extremely large the use of computers
becomes a practical necessity if the search is to be conducted in
a reasonable time. A comparison of X-ray search-match approaches
with each other and with search-match capabilities existing in
other fields may lead to better X-ray phase identification
techniques.

BACKGROUND

In the simplest form of search/match, a file of identified patterns is used along with a method of determining whether one or more
of these "standard" patterns matches, or is a subset of, the unknown
pattern. Complications arise from a number of factors. In cases
where the pattern is a composite of sub-patterns, it is possible to
end up with an ambiguous answer. As an example, consider the composite word pattern "bdornye". Two possible two word sub-patterns,

namely born + dye or bone + dry, result from this composite word
pattern without necessitating a change in the order of the letters
in the unknown pattern.

Further complications arise because of errors in the measurements of the pattern. This is true for standards and unknowns alike.
The presence of experimental errors in our word pattern above might
cause it to appear as a-c, c-e, n-p, q-s, m-o, x-z, d-f. In similar
fashion, the patterns in the standard word pattern file have error
limits. Born, for example, could have been represented by a-c, n-p,
q-s, m-o. It is easily seen, then, that the general solution of a
composite pattern is not simple.

COMPARISONS

Not all areas in which search/match procedures are utilized
must contend with the complications mentioned above. It is unlikely
that a fingerprint identification would be attempted if two or more
prints completely overlapped. In similar manner, signature and
voice verification treat single patterns only. Single pattern
identification by reference to a standard file is so much simpler
than identification of an individual pattern in a composite pattern
that, for many spectroscopy techniques, attempts are made to simplify
the problem by performing separation of phases in the mixture before
any experimental patterns are produced. This reduces the number of
unknowns in any one experimentally collected pattern, to a series of
single unknowns. The GCMS (gas chromatograph - mass spectrometer)
is an example of an instrumented version of such a separation.
Researchers using IR, UV etc. also attempt separation of materials
prior to producing an unknown pattern much more often than is done
in X-ray diffraction analysis. Separations based on magnetic,
optical, density, cleavage, or other physical properties might be
used to greater advantage by X-ray analysts. In addition, micro
techniquies offer the ability to investigate selected areas of a
sample. Even a partial separation would allow the analyst to
subtract the altered unknown from the origninal unknown pattern to
assist in the identification of one or more phases. This capability
is available today using programs developed for diffraction
automation (13).

Many other characterization techniques differ from X-ray powder
diffraction analysis in another important way. These other techniques give valuable fragment information about the individual
components present in a complex mixture, even if those components
cannot be readily separated or identified. In mass spectroscopy,
the masses of fragments as well as elements that make up such
fragements are identifiable. For infrared analysis, the absorption
bands yield information of chemical functionality even if the resulting pattern of bands cannot be correlated with a standard compound

or compounds present in an IR data base. Unlike the mass spectrographic and infared cases, the individual d-I data from multiphase
X-ray powder diffraction patterns does not a-priori provide any
direct insight into the identity or properties of any phase present.

In spite of these difference, some definite similarities
exist among the various search-match characterization techniques.
One similarity is the number of features or attributes of a given
pattern chosen as characteristic, when rapid search-matching of a
file of standards is to be used. In the case of several X-ray
search-match, as well as mass spectra (7) programs the eight
strongest maxima from the standard file pattern are compared with
the sixteen strongest maxima from the unknown to determine whether
further, more detailed, matching should be undertaken. The magic
number 8 is also utilized for fingerprint identification where
eight, two digit numbers are used as identifiers (6).

In addition to this similarity, almost all search-match programs
are designed to attempt to avoid searching the entire file and to try
to use the best possible matching (discriminating) algorithm that can
be devised. These represent the broad areas of similarity and difference between the X-ray powder diffraction search/match problem and
those in other areas. A number of differences also exist in the different approachs to the X-ray search-match problem.

X-RAY SEARCH-MATCH STRATEGIES

There are four principal approaches to the X-ray search-match
area at the present time.

1. DOW/Frevel (1)

2. Penn State/Johnson (2)

3. SEARCH/Nichols - Burfield (3,4)

4. NIH - EPA (CIS System)/JCPDS-Marquart (11)

These programs are similar in a number of respects but their
differences are worth discussing. The most important differences
between the programs occur in the areas of "error" windows, file
search methods, and the need for some of the programs to have
supplemental chemical information provided as input.

Error windows are the limits placed around the experimental data
points (d's and/or I's). A match occurs when two potentialy matching
values agree within these limits. Smaller windows do permit a great
restriction of the number of possible answers retrieved by the search-
match programs. In the DOW program, the smallest possible d spacing

windows are utilized first; if matching patterns are not found then error limits are increased. Both the Penn State and CIS programs use small d windows in order to limit the number of possible answers. The SEARCH program uses large windows and, in a more recent modification, even overlapping windows (4). The designation of error windows is not entirely correct since these acceptance limits can also be used to allow identification of possible solid solution phases. Since the form of this "error" window is not the same for experimental errors and solid solution shifts (especially at high 2θ) it is important for the windows to be "wide" if solid solution possibilities are not to be excluded.

In every program except for the DOW program, all of the patterns in the file which could best match the unknown are considered prior to any alteration of the unknown pattern by subtraction. It is clear that the use of an ordered file with the very efficient inverse search/match (14) will be required in the future as the size of the file continues to increased. Table I summarizes how these programs conduct a search.

There is much controversy over the need for chemical information in order to conduct an effective search/match of the X-ray

TABLE I

COMPARISON OF SEARCH/MATCH PROGRAMS

Program	Error Windows	Search Method	Chemical Information	References
DOW	Small→large	Direct search of ordered file	Required	(1)
Johnson/Vand	Small	Entire file searched sequentially	Greatly assists in limiting answers	(2)
SEARCH	Large (over lapping)	Inverse search/match of ordered file	Not usually required. Used as last resort.	(3,4)
CIS	Small	Direct pre search; inverted match	Required in practice to limit answers	(11)

powder diffraction file. Table I shows where the various programs
stand in this regard. If a program contains a good matching
algorithm, it is possible to match isostructural compounds of
differing chemistry provided chemical restrictions are not employed.
Thus, restricting the use of chemical information to the final
step in an analyses is a definite advantage.

PRESENT AND FUTURE STATUS

      All of the X-ray Search-Match programs have been used success-
fully to identify phases present in complex mixtures. The improved
accuracy that should result from increased automation and the
proposed publication standards for new powder data (12) will
ultimately mean better performance for all X-ray search-match
programs. Active research of X-ray search-match methods by a
number of investigators is expected to lead to further search-match
improvements. Approaches used in other areas suggest also that
fresh look should be taken at the use of conventional, or novel
separation techniques prior to obtaining a pattern. Further
efforts in this area will bear substantial rewards.

REFERENCES

1. L. K. Frevel, Quantitative Matching of Powder Diffraction
        Patterns, in: "Advances in X-ray Analysis, Vol. 20," Plenum
        Press, NY, (1976).
2. G. G. Johnson, Resolution of Powder Patterns, in: "Laboratory
        Systems and Spectroscopy," Marcel Dekker, NY, (1977).
3. M. C. Nichols, A FORTRAN II Program for the Identification of
        X-Ray Powder Diffraction Patterns, UCRL-70078, Lawrence
        Livermore Laboratory, October 1966.
4. M. Burfield, P. L. Hauff, G. Jr. Van Trump M. C, Nichols,
        A New Algorithm for Computer Search-Match Identification
        Techniques, 28th Annual Denver Conf., Denver, CO, 1979.
5. E. Bunge, Automatic Speaker Recognition by Computers, Page 23
        Proceedings of the 1977 Carnahan Conference on Crime
        Countermeasure Univ. of Kentucky.
6. J. H. Wegstein, The Automated Classification and Identificaiton
        of Fingerprints, National Bureau of Standards PB-255-189
        (1974).
7. S. Sokolow, J. Karnofsky, E. P. Gustafson, The Finnigan
        Library Search Program, Finnigan Application Report #2, July
        1978.
8. H. B. Woodruff, M. E. Mank, Computer-Assisted Infared Spectral
        Interpretation p 34, Research and Development, August 1977.
9. J. D. Hanawalt, Phase Identification By X-Ray Powder Diffraction,
        p. 63, Vol. 20 Advances In X-Ray Analysis (1977).

10. M. Martens et al, Systematic Identification of Unknown Drugs
      In Powder Form By Means of Ultraviolet Spectrophotometry
      in Forensic Technology, Anal. Chem. 47 #3 p 458 (1975).
11. R. G. Marquart, et al., Computerized Search-Match Systems for
      Powder Diffraction Patterns CHIF 63, Analysis, 177th
      Natl. ACS. Mtg., Honolulu, HA, April 1-6, 1979.
12. L. D. Calvert, J. L. Flippen-Anderson, C. R. Hubbard, Q. C.
      Johnson, P. G., Lenhert, M. C. Nichols, W. Parrish, D. K.
      Smith, G. S. Smith, R. L. Snyder R. A. Young, Standards
      for the Publication of Powder Patterns, Symposium on
      Accuracy in Powder Diffraction, National Bureau of Standards,
      Washington, D.C. June 1979.
13. R. P. Goehner, L. D. Stang and W. T. Hatfield, "A FORTRAN
      Plotting Package for 4010," GE Report 78CRD068, April
      1978.
14. M. C. Nichols, Performance of the SEARCH Program (1966-1978-...)
      Past-Present-Future, American Cystallographic Association
      Meeting, Honolulu, HA, (1979).

Work supported by U.S. Department of Energy, DOE, under Contract
DE-AC04-76DP00789 and Contract W-7405-ENG-48.

A COMPUTER AIDED SEARCH/MATCH SYSTEM

FOR QUALITATIVE POWDER DIFFRACTOMETRY

Ron Jenkins, Y. Hahm, S. Pearlman and W.N. Schreiner

Philips Electronic Instruments, Inc.

Mahwah, New Jersey 07430

INTRODUCTION

The use of the computer controlled powder diffractometer for the qualitative and quantitative analysis of multi phase powder mixtures is becoming relatively common place, and we ourselves have described in some detail the hardware and software components of typical systems 1,2) of which the APD 3600 is the latest version 3). Although great success has been achieved in the use of the minicomputer for machine control, data acquisition and display, plus some degree of data workup, less success has been gained in search/ matching of unknown powder patterns. Although computer programs have been available for a number of years which utilize relatively large computers for automatic search/matching e.g. 4,5) the success with which these programs have been applied has been somewhat dis-appointing. Much of the difficulty arises from the marginal quality of data in the JCPDS index and from uncertainties in the accuracy of the experimental patterns. Round Robin tests 6,7) have indicated that computer search/matching is very sensitive to the quality of "d" values and acceptable in terms of $\Delta d/d$ are probably of the order of 1/1000. Much is being done to improve the quality of the stand-ard data, and this along with more advanced search/match procedures 8) will doubtless improve the situation during the course of the next decade. Although somewhat tedious, the manual search/matching methods are routinely being applied with a relatively high degree of success, presumably because the eye is well adapted to pattern recognition and is able to accept those minor errors in data which can prove fatal to the computer in its line by line search. Also a manual search system is by definition interactive and the operator can make on the spot judgements about e.g. orientation and line widths, and hence modify his searching strategy almost at will.

During the conception of the APD 3600 we decided to investigate several approaches towards the problem of qualitative analysis of mixtures and essentially three different approaches are included in the overall software package. The first of these supports the CIS system 9) and provides the means through peak hunting, background subtraction and internal calibration, of rapidly converting a measured diffractogram into a CIS format file consisting of a set of "d's" and "I's" with appropriate control and title lines. The second system, called "SANDMAN", provides a completely local computer search/matching system utilizing the whole or subsets of the JCPDS data base stored on a 5 megabyte firm disk. The third system, which forms the subject of the remainder of this paper, is called "CASSAM" and is a computer aided, visual pattern recognition systems, again using the whole or portions of the JCPDS data base.

USE OF DATA FILES

Like the SANDMAN program, CASSAM utilizes the whole of the JCPDS data base which currently contains about 34,000 patterns. A unique packing scheme has been adopted whereby all of the patterns can be stored on a 5 megabyte removable disk pack and from which individual patterns can be rapidly retrieved via use of the JCPDS number. Typical pattern retrieval times are on the order of 0.1 second. The SANDMAN program includes on-line routines to build subfiles from the main file either in terms of pattern numbers or by element or functional group type. A running count is kept of the number of times each individual pattern is called, and this allows each user to build his own most frequently encountered file.

THE BASIC CASSAM SYSTEM

Figure 1 shows the overall scheme of the CASSAM program. This consists of an interactive program which is able to call experimentally recorded patterns, or patterns from the JCPDS data base and display them individually or simultaneously on a high resolution graphics display terminal. This monitor program is also able to call a variety of subprograms which allow spectral subtraction, background stripping, $\alpha 2$ stripping, and recalculation of intensities in terms of different fixed or variable slit widths. Essentially diffraction patterns are acquired and stored in real time after which they can be recalled and examined using CASSAM. Where the computer supports foreground/background, previously acquired patterns can be examined at the same time as new patterns are being acquired.

Figure 2 gives an example of the use of the pattern display and compare option. The upper diffractogram is of Round Robin 7A 7) which contains Si (5-565), KI (5-664), BaCl2.2H20 (24-94), and ZnO (4-471).

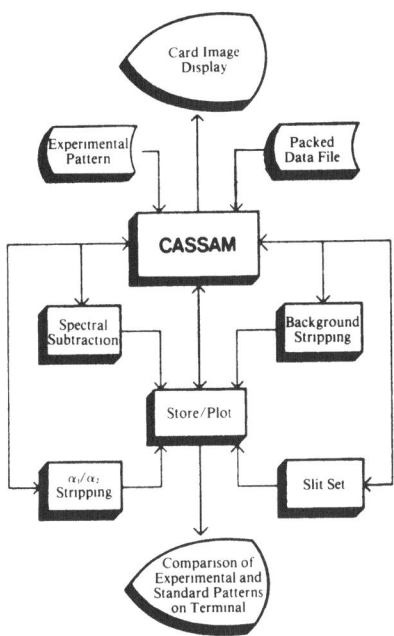

Figure 1.   Command options available in CASSAM

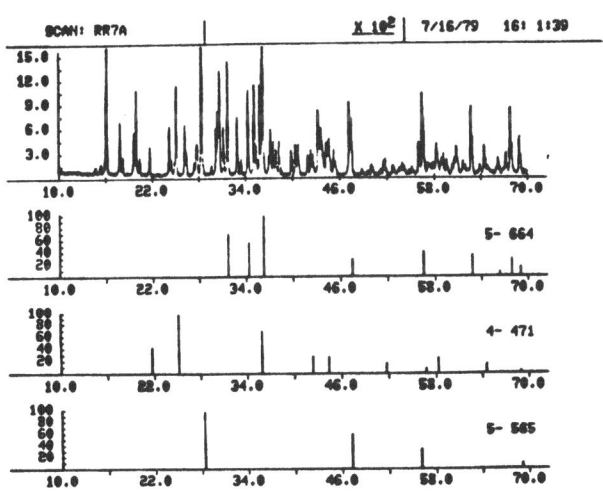

Figure 2.   Example of the use of CASSAM

Three of the patterns are displayed as stick figures and cross hair cursors on the terminal are used to match standard lines with the original pattern. Subtraction and smoothing routines are also available for improving the quality of the original pattern. As an example, Figure 3 shows the same Round Robin 7A specimen before and after smoothing and background subtraction. The change in these patterns is not particularly pronounced because the background is

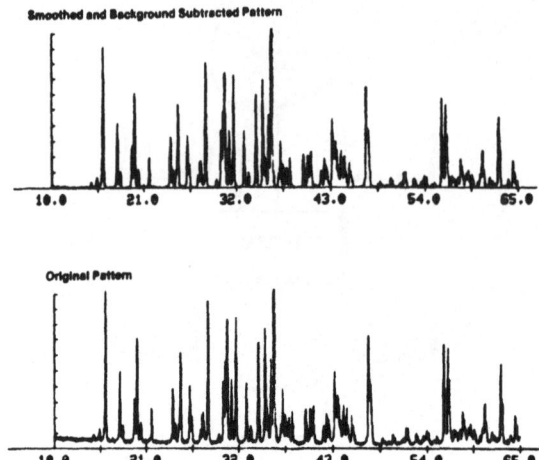

Figure 3.   Example of the use of the background subtraction
             routine using Round Robin 7A sample

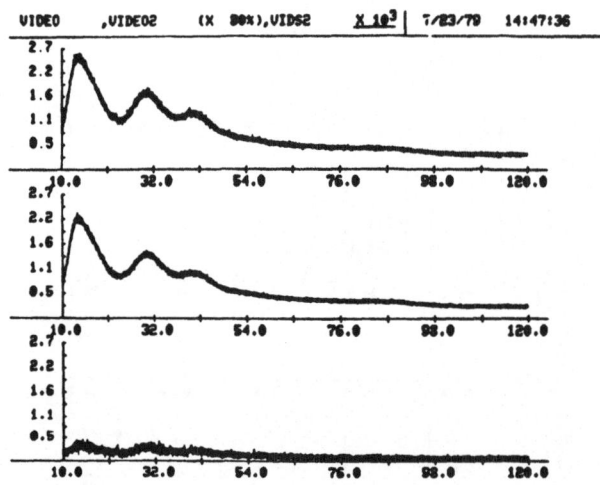

Figure 4.   Use of the subtraction option

fairly low and the counting rate is high enough to give only minor
fluctuations.

     Figure 4 shows a more dramatic example of the background sub-
traction. The upper pattern was obtained from a specimen of an
optical video disk which comprised a 300 Å layer of tellurium on a
plastic substrate. A few very weak lines are just discernible.
The second pattern was obtained from the reverse uncoated side of
the disk. The lower pattern shows the residue after subtraction
which after smoothing gives the upper pattern shown in Figure 5.

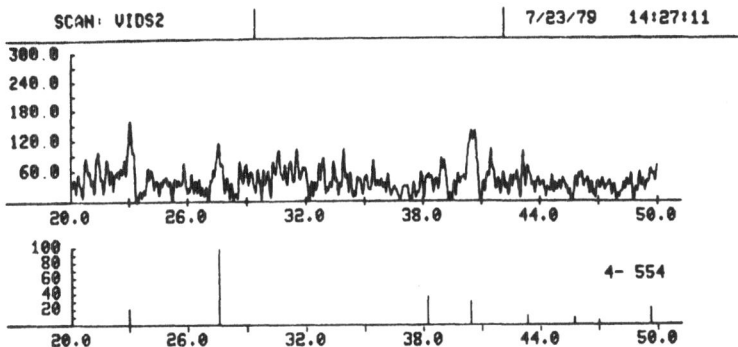

Figure 5.   Matching of the background subtracted pattern

Several weak lines are visible in the pattern which match well with
pattern number 4-554 for tellurium.   It is apparent that the thin
tellurium layer is highly oriented.   The original data from the
same specimen were run in the peak search mode without background
subtraction, and five of the six strongest lines of tellurium were
confirmed 10).

SPECTRAL SUBTRACTION

     Subtraction techniques can also be used for the removal of $\alpha2$
lines from the diagram and for the subtraction of identified phases
from the pattern.   One of the first difficulties to overcome in
subtraction is that whereas the full profile distribution is available for each line in a meassure diffractogram, only d/I data are
available in the JCPDS file.   There are two ways of avoiding the
problem, either the conversion of the measured pattern to a stick
pattern by peak hunting, or the conversion of the JCPDS stick pattern to a synthetic full pattern using a suitable algorithm to artificially broaden each stick to a profile.   We have investigated both
of these methods and conversion of the experimental pattern to a
stick pattern is by far the most attractive.   Figure 6 shows an
example of the profile subtraction method in which the JCPDS stick
pattern for Si was artificially broadened and subtracted from the
pattern RR7B, which is one of the Round Robin specimens referred
to earlier.   The residue pattern is shown below.   The problem arises
in that unless the profile shape is accurately reproduced, only partial subtraction will result.   Figure 7 shows an example of the preferred method in which the full diffractogram from a mixture of
$MgSO_4$, KNO_3 and ZrO_2 has been converted to a stick pattern by peak
hunting.   This pattern contains 53 lines.   The center pattern is of
$MgSO_4$, and the lower pattern the residue of the original pattern
after subtraction, and contains 32 lines.   If this process is

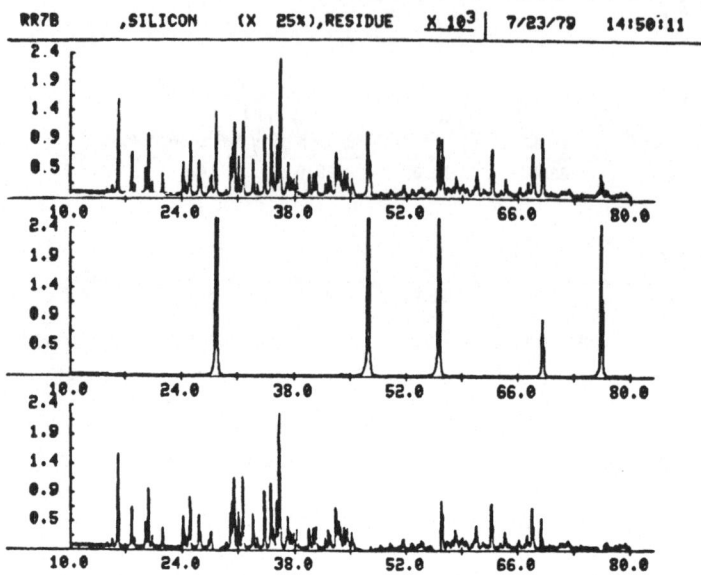

Figure 6.   Use of the profile subtraction method

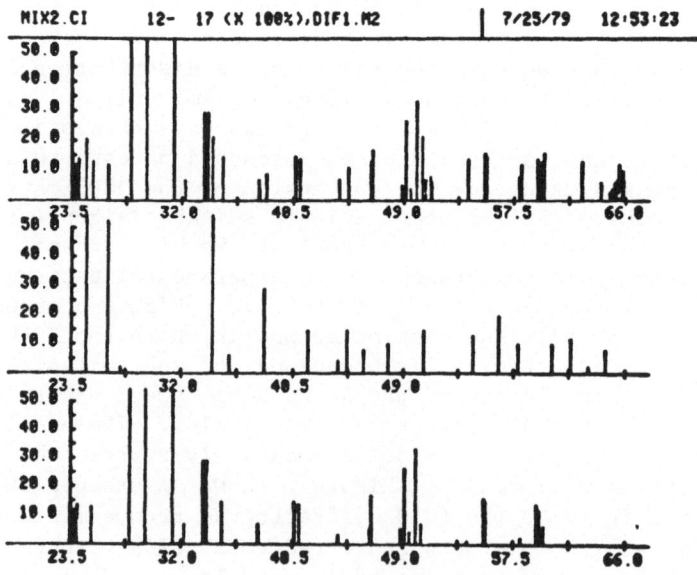

Figure 7.   An example of the use of the stick subtraction method

repeated with the patterns for ZrO_2 and KNO_3 the final residue contains 7 lines of which 5 are ZrO_2 lines left because of the preferred orientation of the ZrO_2. Conversion of all data to stick patterns is also attractive since it greatly simplifies the $\alpha2$ stripping sequence and also allows easy correction of "d" data where internal standards have been used. It also allows recalculation of the original (or standard) diffraction in terms of different fixed or variable slit widths.

CONCLUSIONS

It appears to us that the use of fully automated search/match procedure will not come to full fruition until the problem of intolerable errors in experimental or standard diffraction patterns is solved. Until this time we believe that computer aided searching techniques based on pattern recognition should be exploited since these reduce much of the tedium associated with manual search/matching and gives the operator better control of his experimental data.

REFERENCES

1. R. Jenkins, D. J. Haas and F. R. Paolini, Norelco Reporter 18:1 (1971).
2. R. Jenkins and R. G. Westberg, Norelco Reporter 22:7 (1975).
3. R. Jenkins, Y. Hahm, and S. Pearlman, Norelco Reporter 26:1 (1979).
4. G. C. Johnson and V. Vand, Ind. Eng. Chem. 59:19 (1967).
5. L. K. Frevel, C. E. Adams and L. R. Ruhberg, J. Appl. Cryst. 9:300 (1976).
6. R. Jenkins, Adv. X-Ray Anal. 20:125 (1976).
7. R. Jenkins and C. R. Hubbard, Adv. X-Ray Anal. 22:133 (1978).
8. R. G. Marquart et al. (in press).
9. S. P. Heller, G. W. A. Milne and R. J. Fedmann, Science 195:253 (1977).
10. W. N. Schreiner and R. Jenkins, Adv. X-Ray Anal. 23 (1980) this work.

A SECOND DERIVATIVE ALGORITHM FOR IDENTIFICATION OF PEAKS

IN POWDER DIFFRACTION PATTERNS

W. N. Schreiner   Ron Jenkins

Philips Laboratories   Philips Electronic Instruments

Briarcliff Manor, NY   Mahwah, NJ

INTRODUCTION

Modern (>1950) data analysis techniques are commonly applied in many scientific fields [1] but their careful application in other fields, including X-Ray diffraction, has been notably lacking or generally confined to the environment of university research. Perhaps this lack of widespread application in XRD can be blamed on the absence of sophisticated computer controlled instrumentation, and, if this is so, the situation should change rapidly, since today's automated diffractometers are driven by very powerful minicomputers with large firm disks and sophisticated operating systems. In such an environment it is entirely practical to implement, on-line, state-of-the-art data analysis techniques, and in this paper we present one such example.

PEAK HUNTING

Fig. 1 shows a typical X-Ray diffraction pattern of quartz from 20° to 100°. Many peaks are observed; below about 40° the α_1 and α_2 peaks are not resolved. There also appear to be bumps (with $I<1$) barely visible even in this figure where full scale is $I \approx 25$. In Fig. 2 we plot the same data on a square root scale with the same full scale. Here we can clearly see the small peaks as well as the large peaks. A logarithmic scale would give similar results, however the square root scale is more desirable since it has the property that the error bars on the data points (due to the statistical counting process) are all the same size. Hence the statistical error bars at a peak are the same size as the "width" of the background line. This is very convenient for judging which fluctuations are statistically significant and which are not.

287

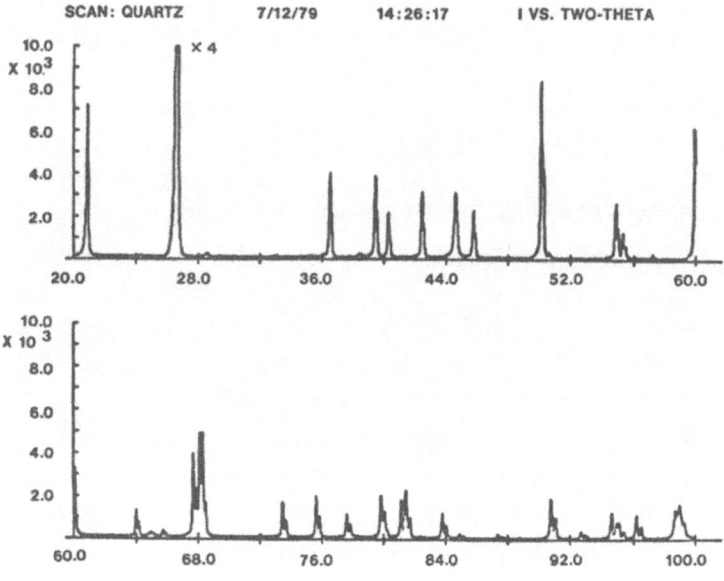

Fig. 1.

    Ideally, a peak searching algorithm should be capable of finding all statistically significant peaks, whether or not they are required in later analysis. This is because all peaks have their origin in some physical process. Fig. 3 shows the highly magnified quartz data just beyond the (101) line. Here we can clearly see, or strongly suspect, 6 small peaks, some barely more than 1 standard deviation

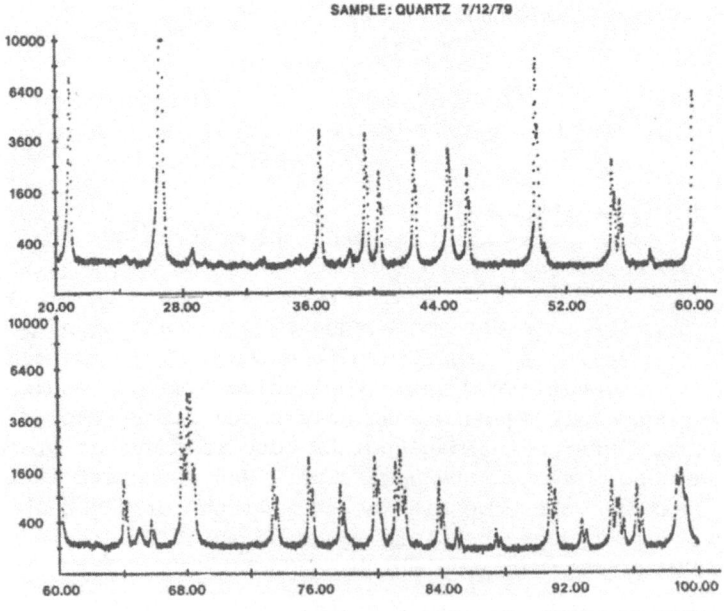

Fig. 2.

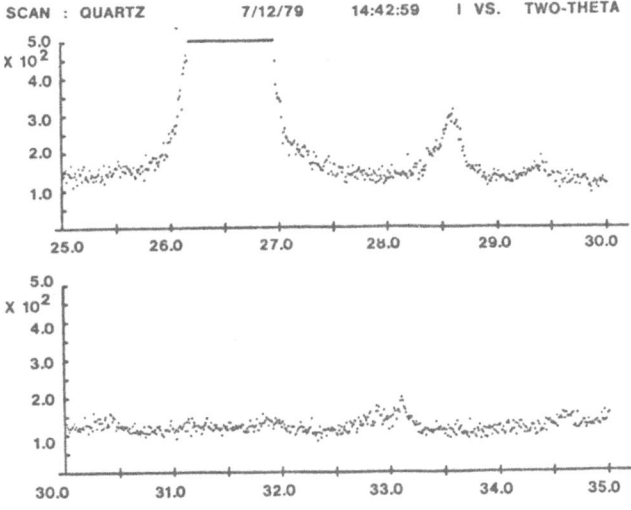

Fig. 3.

above background. The reason the human eye can recognize them is
that it integrates the area under the peaks. Thus the 1 standard
deviation bump at 31.9°, for example, is really more like a 15 standard deviation peak since it was, in effect, measured 15 times.

Fig. 4 shows the (101) line of quartz. Although the α_2 is not
readily visible, if the trailing edge of the peak is viewed at a
glancing angle, a slight deviation from a smooth fall-off may be noted
compared with the leading edge. Small discontinuities are due to the
resolution of the plotting device used to make the figure, nonethe-
less the overall shape is real. Here the eye is perceiving subtle
changes in the slope of the curve which are large compared to the
statistical fluctuations in the data. Again a good peak hunting algorithm might be expected to find this.

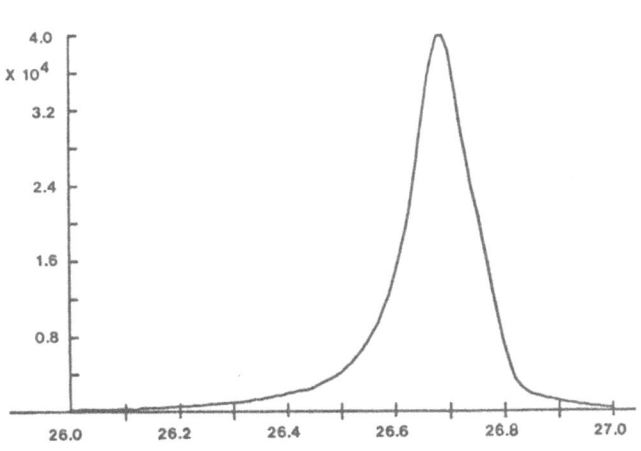

Fig. 4.

2nd DERIVATIVE METHOD

The key to writing such an algorithm is to note how the eye is
able to distinguish statistically significant bumps - changing slopes
and peak integration in the two cases above. Figure 5a shows a
Lorentzian peak, 5b its first derivative and 5c its second derivative. The 1st derivative crosses zero with a negative slope at the
peak. This property could be used to identify peaks, however, it
does not lend itself well to integration. The 2nd derivative is
negative around the peak and positive elsewhere, hence one can use
the existence of a region of negative 2nd derivative to establish
the existence of a peak, and integrate the negative region to determine its statistical significance. Furthermore, the width of the
negative region provides an estimate of the width of the original
peak.

When peaks overlap, the 2nd derivative provides a powerful separation technique. In Fig. 6a-c, we see two lorentzians (α_1 and α_2)
are barely resolved when separated by one FWHM. An algorithm based
on the first derivative could also be devised for this case. When
the peaks are separated by one half the distance, Fig. 6d-f (as is
the case with the 26.66° line of quartz) the second derivative still
shows two distinct negative regions suitable for peak searching, but
the first derivative can no longer be used.

Based on these properties we devised a second derivative algorithm. Although the 2nd derivative has nice mathematical properties,
it is extremely sensitive to statistical fluctuations in the data.
Hence great care must be exercised in computing the value of the 2nd
derivative and in a proper statistical treatment of all intermediate

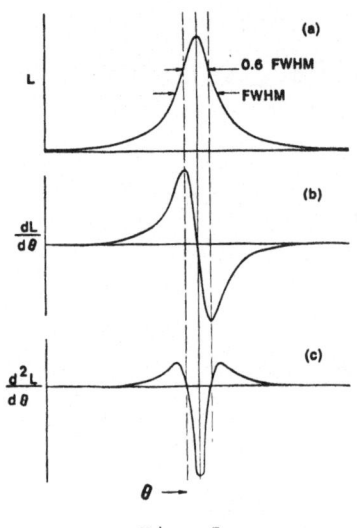

Fig. 5.

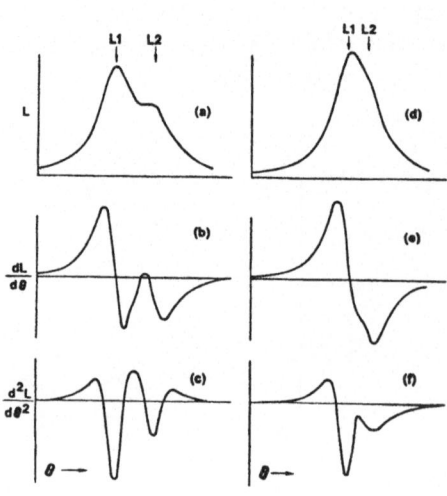

Fig. 6.

and derived quantities. Our algorithm employs a multiplicity of techniques to overcome these problems. These include smoothing, variable
width windows, least squares fitting, χ^2 and other heuristic tests.

Fig. 7 shows a portion of the output from the algorithm for the
quartz data. It can be seen that the 26.66° line is resolved into
α_1 and α_2. The α_1 is found at 26.68 which is .04° high, corresponding
to a specimen displacement error of 75 microns. The α_1/α_2 separation
is .08°, very close to the theoretical value of .07°. All of the
small peaks between 28° and 35° observed in Fig. 3 are found reliably
and no noise peaks were found. When the background of ~110 cts is
subtracted, it is seen that the smallest peaks have intensities of
~20 cts. Thus peaks with normalized intensities of I<0.1 can be reliably identified. In statistical terms this means that peaks
~1.5 S.D. above background can be retrieved. On a Nova 4 minicomputer
our algorithm requires only 1700 words and it processes 2500 data
points/min.

With such a powerful analytic technique, the range of problems
that can be studied with a diffractometer is significantly extended.
Three obvious applications are a) more rapid data collection (the
sensitivity of the algorithm can be traded for smaller count times),
b) improved low limit of detection and c) ability to work with sample
with a high background. Figs. 8-11 are an example of the latter two
applications.

Fig. 8 shows a "rapid" scan (1 sec/point, 0.025° step size) of
a 300 Å layer of Tellurium vacuum deposited on a 1 mm thick plastic
substrate.[2] It was desired to determine if Te exhibited preferred
orientation in the deposition process. One's initial reaction would
be that this could not be done with X-ray diffraction since the Te
layer was so thin and any crystalline peaks would be overwhelmed by
the amorphous background from the plastic. Indeed the rapid scan
showed this to be the case, however, the peak searching algorithm
surprisingly found 3 peaks (Fig. 9). An examination of the JCPDS
card file confirmed that these were indeed crystalline Te peaks.

2-THETA (DEG)	APPROX WIDTH	INTENSITY	D SPACING	NORMALIZED INTENSITY
20.91	0.11	7123.	4.244	18.19
24.40	0.25	200	3.644	0.51
24.96	0.11	180	3.565	0.46
26.68	0.09	39160.	3.339	100.00
26.76	0.05	15750.	3.328	40.22
28.60	0.18	298	3.118	0.76
29.37	0.24	164.	3.039	0.42
30.40	0.26	137.	2.938	0.35
31.90	0.24	141	2.803	0.36
33.12	0.15	192.	2.703	0.49
34.59	0.24	153	2.591	0.39
35.32	0.06	211	2.539	0.54
36.57	0.09	3830.	2.455	9.78
36.68	0.06	1833	2.448	4.68

Fig. 7.

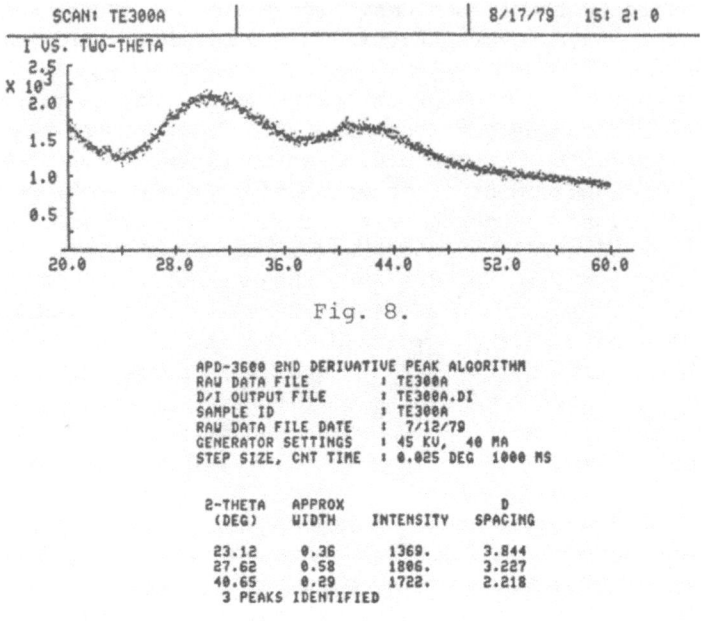

Fig. 8.

Fig. 9.

Encouraged by these results, a second run was made with 9 sec/point
and 0.025° step size (Fig. 10). The original 3 peaks showed up
clearly and the peak algorithm found two additional peaks (Fig. 11).
After background subtraction and peak normalization, the intensities
were determined to approximately 10%, sufficient to answer the question of preferred orientation.

CONCLUSIONS

    Experimental measurements contain a great deal of information -
generally much more than is routinely extracted. Although the residual information is often unimportant, there are freqently occasions
where it is crucial and one must have the capability of recovering it.
By applying good data analysis techniques, such as proper statistical

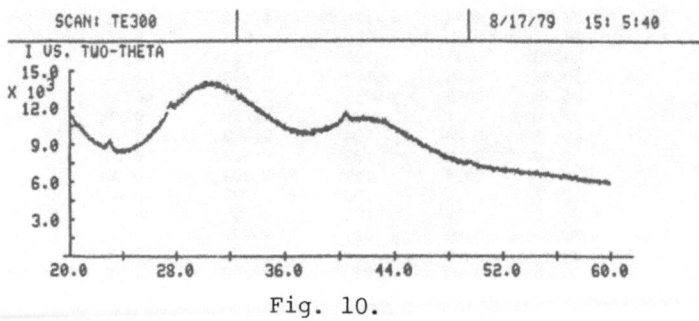

Fig. 10.

```
APD-3600 2ND DERIVATIVE PEAK ALGORITHM
RAW DATA FILE          : TE300
D/I OUTPUT FILE        : TE300.DI
SAMPLE ID              : TE300
RAW DATA FILE DATE     :  7/11/79
GENERATOR SETTINGS     : 45 KV,  40 MA
STEP SIZE, CNT TIME    : 0.025 DEG  9000 MS


  2-THETA   APPROX                        D
   (DEG)    WIDTH    INTENSITY     SPACING

   23.07    0.41       9293.        3.852
   27.60    0.43      12232.        3.230
   40.57    0.51      11428.        2.222
   43.35    0.69      10816.        2.086
   49.77    0.32       7586.        1.830
   5 PEAKS IDENTIFIED
```

Fig. 11.

treatment in a peak hunting algorithm, important but subtle effects
can be measured. Such capability can significantly extend the appli-
cability of an instrument.

REFERENCES

1. A random selection of books and articles found in our library
    yielded the following examples:
    NMR - "Magnetic Resonance Spectroscopy", Hecht (1966) J. Wiley.
    IR,ED - "Molecular Structures and Vibrations", S. J. Sven (1972)
            Elsevier
    γ-Spect - "Computer Program for Multicomponent Spectral Analysis
            Using Least Squares", J. Blackburn Anal.Chem. <u>37</u>,
            1000 (Jl.1965).
2. "Optical Disc Replaces 25 Magnetic Disks", Kenney, Lou, McFarlane,
        Chan, Nadan, Kohler, Wagner, Zernike, IEEE Spectrum (Fb.1979).

ADVANCES IN THE COMPUTER INDEXING OF POWDER PATTERNS

Gordon S. Smith

Chemistry and Materials Science Department
Lawrence Livermore Laboratory, University of California
Livermore, California 94550

ABSTRACT

Indexing of powder-diffraction patterns by computer techniques has advanced to the state that it is now often possible to
determine unit-cell dimensions and crystal system for an unknown
material solely from its powder-diffraction data. This indexing
is fully automated, proceeding directly from positions of observed
diffraction lines as input, with decision-making steps being made
by a computer. Ease of indexing depends on quality of data (accuracy and completeness), volume of the unit cell, and symmetry of
the crystal system. In general, a powder pattern of a triclinic
compound with a large unit cell requires a more accurate and complete data-set for successful indexing than does a cubic material
having a small unit cell. Fortunately, data from a well-aligned
diffractometer or Guinier camera ordinarily suffices for computer
indexing. Because of systematic errors in the low-lying diffraction lines, data from the Debye-Scherrer technique usually are not
adequate for computer indexing (except for the simpler cases). A
brief review of the strategies/algorithms of some of the computer
indexing codes now available is given. Criteria for assessing the
reliability of a particular computer-assisted indexing are discussed. Finally, attention is directed toward future developments
such as by automating the collection of powder-diffraction data,
analyzing data by computer data processing, and increasing the
speed and reliability of computer indexing.

INTRODUCTION

Many materials readily exist in polycrystalline form. Furthermore, it is generally easy experimentally to obtain the powder-
diffraction pattern of a substance. Thus, the unraveling of the
powder pattern of previously uncharacterized materials in terms of
unit-cell dimensions and crystal system has long been pursued by
diffractionists. Indeed, over the years, a number of graphical and
analytical methods have been devised for that purpose (see, for
example, Ref. 1). Today much of this effort is done by computer,
and at this stage of development it often is possible by means of
computer indexing to obtain the unit-cell information of an unknown
solely from its powder-diffraction data.

This paper reviews briefly developments in computer indexing:
the differing strategies involved, the quality of data normally
required, and advances in assessing the reliability of the solutions obtained. Possible future developments based on laboratory
automation and data processing also are discussed.

MEANING OF COMPUTER INDEXING

Computer indexing in the present paper is taken to mean that
the crystallographic characterization of a material from its powder-
diffraction pattern is carried out by the computer in a fully automated manner. (We exclude from consideration computer programs
that, from known lattice parameters, generate Miller indices for the
diffraction lines in a pattern.) The computer not only assists the
user in carrying out the large number of calculations usually necessary to obtain a solution but also takes charge of decision-making
steps in the process.

The amount of input to a modern computer-indexing program can
be as small as a title-card plus a list of the positions of the
lines measured for the pattern. Questions related to, for example,
the radiation used and experimental errors in the line positions
can be handled by default options. There also has been a proposal
(2) to standardize the input so that the same data-set can be put
through several indexing programs without reformatting of the data.

Output typically consists of a small number of possible solutions. Each solution normally gives the unit-cell information and
a line-by-line listing of (a) the Miller indices hkℓ and (b) the
agreement between observed and calculated values of $Q = \lambda^2/$
$\left[4 \sin^2 (2\theta/2) \right]$ or 2θ.

Once a solution or a group of solutions is obtained, the user
must assess which, if any, is correct. In the past, few programs
gave any help in this matter. Instead, the user generally em-

ployed intuitive notions, such as whether the agreement between observed and calculated line-positions looks reasonable. Today, important strides have been made in developing guidelines to help the user assess the reliability of the unit cell(s) obtained by computer indexing. This point will be discussed later.

STRATEGIES OF COMPUTER INDEXING

The basic equation to be solved is:

$$Q_i = 1/d_i^2 = h_i^2 a^{*2} + k_i^2 b^{*2} + \ell_i^2 c^{*2}$$

$$+ 2 k_i \ell_i b^* c^* \cos \alpha^*$$

$$+ 2 h_i \ell_i a^* c^* \cos \beta^*$$

$$+ 2 h_i k_i a^* b^* \cos \gamma^*, \tag{1}$$

where h_i k_i ℓ_i are the Miller indices for each of the N observed diffraction lines and a^*, b^*, c^*, α^*, β^*, and γ^* are reciprocal cell parameters.

Almost all computer-indexing schemes solve this equation by assigning provisional indices to the low-angle lines, obtaining trial lattice parameters, and then seeing if the remaining lines can be indexed. If the entire pattern cannot be indexed, the solution is rejected, and another attempt is made using the above cycle. Successful solutions are stored for later output to the user. For the beginning lines, the range of the absolute value of each of the individual hkℓ indices is usually ≤ 4. In this way, the number of possible solutions may be large but the solutions are denumerable.

Strategies in computer indexing divide sharply as to how the effect of crystal symmetry is treated. In Type 1, solutions are sought in particular crystal systems, usually in the highest lattice symmetry, cubic, and proceeding through tetragonal, hexagonal, etc., as required. An example of this approach is the program of Goebel and Wilson (3), which treats cubic down through orthorhombic symmetry. In Type 2, solutions are obtained in the lower crystal systems, triclinic or monoclinic, and then in the correct system, if necessary, through a cell-reduction algorithm. Examples of this approach include the programs of Visser (4), Roof (5), and Smith and Kahara (6).[†]

[†]This listing of computer-indexing codes is not intended to be exhaustive, but includes those that the author has used. A full review of codes now available is being prepared by R. Shirley.

The user should be aware that some codes tolerate "impurity lines" and some do not. ("Impurity lines" can result from chemical or wavelength impurities or from mistakes in data reduction and data input.) Some codes establish maximum discrepancies allowed in a solution internally, or have default (but overridable) error settings, and some require the user to input his own estimate of errors. The author also has seen a perhaps 20:1 variation between programs in speed of execution on the same data-set.

As a specific example of the strategies involved in computer-indexing, we consider the program of Smith and Kahara (6). This program is designed primarily to treat the monoclinic system but is applicable to orthorhombic and higher symmetry, since these are special cases of monoclinic as far as Eq. (1) is concerned.

A key part of their strategy is to determine the 020 reflection, which occurs far more frequently in monoclinic patterns than does the 010 reflection. The relations that serve as 020 detectors are

$$2\,Q_{020} + Q_{h10} = Q_{h30} \qquad\qquad\qquad (2)$$

and

$$3\,Q_{020} + Q_{h20} = Q_{h40} \, . \qquad\qquad\qquad (3)$$

The above relations are equally true if ℓ rather than h is the running index. Low 2θ-reflections are tried as the 020 reflection in combination with all lines as h10 and h20 (or 01ℓ and 02ℓ), and coincidences (within a prescribed ΔQ limit) are looked for between the generated values $2Q_j + Q_i$ and $3Q_j + Q_i$ and observed Q's. A coincidence in row p and columns q and r in the $2Q_j + Q_i$ table means, for example, that reflection p is potentially 020 and q and r are potentially h10 and 01ℓ. (Values of h and ℓ are assumed to be ≤ 2.) Trial values of b*, a*, and c* can be derived. All coincidences are investigated, and all possible hk0 and $0k\ell$ zones are formed.

For a given combination of hk0 and $0k\ell$, the computer attempts to find the monoclinic angle β* by trying various simple h0ℓ and hkℓ values for those reflections not already indexed as belonging to the orthogonal hk0 and $0k\ell$ zones. The various unit cells obtained are ordered according to magnitude of the volumes, and indexing of the entire pattern is attempted starting with the smallest cell. Output consists of least-squares-refined cell constants and Miller indices for (up to) the ten smallest cells.

PROGRESS TOWARD FAST COMPUTER INDEXING

When only one or even a small number of patterns is to be indexed, time spent on the computer ordinarily would not be a concern. However, suppose one wanted to index the entire Powder Dif-

fraction File of more than 30,000 patterns available now on magnetic tape. In addressing this problem, Smith and Kahara (7)
pointed out that the computer-indexing code not only must be fast
per pattern but also must cope with wide variations in pattern accuracy and must spare the user from the time-consuming process of
evaluating the results of each indexing. Smith and Kahara employed a rapid permutation of the Miller indices of the beginning
lines; error settings and evaluations of the results via an empirical figure of merit also were done by the computer. Tests showed
that for cubic patterns, computer indexing could be done at the
rate of about 1600 patterns per minute (CDC 7600 computer) and the
correctness of the indexings was about 95%. For hexagonal and tetragonal patterns, the speed fell to about 50 patterns per minute,
and the correctness of the solutions obtained was about 90%.
These experiences suggest that for noncubic patterns the maximum
rate of computer indexing possible today might be about 50 patterns per minute. Hence, it would be rather expensive to attempt
to index the entire Powder Diffraction File. A possible alternative is to put the File indexing problem on a minicomputer as long-
term background work.

ASSESSING THE RELIABILITY OF INDEXING

Perhaps the most important part of the computer indexing of
an unknown compound is the assessment of which solution, if any,
of those obtained is correct. The chance of a successful indexing
naturally depends on the accuracy with which the positions of the
lines are measured and on the completeness with which the lines
are recorded. Size of the unit cell and crystal symmetry also
play a part. A triclinic compound having a large unit cell ordinarily requires more accurate data than a cubic compound with a
small unit cell.

Two schemes that combine the concepts of accuracy and pattern
completeness have been proposed to give numerical criteria for assessing the reliability of a proposed indexing. These are the M_{20}
figure of merit of de Wolff (8), defined as

$$M_{20} = \frac{Q_{20}}{2\overline{|\Delta Q|}} \; \frac{1}{N_{poss}} , \qquad (4)$$

and the F_N figure of merit of Smith and Snyder (9), defined as

$$F_N = \frac{1}{\overline{|\Delta 2\theta|}} \; \frac{N}{N_{poss}} . \qquad (5)$$

In Eq. (4), N_{poss} is the number of possible diffraction lines
up to the 20th observed line, Q_{20} is the value of Q for the
20th observed line, and $|\overline{\Delta Q}|$ is the average absolute discrepancy
between observed and calculated Q-values for these 20 lines.   In
Eq. (5), N_{poss} is the number of possible diffraction lines up
to the Nth observed line, and $|\overline{\Delta 2\theta}|$ is the average absolute
discrepancy between observed and calculated 2θ values for the N
lines.

Experience has shown that an $M_{20} > 10$ and an $F_{20} > 20$
means the indexing is almost certainly correct. For a pattern
completeness factor N/N_{poss} of 0.7 (a value typical of good-
quality data), an $F_{20} > 20$ means $|\overline{\Delta 2\theta}|$ is < 0.035° 2θ. No
such easy conversion exists for M_{20}, but in test cases one usu-
ally experiences difficulties in obtaining the correct solution if
any $|\overline{\Delta Q}|$ is > 0.0008 $\mathring{A}^{-2}$.

Conversely, an indexing having $M_{20} < 3$ and $F_{20} < 7$ is
almost certainly incorrect. Incorrect solutions often are characterized also by excessively large unit-cell volumes with a prim-
itive lattice-type displaying no systematic absences. Both of
these factors lead to low values of N/N_{poss}, and hence to low
values of M_{20} and F_{20}.

Values of M_{20} and F_{20} between 3 and 10 and 7 and 20, respectively, represent gray areas of reliability. Some of the indexings may be correct, but are not necessarily so. The pattern
should be remeasured or the data retaken, with increased accuracy, so that an indexing can be totally convincing.

An advance knowledge of the magnitude of the unit-cell volume
V would be extremely valuable when assessing the results of a particular computer indexing. It was noted by Smith (10,11,12) that
many powder patterns of high quality contain information concerning the magnitude of the unit-cell volume, and that this information can be brought out by a graphical plot directly from the d
spacings or by a simple calculation of the type $V \sim k_N\ d_N^3$.
Of course, if the density and formula weight of the substance are
known, the unit-cell volume obtained from a particular indexing
can be checked more definitively by ascertaining whether the calculated and experimental densities agree.

As an example of the use of F_{20} in a reliability assessment, we choose powder data for PbV_2O_7; two patterns are available, PDF 15-680 and PDF 18-708. When we attempted to index the
pattern on PDF 15-680, a monoclinic cell was obtained (Table 1)
that indexed 19 out of the first 20 lines and had a low value of
both N/N_{poss} (20/91) and F_{20} (5.3). According to the above

Table 1. An example of F_{20} reliability assessment. PbV_2O_7: two patterns available, PDF 15-680 and 18-708.

Solution	Cell Parameters					N/N$_{poss}$	F_{20}
	a	b	c	β	V		
Computer-obtained[a]	9.71	9.69	10.06 Å	101.6°	927 Å^3	20/91	5.3	incorrect
PDF 15-680	13.47	7.326	6.956 Å	107.4°	655 Å^3	20/52	1.8	incorrect
PDF 18-708	13.30	7.14	7.08 Å	106°	646 Å^3	20/29	21	correct

[a]One line not indexed; d values from PDF 15-680

criterion for F_{20}, this cell is almost certainly incorrect. Next, the patterns themselves were evaluated using the lattice parameters listed on the PDF cards. The value of F_{20} for this calculation of PDF 15-680 was extremely poor, namely, 1.8. Clearly, the solution given on the card is in error. On the other hand, evaluation of the data listed on PDF 18-708 gives satisfactory values for N/N$_{poss}$ and F_{20}. Hence, we conclude that the solution is almost certainly correct. Note that the computer-obtained unit-cell volume is almost 50% too large. Note also that unit cells on the two PDF cards are not very different; however, the F_{20} criterion readily points out problems with the cell listed on PDF 15-680. [Independent of these considerations, this pattern already has been deleted from the File.]

PROSPECTS FOR DEVELOPMENT IN COMPUTER INDEXING

For computer indexing, the data must be good, but a Guinier camera or a well-aligned diffractometer usually provides adequate data. Moreover, the basic technology for these two instruments has been well-established for more than two decades. A greater impact on development, we believe, will come from the automation of powder diffraction and from computer data processing. In this regard, we can foresee an increased _ease_ of measuring line positions and recording them in a computer-ready format through automation. There also will be increased _accuracy_ in the measurement of line positions through computer processing of the data, e.g., location of peak positions by polynomial fitting. Higher values of the pattern-completeness factor will result from deconvolution of complex spectra through computer-profile fitting, e.g., the method of Parrish, Huang, and Ayers (13). Thus, instead of the step-by-step process in use today for obtaining and preparing data for computer indexing, it should be possible to develop a continuous, one-step operation with all of the work being done by the computer.

ACKNOWLEDGMENTS

This work was performed under the auspices of the U.S. Department of Energy by Lawrence Livermore Laboratory under contract No.
W-7405-Eng-48.

REFERENCES

1.   H. P. Klug and L. E. Alexander, X-ray Diffraction Procedures
     for Polycrystalline and Amorphous Materials, J. Wiley and Sons
     (1974).

2.   R. Shirley, Personal communication (1977).

3.   J. B. Goebel and A. S. Wilson, "INDEX, U.S. Atomic Energy
     Commission, Research and Development Report," BNWL-22 (1965).

4.   J. W. Visser, "A Fully Automatic Program for Finding The Unit
     Cell from Powder Data," J. Appl. Cryst. 2, 89-95 (1969).

5.   R. B. Roof, Jr., "INDEX, A Computer Program to Aid in the
     Indexing of X-ray Powder Patterns of Crystal Structures of
     Unknown Symmetry," Los Alamos Scientific Laboratory Report,
     LA-3920 (1968).

6.   G. S. Smith and E. Kahara, "Automated Computer Indexing of
     Powder Patterns: the Monoclinic Case," J. Appl. Cryst. 8,
     681-683 (1975).

7.   G. S. Smith and E. Kahara, "Progress Toward Fast Indexing of
     Powder Patterns: High Symmetry Cases," Paper PA20, American
     Crystallographic Association Winter Meeting, Norman, Oklahoma
     (1978).

8.   P. M. de Wolff, "A Simplified Criterion for the Reliability of
     a Powder Pattern Indexing," J. Appl. Cryst. 1, 108-113 (1968).

9.   G. S Smith and R. L. Snyder, "F_N: A Criterion for Rating
     Powder Diffraction Patterns and Evaluating the Reliability of
     Powder-Pattern Indexing," J. Appl. Cryst. 12, 60-65 (1979).

10. G. S. Smith, "Estimating Unit-Cell Volumes from Powder
     Diffraction Data: the Triclinic Case," J. Appl. Cryst. 9,
     424-428 (1976).

11. G. S. Smith, "Estimating the Unit-Cell Volume from One Line in
     a Powder Diffraction Pattern: the Triclinic Case," J. Appl.
     Cryst. 10, 252-255 (1977).

12. G. S. Smith, "Estimating Unit-Cell Volumes from Powder
     Diffraction Data: the Monoclinic and Orthorhombic Cases,"
     unpublished.

13. W. Parrish, T. C. Huang, and G. L. Ayers, "Profile Fitting: A
     Powerful Method of Computer X-ray Instrumentation and
     Analysis," Transactions of the American Crystallographic
     Association 12, 55-73 (1976).

NOTICE

SPECPLOT - AN INTERACTIVE DATA REDUCTION AND DISPLAY PROGRAM FOR SPECTRAL DATA

Raymond P. Goehner

General Electric Corporate Research and Development

Schenectady, New York 12301

The automation of analytical equipment is proceeding at a rapid pace, particularly since the introduction of inexpensive microcomputer systems. Most of this equipment has one characteristic in common, that is, they produce digital spectral data. The usual method of recording spectral data has been the strip chart recorder. Strip charts require the hand encoding of position and intensities of the spectral lines. This requires that all of the lines be on scale or that the sample be run several times in order to amplify weaker lines. This problem is eliminated by recording the data digitally. Digital data can then be rapidly plotted on a cathode ray terminal to any desired scale. The user of digital data has access to a great variety of automatic data reduction programs.

SPECPLOT is a modular FORTRAN program which was written to incorporate a great variety of data reduction and interactive plotting subroutines. The program has been applied to the analysis of spectral data from nuclear magnetic resonance, differential scanning calorimeter, liquid chromatography, gel permeation chromatography, as well as x-ray fluorescence and diffraction data. The diffraction data analysis which will be discussed here comes from an automated diffractometer and a microdensitometer. The program plots on a Tektronix 4010 terminal with x & y cursor controls. The 4010 terminal has 1024 by 780 addressable points. When plotted full screen, the actual spectrum is 936 by 630. The rest of the space is used for labeling, sample identification, title, running conditions, and results returned from interactions with the data. The plotting routines are all written in FORTRAN and contained with the program.

Figure 1 is an example of the questions asked by the program.
The first question is what type of display is requested. There
are five types of displays available: full screen, split screen,
overlay, addition and subtraction. The split screen plots two
requested data files, with the first file plotted on the bottom
portion of the screen and the 2nd file at the top. The overlay
plots the two requested spectra, with the 2nd spectrum displaced
slightly in the y direction. Figure 2 is an example of split
screen and overlay displays. The results of an addition or subtraction of two data files can be displayed full screen. In this
case, in addition to requesting a 2nd data file, the program will
also request the portion of the second data file to be added or
subtracted from the first. If a carriage return (default) is
given in response to the first question asked by the program, the
plot will be done full screen. The digital data file is then
requested. Once this data filename is read, the sample number,
date, instrument type, title information, and running conditions
are read from the first two lines of the file and displayed. The
program then reads the binary data file and returns the number of
points contained within it.

The next set of questions consists of the data reduction
portion of the program. The smoothing algorithm used is a least
square moving average which was popularized by Savitzky and
Golay (1967). The subroutine allows smoothing using 5, 7, 9, 11,
15, 17, or 25 points. Figure 3 is an example using the 25 point
smooth. The top is the actual spectrum and the bottom has been
smoothed. The background subtraction algorithm is an iterating

SPLIT SCREEN-1 OVERLAY-9 SUBTRACT-8 ADD-7? 7
DATA FILE? KU700002
2ND DATA FILE? K9700001
   2 11-30-77 DIF
24407 10831-91-BO BOEHMITE SLURRY A. FACTOR .3X. 1

START ANGLE= 12.000 DELTA 2THETA= 0.050 TIME INTERVAL= 4.0
NUMBER OF POINTS= 1121

SMOOTH-00,05, 07, 09, 11, 15, 17, 25?

   FRACTION OF 2ND FILE? 1.0

BKGR-1?
SEARCH-1?
XMIN= 13.0 XMAX= 67.9
XMIN=
XMAX=?
YMIN=   9.5 YMAX= 3551.3
YMIN=?
YMAX=?

Fig. 1. Example of how data are entered into the
        program

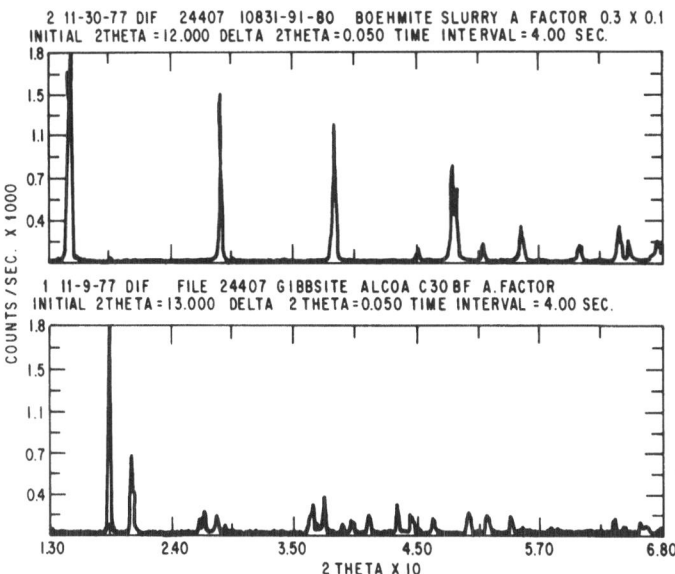

Figure 2a.   Split screen display

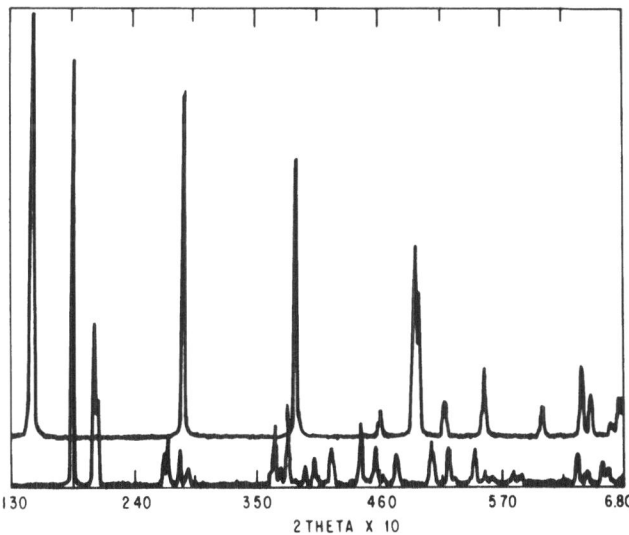

Figure 2b.   Overlay display

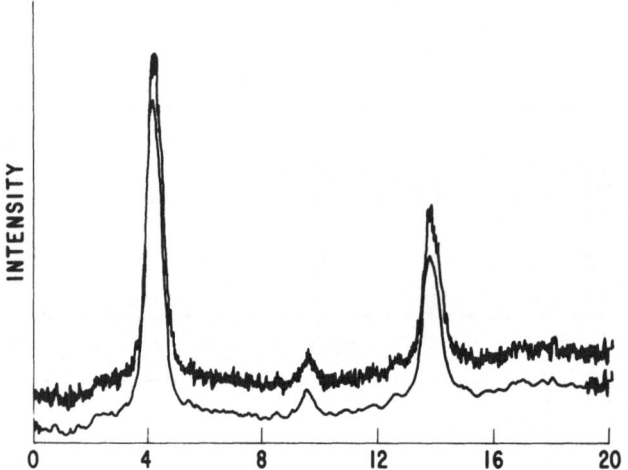

Figure 3.   25 point smooth.   The upper spectrum
is the actual digital data and the bottom is
the same data smoothed.

linear least square subtraction developed by Goehner (1978).   The
background can be fitted over 50 point blocks with an overlap, if
desired.   The background of the entire spectrum is fit linearly
if a carriage return is entered.   The Rachinger (1948) method is
used for $K\alpha_2$ elimination.   The user can now search his data for
peaks using a 2nd difference peak finding algorithm similar to the
ones described by Marescoth (1967), Mills (1970), and Sommerweld
and Visser (1974).   The subroutine calculates the 2nd difference
over a specified number of data points using a moving least square
procedure.   The most negative point of the 2nd difference then
corresponds to the peak position.   Figure 4 shows the tabulation
of the peak search from the inputs shown in Figure 1.   Once the
tabulation is done the program will plot out the spectrum on the
screen of the terminal (Figure 5).   Each peak position is labeled
in the same manner as in the tabulation.

When the plotting is completed, the faint x & y cursors appear on the screen.   The program is now in the interactive mode
of operation.   This mode could have also been achieved by simply
defaulting (carriage return) all questions except data filename
asked by the program.   The x & y cursors can be moved to any point
on the screen by moving the thumb wheels on the terminal.   A series
of one letter commands can be entered.   The program will read the
x & y position of the cursor and the terminal key.   Control is
then passed to the appropriate subroutines based on the letter
entered.

66 3/10/78 DIF 24590 LEN NEIDRACH ZRO2 POWD:
11062-1200
INITIAL 2THETA= 10.000 DELTA ANGLE= 0.050
DELTA TIME= 5.00SEC
NO2= 9 DR2L= 0.5 NUM= 1 IPF=1

M	2THETA	DSPACE	HEIGHT	WIDTH
1	17.437	5.0819	4.3	0.458
2	24.108	3.6885	16.6	0.308
3	24.463	3.6359	10.5	0.255
4	28.238	3.1578	100.0	0.344
5	30.481	2.9304	3.2	0.686
6	31.504	2.8374	65.9	0.386
7	34.165	2.6223	21.1	0.311
8	34.497	2.5978	13.4	0.252
9	35.294	2.5410	14.7	0.396
10	35.837	2.5037	3.1	0.351
11	38.662	2.3270	4.8	0.401
12	39.492	2.2800	1.3	0.333
13	40.778	2.2110	11.0	0.431
14	41.423	2.1781	5.8	0.366
15	44.857	2.0190	6.5	0.407
16	45.531	1.9906	5.8	0.477
17	49.347	1.8453	17.6	0.427
18	50.212	1.8155	25.2	0.419
19	50.195	1.8161	14.9	0.250
20	51.281	1.7801	6.2	0.310
21	54.053	1.6952	9.6	0.488
22	55.451	1.6557	13.5	0.506
23	56.038	1.6398	7.8	0.377
24	57.174	1.6088	6.6	0.496
25	58.296	1.5815	5.8	0.750
26	58.993	1.5408	11.6	0.537
27	61.415	1.5085	5.5	0.506
28	61.928	1.4972	5.6	0.588
29	62.611	1.4783	7.7	0.499
30	64.430	1.4450	2.8	0.865
31	65.742	1.4193	6.6	0.660
32	68.069	1.3588	1.8	0.432
33	71.214	1.3230	4.2	0.548
34	74.690	1.2698	1.9	0.318
35	75.353	1.2603	3.1	0.561
36	76.369	1.2460	2.2	0.566
37	77.960	1.2245	1.7	0.792
38	79.277	1.2075	2.1	0.578

Figure 4. Tabulation of peak search from
inputs shown in Figure 1

66 3-10-78 DIF   24590 LEN NIEDRACH ZRO2 POWDER 11062-1200
INITIAL 2THETA 10.000 DELTA 2THETA = 0.050 TIME INTERVAL = 5.00 SEC.

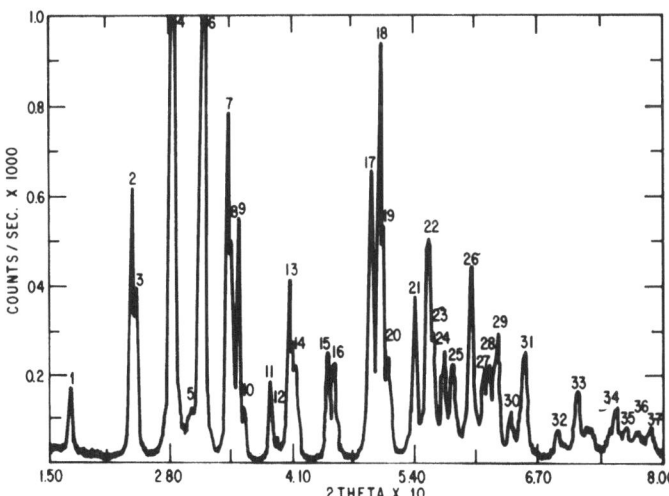

Figure 5. The spectrum obtained from inputs shown
in Figure 1 with the peaks labeled

As an example, if the cursors are put just above a peak on
the screen and a G entered, the d spacing of the x cursor is calculated. A number based on how many commands were entered is
printed above the peak and the resulting d spacing printed above
the spectrum. Figure 6 shows this for two small peaks, as well as
an example of integrating a large amorphous peak with a user
specified background. The background is supplied by the user using
the cursor controls. The output from this command consists of
the position of the highest point, integral, peak height, integral
half width, and centroid of the peak.

Another example of the cursor control is the replotting of a
windowed region of the spectrum full screen. The x & y cursor
of the lower left hand corner and upper right corner of the region
of interest is entered using the cursor command E. This portion
of the spectrum is then plotted full screen. The cursor then
reappears and the process can be repeated. If the original
spectrum is needed, the cursor command R is entered and the entire
spectrum is replotted.

SPECPLOT is a generalized data reduction and plotting program
which allows the user to interact directly with spectral data.
The program is written using a modular procedure in FORTRAN and
will plot on a Tektronix CRT terminal with cursor controls. The

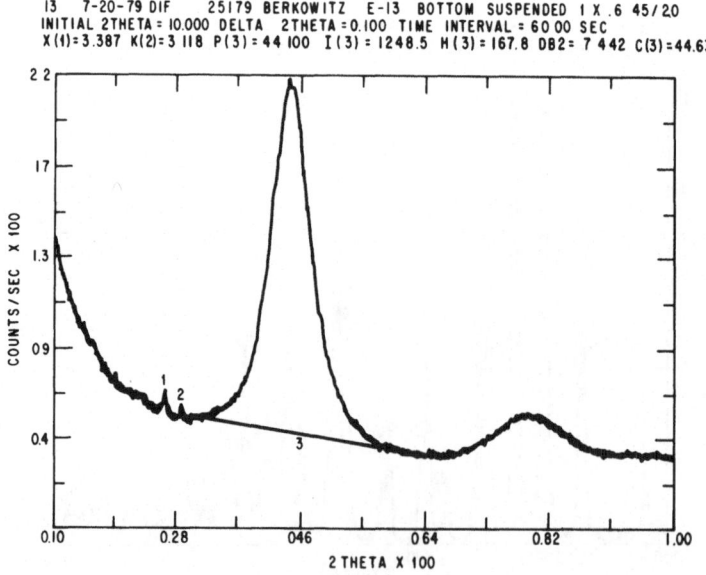

Figure 6.   Integrated amorphous peak and small peaks
and corresponding command data

program can be extensively overlayed and occupies about 25K bytes
of core, 8.1K bytes of which is buffered data. This much data is
buffered in order to give quick access for data reduction, display
and interaction, with up to 4096 data points. The program is
operational at several laboratories and a manual complete with a
listing can be obtained from the author.

REFERENCES

1. Savitzhy, A. and Golay, M. J., Analytical Chemistry, 36,
    pp. 1627-1639 (1964).
2. Goehner, R. P., Analytical Chemistry, 50, pp. 1223-1225
    (1978).
3. Rachinger, W. A., J. Sci. Instrum., 25, pp. 254-255
    (1948).
4. Mariscoth, M. A., Nuclear Instruments and Methods, 50,
    pp. 309-320.
5. Mills, S. J., Nuclear Instruments and Methods, 81, pp.
    217-219 (1970).
6. Sommerveld, E. J. and Visser, J. W., J. Appl. Crystl.
    8, pp. 1-7 (1974).

A MINICOMPUTER AND METHODOLOGY FOR X-RAY ANALYSIS

W. Parrish, G. L. Ayers and T. C. Huang

IBM Research Laboratory, 5600 Cottle Road

San Jose, California   95193

ABSTRACT

This paper outlines the use of an IBM Series/1 small computer
for instrument automation and data reduction for X-ray
polycrystalline diffractometry and wavelength dispersive X-ray
fluorescence spectrometry. The profile fitting method is used
to determine 2θ, d and relative peak and integrated intensities
in diffraction, and the fundamental parameters method (LAMA
program) is used for quantitative analysis of bulk and thin film
samples. The methods are precise and rapid.

INTRODUCTION

The rate of growth in the use of computers for X-ray analysis
is accelerating at a rapid rate. Among the factors contributing
to this trend is the availability of small, relatively low cost
computers and programs which greatly simplify automatic instrument
control, data collection and data reduction. This paper outlines
the use of an IBM Series/1 small (mini) computer for automation,
X-ray polycrystalline diffractometry and wavelength dispersive
fluorescence analysis. The programs and methods were developed
over several years using a large computer system – an IBM System/7
for on-line instrument control and a IBM 370/168 host computer
for data reduction (1). The small computer gives the same
precision as the large system in acceptable turn-around times.
The methods have been developed to permit use by operators with
little or no computer experience.

MINICOMPUTER X-RAY SYSTEM

The Series/1 requires a minimum of 64 kB memory to run our programs for X-ray analysis. It is a stand-alone (in contrast to an embedded mini) with EDX (Event Driven Executive) and FORTRAN capabilities. It can control the simultaneous operation of several instruments as shown in Fig. 1. An instrument control card containing the interface, scaler-timer and stepper motor control is mounted in the computer and handles one stepper motor; four cards can be used with the minimum version. Automating a diffractometer only requires mounting the stepper motor to directly drive the worm gear and connecting it through a translator to the card; the other connection from the output of the pulse height analyser feeds the intensity to the card. Retrofitting existing equipment is thus a simple task.

Operation

The use of Question and Answer formats which appear on the terminal screen to prompt the user in entering the desired experimental parameters simplifies the operation. Two types of parameters are required: default values which are characteristic of the particular instrument, and experimental run values selected for a particular run. The computer uses these values to generate the electronic and timing signals which operate the motor and track the angles.

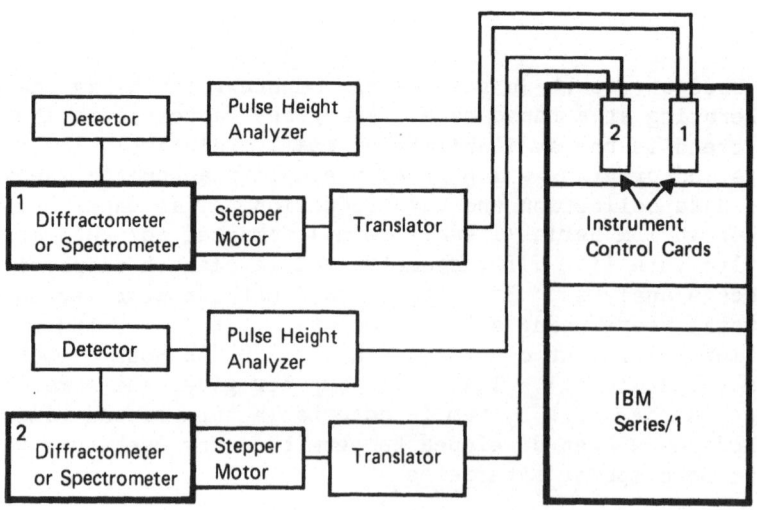

Figure 1. Series/1 connections for simultaneous operation of two instruments.

Default values include the minimum 2θ angle step generated
by one pulse on the stepper motor, the dead time of the detector
system, the maximum and minimum 2θ angles permitted on the
instrument, and others. Experimental run parameters include the
present 2θ angle to set a fiducial point, the start and end 2θ's,
the angular increment $\Delta 2\theta$ and counting time t for the steps. Data
may be collected by step scanning at selected $\Delta 2\theta$ and t,
continuous scanning with the motor operating at a fixed high
speed and data read out on the fly, and point-to-point in which
particular 2θ's and t's can be selected for up to 60 points.
Repeat runs or a number of runs with different parameters can be
requested prior to beginning a run.

POLYCRYSTALLINE DIFFRACTOMETRY

After the experimental parameters are entered the data
collection and reduction are automatic. The profile fitting
method determines the exact peak shapes and resolves overlapping
peaks with high resolution (2). The program accurately determines
2θ's, d's, relative integrated and peak intensities, and the
differences in width between the experimental profile and a set
of stored standard profiles. The experimental points of the
pattern and the fitted profiles are also displayed on a terminal
screen.

To avoid excessive computer time the weak peaks may be
required to exceed the background by say 3σ and be greater than
1% of the highest peak to be included; the numbers are selectable.
The precision of peak angles can be better than $0.001°$ (excluding
systematic errors) if the counting statistical accuracy and X-ray
technique are adequate (3). In routine analyses the data can be
collected at high speeds and the precision is better than $0.01°$
and 1-2% in intensities (4). Data reduction time varies from 1
to 15 minutes depending on the complexity and range of the
pattern.

The method is applicable to all diffractometer geometries.
Unlike the position sensitive proportional counter and energy
dispersive diffraction methods the data can be collected with a
diffracted beam monochromator thereby greatly increasing the
peak-to-background ratio, and has higher resolution.

FLUORESCENCE ANALYSIS

The computer control method is used to collect data and
operator interaction is required for peak selection and
qualitative analysis. After the intensities are entered,
quantitative analysis proceeds automatically. The LAMA program
based on the fundamental parameters method (5) was modified to
use linear approximations for faster convergence. Only pure

element bulk standards are required for bulk and thin film
analyses and multi-element standards may be used if available.
The precision is about the same as the electron microprobe and
atomic absorption spectroscopy, and the mass thickness of thin
films is determined with a precision equal to conventional
techniques (6). The computer time varies from about 1/2 minute
for binary element specimens to 10-12 minutes for a ten element
analysis. Another option in the program is that it can be used
in reverse to calculate the intensities from input concentrations.

REFERENCES

1. W. Parrish, T. C. Huang and G. L. Ayers, "Profile Fitting:
    A Powerful Method of Computer X-Ray Instrumentation and
    Analysis," Trans. Am. Cryst. Assoc. 12, 55-73 (1976).

2. T. C. Huang and W. Parrish, "Qualitative Analysis of
    Complicated Mixtures by Profile Fitting X-Ray Diffractometer
    Patterns," in C. S. Barrett, D. E. Leyden, J. B. Newkirk and
    C. O. Ruud, Editors, Advances in X-Ray Analysis, Vol. 21,
    p. 275-288 (1978).

3. W. Parrish and T. C. Huang, "Accuracy of the Profile Fitting
    Method for X-Ray Polycrystalline Diffractometry," Proc. Symp.
    on Accuracy in Powder Diffraction, Nat. Bur. Standards,
    Washington, (1979).

4. G. L. Ayers, T. C. Huang and W. Parrish, "High-Speed X-Ray
    Analysis," J. Appl. Cryst. 11, 229-233 (1978).

5. D. Laguitton and M. Mantler, "LAMA I - A General Fortran
    Program for Quantitative X-Ray Fluorescence Analysis," in H.
    F. McMurdie, C. S. Barrett, J. B. Newkirk and C. O. Ruud,
    Editors, Advances in X-Ray Analysis, Vol. 20, p. 515-528
    (1977).

6. D. Laguitton and W. Parrish, "Simultaneous Determination of
    Composition and Mass-Thickness of Thin Films by Quantitative
    X-Ray Fluoresnce Analysis," Anal. Chem. 49, 1152-1156 (1977).

FRACTURE SURFACE ANALYSIS OF BALL BEARING

STEEL BY X-RAY RESIDUAL STRESS MEASUREMENT

Hirokazu Nakashima, Noriyuki Tsushima,
and Hiroshi Muro

NTN Toyo Bearing Co. Ltd.

511 Kuwana, Japan

INTRODUCTION

Plastic deformation necessarily accompanies fracture even in high hardness steels such as ball bearing steel, though it is within a very shallow layer just under the fracture surface. The depth of the plastic deformation zone can be determined by X-ray measurement of half value breadth.[1,2] However, in the case of high hardness steel, the half value breadth is not changed significantly by this kind of plastic deformation. On the contrary, residual stress on the fractured surface was found to change remarkably depending on the mode of fracture such as static, fatigue or delayed types.[3]

Measurement of residual stress on fractured surfaces of ball bearing steel was conducted with various specimens used in fracture toughness tests, in order to obtain the quantitative relations between X-ray measurement parameters and fracture mechanics parameters such as the toughness factor K_{IC}.

EXPERIMENTAL PROCEDURE

The notch bend specimen used in the fracture toughness test is shown in Fig. 1. Specimens were made of SUJ 3 steel. After heat treatment, loads of 15 to 20 tons were repeatedly (5 to 20 times) applied at right angle to the notch to produce a precrack of about 1 mm depth at the bottom of the notch. The value of K_{IC} was obtained by the three point bending test, using the following equation:[4]

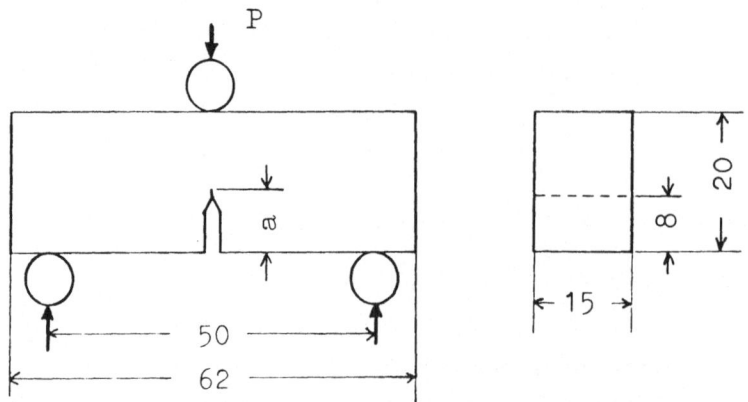

Fig. 1. Three point bending specimen with precrack
for K_{IC} and fatigue tests

$$K_{IC} = \frac{PL\sqrt{a}}{BW^2} \{5.8 - 9.2(\tfrac{a}{W}) + 43.6(\tfrac{a}{W})^2 - 75.3(\tfrac{a}{W})^3 + 77.5(\tfrac{a}{W})^4 \}$$

where P is the fracture load, L half of the distance between the
two supporting points, B the width of the specimen, W the height of
the specimen, a the depth of the crack and C the depth of the
notch.

    The residual stress in the direction of crack propagation on
and under the fracture surface was measured by X-rays.

RESULTS

Influence of Various Factors on K_{IC}

    Tempering temperature. K_{IC} values were obtained from speci-
mens quenched from the normal quenching temperature and tempered
at various temperatures for 1 hour. Fig. 2 shows the relation
between K_{IC} and tempered hardness. The K_{IC} value of the specimen
subjected to normal heat treatment is about 50 kg$\sqrt{mm}$/mm^2. The K_{IC}
value increases remarkably when the Rockwell C hardness decreases
below HRC 50.

    Inhomogeneous structure. Bainite structure is known to be
tougher than full martensite structure of the same hardness.[5]
K_{IC} values of high hardness materials containing various amounts
of bainite were investigated. Results are shown in Table 1. The
K_{IC} value varies from 44 to 60 kg$\sqrt{mm}$mm^2 with the amount of bainite
within the range of hardness over HRC 60. Comparing K_{IC} values of
bainite containing structures (Table 1) with that of the full
martensitic one tempered to the same hardness (Fig. 2), it is

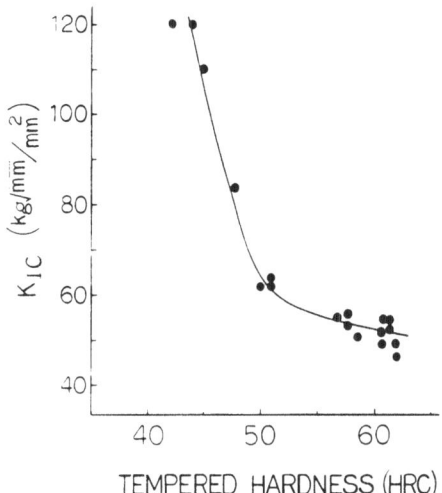

Figure 2.   Relation between K_{IC} and
tempered hardness of ball bearing steel

Table 1

Influence of heat treatment upon crack toughness K_{IC}

Heat treatment 820°C OQ, 170°C Temper	HRC	$K_{IC} kg\sqrt{mm}/mm^2$	Residual stress
Full martensite structure	63.2	43.8 44.9	10 kg/mm^2
A few % bainite structure	63.0	45.6 49.4	10
10% bainite structure	62.0	57.9	35
50% bainite structure	59.6	60.1	70

found that the inhomogeneous structure shows higher K_{IC} value than
the homogeneous martensitic structure.

Residual Stress on the Fractured Surface

When a crack grows, plastic deformation usually takes place

around the crack tip. Therefore residual stress is produced on the
fractured surface. Fig. 3 shows several examples of residual stress
distributions measured on the fractured surfaces of the specimens
fractured by static bending and bending fatigue. Specimens were
quenched normally and tempered at various temperatures. From one
specimen were obtained the four curves of each group in Fig. 3a,
the three curves in each group of Fig. 3b and the two of each group
in Fig. 3c. The measured position for the curves with circles differ from the position of those with triangles. Open and closed
circles are from positions on the fracture surface where there had
been a crack in an early stage of its propagation, open and closed
triangles were from areas corresponding to the last stage of propagation. Crosses indicated data from a surface rapidly fractured
in bending fatigue. The measuring direction for the closed circles
and triangles differs from that with the open circles and triangles.
Closed circles and triangles were measurements perpendicular to the
crack propagation direction, other symbols were measurements parallel to the propagation direction.

Residual stress on and under the fracture surface is tensile
and largest at the surface. Higher residual stress is observed in
statically fractured surfaces than in fatigued ones. The residual stress below the surface is so small that only surface residual stress needs to be considered. The relation between surface
residual stress and K_{IC} is plotted, from the data of Fig. 3c and
some additional specimens, in Fig. 4. A linear correlation exists
up to the value of K_{IC} of 70 kg$\sqrt{mm}$/mm^2. Therefore, the K_{IC} value
can be assumed from residual stress values on a fracture surface
provided that the K_{IC} of the material is less than 70 kg$\sqrt{mm}$/mm^2.

APPLICATION TO FAILURE ANALYSIS

Fig. 5 is an example of residual stress measurement on the
fracture surface of a spherical roller bearing inner ring fractured in a practical use. The stresses indicated in Fig. 5 are
principal stresses parallel to the crack direction and parallel
to the crack plane. From the residual stress data, it is assumed
that the crack with its origin at the center of the raceway propa-
gated gradually to the size a = 16 mm, 2c = 69 mm before the fast
and final crack propagation to through-fracture occurred.
Residual stress on the fast fractured surface indicates that K_{IC}
of the material is about 70 kg$\sqrt{mm}$/mm^2. In this case the fracture
was caused by hoop tension due to tight fitting. The condition
for a fast fracture is expressed as follows.[7]

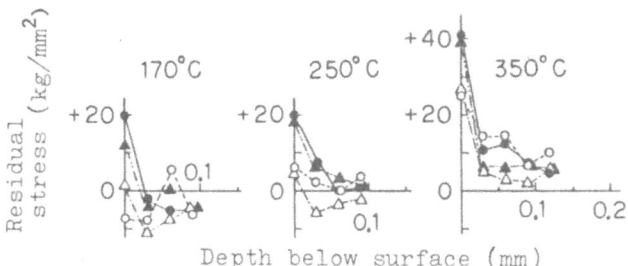

Fig. 3(a). Residual stress distributions on fracture surfaces of K_{IC} specimens, quenched normally and tempered at various temperatures.

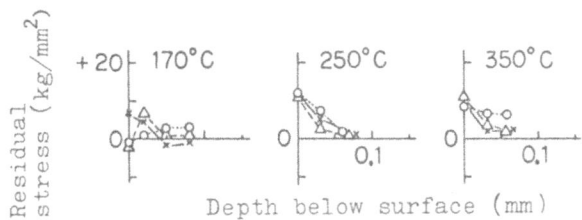

Fig. 3(b). Residual stress distributions on fracture surfaces of fatigue specimens, quenched normally and tempered at various temperatures.

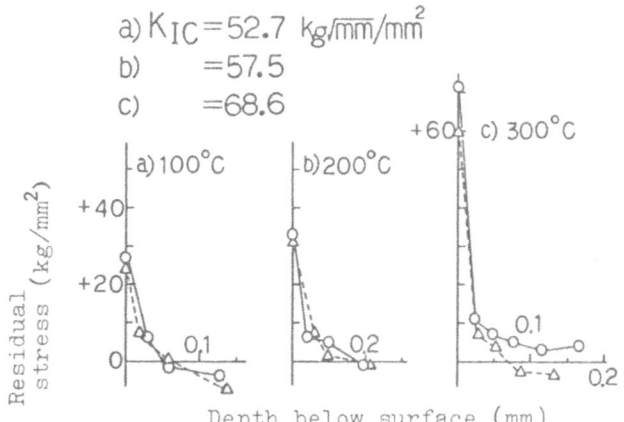

Fig. 3(c). Residual stress distributions on fracture surfaces of K_{IC} specimens, fractured at high temperatures.

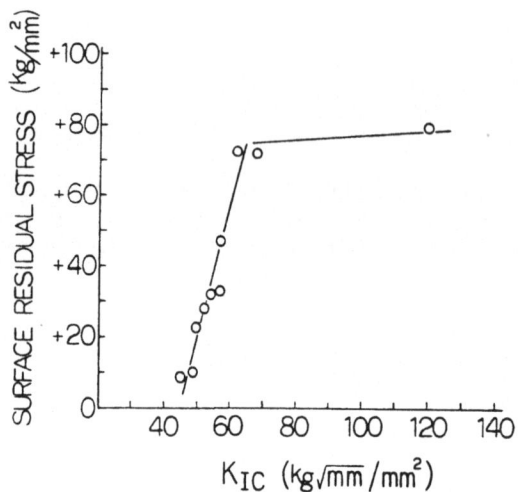

Fig. 4.   Relation between surface residual stress
          and K_{IC} .

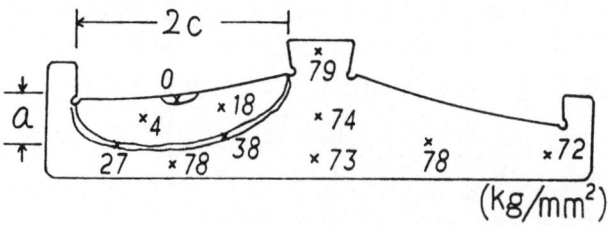

Fig. 5.   Residual stress distribution on the fracture
          surface of a spherical roller bearing inner
          ring.

$$K^2_{IC} = \frac{1.21 \; \sigma^2 \; \pi a}{\phi^2 - 0.212(\frac{\sigma}{\sigma_y})^2}$$

$$\phi = \int_0^{\pi/2} \sqrt{1 - \frac{c^2 - a^2}{c^2} \; \sin^2\theta} \; d\theta$$

Substituting $a = 16$ mm, $c = 34.5$ mm, $K_{IC} = 70 \; kg\sqrt{mm}/mm^2$ and $\sigma_y = 180 \; kg/mm^2$ (= yield stress) in the equation, we obtain $\sigma = 11 \; kg/mm^2$. It is presumed that the hoop tension was $11 \; kg/mm^2$.

REFERENCES

1. S. Taira and K. Hayashi, Trans. of the Iron and Steel Institute of Japan, 8:20 (1968).
2. H. Kitagawa and T. Matsumoto, Trans. of the Japan Society of Mechanical Engineers, 41:22 (1975).
3. N. Tsushima, H. Nakashima, and H. Muro, J. of the Japanese Society for Strength and Fracture of Materials, 10:91 (1975).
4. Fracture Toughness, ISI Publication 121, p. 55.
5. F. Borik and R. D. Chapman, Trans. ASM, 53:447 (1961).
6. M. L. Williams, J. Appl. Mech., 24:109 (1957).
7. Fracture Toughness Testing and Its Application, ASTM STP 381, p. 190 (1965).

A POSITION-SENSITIVE PROPORTIONAL COUNTER FOR RESIDUAL STRESS MEASUREMENT BY MEANS OF MICROBEAM X-RAYS

Yasuo Yoshioka
Musashi Institute of Technology
1, Tamazutsumi, Setagaya, Tokyo 158, Japan

Ken-ichi Hasegawa and Koh-ichi Mochiki
Dept. of Nuclear Engg., University of Tokyo
7, Hongoh, Bunkyo, Tokyo 113, Japan

ABSTRACT

A position-sensitive proportional counter (PSPC) with high counting efficiency has been made for stress analysis with low intensity X-rays such as microbeam X-rays.

This PSPC system has made it possible to measure the residual stress in a small region such as a fatigue crack tip in very short time compared with the measurement by standard diffractometer or film methods.

INTRODUCTION

It is very important to know the behaviour of residual stress around a fatigue crack tip for discussing the crack growth mechanism. Residual stresses in the vicinity of crack tip are often measured by the X-ray film method;[1] however this process requires much time, and the precision of the stress values is not sufficient for quantitative discussion. If the position-sensitive proportional counter (PSPC) is applied to this field, the time required for data accumulation will be very much reduced and it also should be possible to obtain precise residual stress values in small regions.

We have reported a PSPC with uniform angular resolution and several examples of stress measurement by this PSPC.[2] The time for stress determination was shown to be 1/10 to 1/30 of the time required for the standard diffractometer method.

325

However, a PSPC with higher counting efficiency is being developed. It is favorable for stress measurement in a small region
such as a crack tip. In the present study, a PSPC suitable for
this purpose was made, and stresses in the vicinity of a fatigue
crack tip were measured.

DETECTOR DESIGN

In the past work,[2] we made a resistive wire cathode type PSPC
which has an angular span of 20° in 2θ at a radius of 200 mm. The
entrance window has the dimensions of 8 mm height and 100 mm width.
Though a PSPC made in the present study has the same specifications,
the window height was widened to 20 mm in order to increase the
counting efficiency. Fig. 1 shows schematically the construction
of the detector. The cathode consists of a 200 turn coil which
consisted of a 27 μm diam. Ni-Cr wire wound with 0.5 mm pitch on
an acrylic resin plate, 110 mm wide by 50 mm high by 2 mm thick.
This cathode was placed parallel to the window at a spacing of 8 mm,
and nine anode wires of 20 μm diam. of gold-plated tungsten were
stretched at a spacing of 4 mm in the middle between the window
and the cathode. The material of the window is a vanadium foil of
10 μm thickness for removing the Cr-Kβ radiation (the Cr-Kα radiation alone was wanted). The detector chamber is filled with PR
gas (Ar:90%, Ch_4:10%) at 1 atm. Each end of the cathode wire is
connected to a charge sensitive preamplifier; we have adopted the
charge division method with an analog divider as the basis of position read out. The output pulses from the divider are processed by
the 512 channels pulse height analyzer, and these data are analyzed
by an on-line micro-computer.

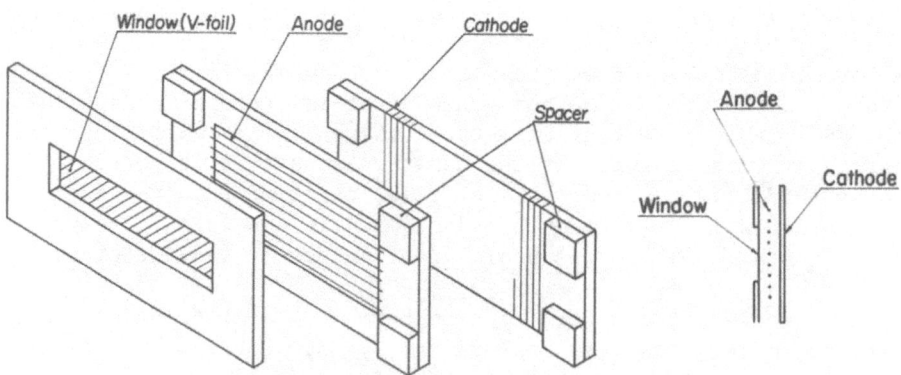

Fig. 1.   Schematic construction of detector

DETECTOR PERFORMANCE

Angular Resolution

The X-ray beam collimated by a 50 μm wide slit was directed into the PSPC with path angle of $\alpha°$ from a point at a distance 200 mm from the PSPC window. As shown in Fig. 2, the minimum angular resolution was about 0.2° in 2θ (FWHM) at $\alpha=0°$ and it slightly increases with increase of path angle α. Fig. 3 shows an $\alpha(211)$ profile from an annealed steel specimen produced by Cr-Kα radiation, the beam size being 2X10 mm^2. The Kα_1 and Kα_2 profiles overlap almost completely in spite of the high angular resolution. This fact shows that the "effective" resolution for the Debye–Scherrer ring is poorer than that obtained by either a "point" or a "line" shaped beam when the window height is widened. As the arc length of the ring into the PSPC increases and thus the height of segment of the arc that enters also increases, the angular resolution is blurred. This segment height is about 0.2° in 2θ at 20 mm height for an X-ray point source. The geometrical blurring and beam size affect the total "effective" resolution, which is less than that for the conventional parallel beam diffractometer method. However, the resolution is sufficient for stress measurement.

The time required for data accumulation was 15 sec. when the window height was limited to 8 mm, equalling the window height of the previous detector.[2] It was reduced to 6 sec. when the height was widened to 20 mm.

The linear response with respect to the path angle α was measured. A plot is obtained that indicates very good linearity in a range of $\alpha = -7.5°$ to $+5°$.

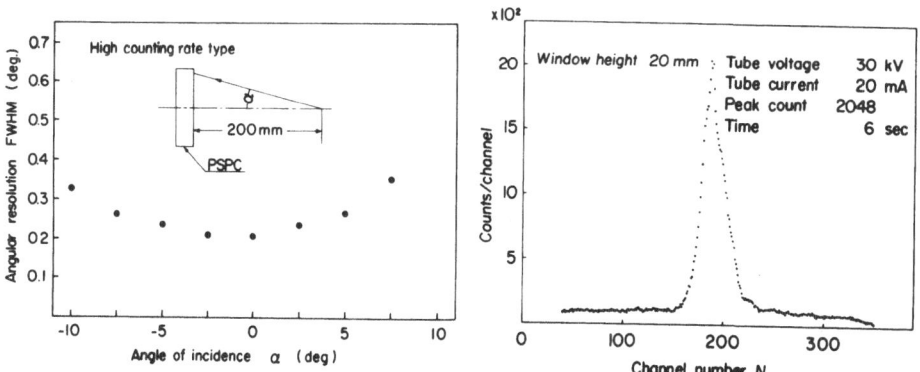

Fig. 2. Angular resolution in 2θ. Fig. 3. Diffraction profile
(50 μm wide slit)   from an annealed car-
                                          bon steel

Linearity and Maximum Allowable Counting Rate

The maximum allowable counting rate was found to be about 30,000 counts per second. This value is sufficient for the stress measurement by means of microbeam X-rays.

EXAMPLES OF STRESS MEASUREMENT

Method and Equipment

An ASTM compact tension type specimen of annealed 0.15% carbon steel was fatigued under pulsating cyclic loads (Pmax = 700 Kg, stress ratio R = 0.05) and the distribution of residual stress, the direction of which is perpendicular to the direction of crack growth, was measured.

Measurements were made using a Cr-target microbeam X-ray generator equipped with this PSPC system. To obtain the residual stress in a very small region, a 200 μm diam. pin-hole slit was placed at the front of a window of the X-ray source. The PSPC was placed at a distance 200 mm from the irradiated position of the X-ray beam on the specimen, and at an angle of 24° to the X-ray beam. The X-ray irradiated diameter on the specimen was about 600 μm. To prevent the disorder of profile, the specimen was oscillated about the irradiated point as a center of oscillation, through a range of ± 5°. A closeup view of the equipment and specimen is shown in Fig. 4.

Experimental Results

Fig. 5 shows an example of a profile obtained from the

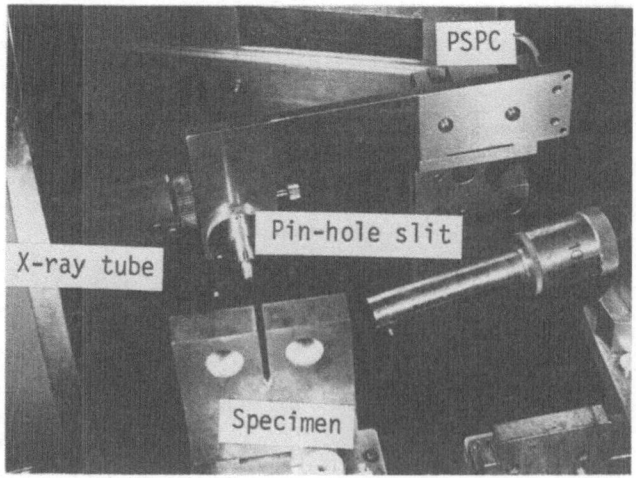

Fig. 4. Closeup view of equipment and specimen

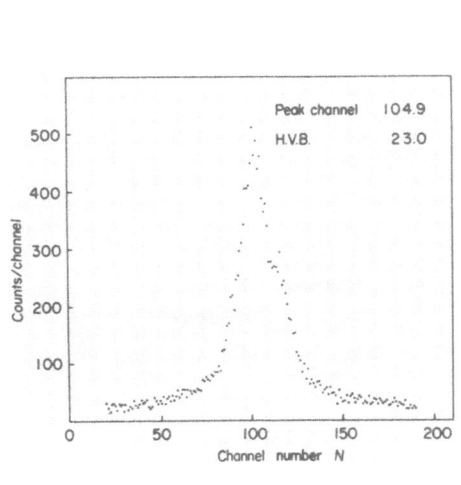

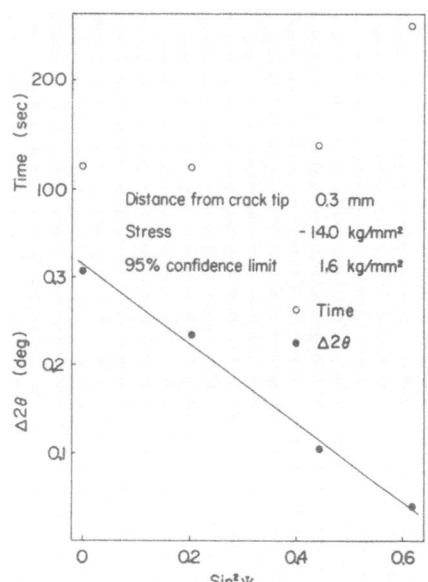

Fig. 5.   An example of profile
          from specimen.   X-ray
          beam was collimated by
          a 200 μm dia. pin-hole   Fig 6.   An example of $\mathrm{Sin}^2\psi$ dia-
          slit   gram and time required

specimen used.   The maximum counts were preset to 512 and the time
taken to reach this count was only 119 seconds with the X-ray source
operating at 30 KV-1.2 mA.   If a Debye-Scherrer pattern is photo-
graphed on an X-ray film under the same X-ray conditions, the exposure time would require more than 2 hours.   Fig. 6 shows an example of a $\mathrm{Sin}^2\psi$ diagram and the time required for forming a profile at every ψ_0 angle.   A good linear relation is found to hold
on the $\mathrm{Sin}^2\psi$ diagram.   The total time for stress determination was
about 20 minutes (ψ_0 = 0°, 0°, 15°, 30°, 40°, 40°).   It was very
hard to measure the residual stress in a very small region by the
diffractometer method or the film method, the diffraction intensity
being very weak.   It is evident that the PSPC system reduces the
time required for measurement very much.

The distribution of residual stress in the vicinity of the
crack tip is shown in Fig. 7, the crack length being 31.9 mm and
the stress intensity factor K being 131.6 $\mathrm{Kg/mm}^{3/2}$.   The compressive residual stress shows a maximum value which exceeds the yield
stress (27.2 $\mathrm{Kg/mm}^2$) at a position 0.5 mm from the crack tip and
it gradually decreases nearly zero at the boundary of the mono-
tonic plastic deformed zone obtained theoretically, indicated by
a dotted line in the figure.

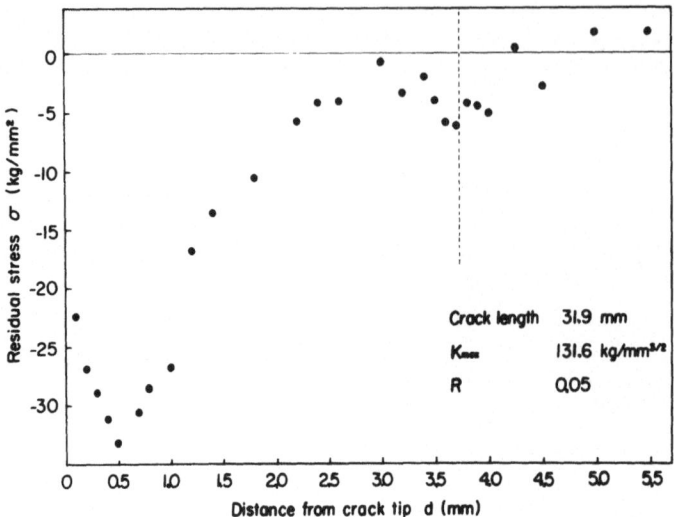

Fig. 7.   Residual stress distribution at a fatigue crack tip

Although the effect of residual stress on the crack growth is
well understood in a qualitative manner, this effect has not been
completely quantified. For example, crack closure was often observed and the crack opening stress was estimated. There are,
however, very few reports of this problem based on residual stress
measurement in the vicinity of the crack tip. The reason is that
to measure the residual stress in a small region was inefficient,
laborious, tedious and probably unsuccessful by the conventional
experimental technique. As the PSPC system makes it possible to
measure such residual stress easily, it will enable one to analyze
the crack closure or the overload effect which affects the crack
growth, quantitatively.

REFERENCES

1.   Y. Yoshioka, T. Kaneko and M. Terasawa, "Fatigue Crack Propa-
     gation and Residual Stress in Tufftrided Steel," J. JSMS,
     26:74 (1976).
2.   Y. Yoshioka, K. Hasegawa and K. Mochiki, "Study on X-Ray Stress
     Analysis Using a New Position-Sensitive Proportional Counter,"
     Advances in X-Ray Analysis, 22:233 (1979).

STRESS ANALYSIS IN GRAPHITE/EPOXY

C. S. Barrett and Paul Predecki

University of Denver Research Institute

Denver, Colorado 10011

X-rays can be used to measure residual as well as applied stresses in polymers and reinforced polymeric composites, by diffracting from filler particles that are embedded before curing.[1-3] We have investigated various fillers that exhibit suitable stress-induced shifts of diffraction angle when embedded in uniaxial graphite fiber/epoxy composites. X-ray measurements of elastic strains in the particles are proportional to the corresponding composite strains, in agreement with the model of H. T. Hahn.[4] Results indicate that the stress sensitivity (change in X-ray diffraction angle per MPa applied to the composite) increases in the order W, CdO, Ni, Ag, Nb, Al. With Al 333 + 511 reflections of $CuK\alpha_1$, the diffraction angle shifts 8.52×10^{-4} deg 2θ per MPa. Theoretical predictions are in reasonable agreement with this figure. The elastic range terminates at a yield point beyond which little or no further shift of the diffracted beams is seen for metallic fillers, but no such yield point has appeared with CdO particles. Oppositely directed shifts are seen after unloading from stress levels that have exceeded the yield point.

Residual stresses in the particles were determined after curing and storage under ambient conditions. Measurements of diffraction angle were made with three different specimen tilts with respect to the incident X-ray beam and combined with measurements of unstressed filler powder so as to determine the three principal strains and three principal stresses. In the fiber direction of the uniaxial composites the residual stresses were always found to be positive and larger than the principal stresses orthogonal to these.

This research is supported by Air Force Office of Scientific Research under Grant #77-3284.

References

1. C. S. Barrett and Paul Predecki, "Stress Measurement in Polymeric Materials by X-Ray Diffraction," Pol. Eng. Sci., $\underline{16}$, 602 (1976).
2. Paul Predecki and Charles S. Barrett, "Stress Measurement in Graphite/Epoxy Composites by X-Ray Diffraction," J. Composite Materials $\underline{13}$, 61 (1979).
3. Charles S. Barrett and Paul Predecki, "Stress Measurement in Graphite/Epoxy Uniaxial Composites," accepted 1979, for publication in Polymer Science and Engineering.
4. H. T. Hahn in "Composite Materials Workbook," AFML-TR-78-33, p. 65, March 1978.

PROBLEMS ASSOCIATED WITH Kα DOUBLET IN RESIDUAL STRESS MEASUREMENTS

S. K. Gupta* and B. D. Cullity**

Department of Metallurgical Engineering and Materials Science, University of Notre Dame, Notre Dame, Indiana 46556

INTRODUCTION

Since the measurement of residual stress by X-ray diffraction techniques is dependent on the difference in angle of a diffraction peak maximum when the sample is examined consecutively with its surface at two different angles to the diffracting planes, it is important that these diffraction angles be obtained precisely, preferably with an accuracy of $\pm$ 0.01 deg. 2θ. Similar accuracy is desired in precise lattice parameter determination. In such measurements, it is imperative that the diffractometer be well-aligned. It is in the context of diffractometer alignment with the aid of a silicon powder standard free of residual stress that the diffraction peak analysis techniques described here have been developed, preparatory to residual stress determinations.

EXPERIMENTAL PROCEDURE

The final step in the alignment of a GE XRD-3 diffractometer is the determination of the center of rotation of the system, for it is usual to assume that the systematic errors present are dominated by those arising from sample displacement δ from this center on the focusing circle. In this procedure a silicon powder pellet (Norelco Standard) was employed with chromium radiation. Kα and Kβ components were included. The maximum of each peak was

---

* Present address: Morris Electronics Limited, Bhosari Industrial Estate, Poona 411026 India
** Deceased

determined by the three[1] and nine point parabola methods from fixed
count technique data corrected by appropriate LPA factors. Lattice
parameters calculated from the d-spacings thus determined are
plotted[2] against $\cos^2\theta/\sin\theta$ in Fig. 1.   Such a plot should be
linear with slope proportional to the magnitude and direction of
sample displacement δ from the focusing circle position; the ordi-
nate intercept is a_o, the best estimate of the true lattice
parameter.

      Least squares straight lines were drawn for both Kα and Kβ
points. The correlation coefficient γ^2 of the lines was also calculated. The intercept corresponding to the precise parameter value
obtained from Kα points was found to be consistently higher than
that for Kβ though with identical slopes. It was clear that the
apparent diffraction peak angles for Kα lines were smaller than their
true values, which forced critical evaluation of methods in such
precise determination of peak maxima from the incompletely resolved
Kα doublet.

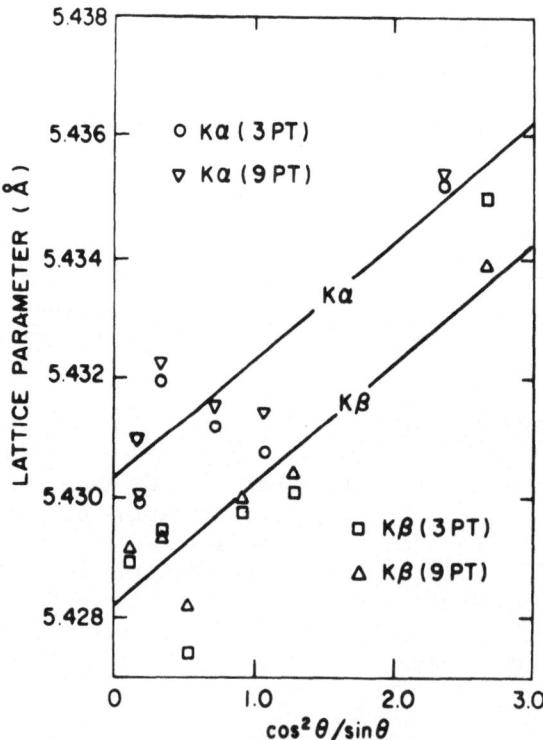

Fig. 1.   Lattice parameter of silicon deduced from Kβ and composite
          Kα peak positions.

The $K\alpha$ doublet has been known to create difficulties in the measurement of various crystal properties by X-ray diffraction. It has been suggested that problems arising from a partially resolved $K\alpha$ doublet may be eliminated by utilizing wider slits and thereby producing a single peak corresponding to the wave-length $\lambda_{K\alpha} = (2\lambda_{K\alpha_1} + \lambda_{K\alpha_2})/3$. This implies that, once the $K\alpha_1$ peak is fixed, the unresolved $K\alpha$ peak center is independent of the shape and width of $K\alpha_1$ and $K\alpha_2$ components. This can be easily demonstrated to be false when any reasonable functional form (Gaussian, Cauchy etc.) for the $K\alpha_1$ and $K\alpha_2$ line shapes is assumed. To circumvent these difficulties it is necessary therefore to completely separate the $K\alpha$ doublet components first.

RACHINGER METHOD

The Rachinger method is most commonly employed to resolve the doublet components.[3] The chief advantages of this method are:

(i)   The $K\alpha_1$ and $K\alpha_2$ profiles are not approximated by any assumed mathematical function.

(ii) The method does not require excessive computational labor.

The chief disadvantages are:

(i)   The time required to measure the complete $K\alpha$ profile is rather long.

(ii) The accuracy of determining the $K\alpha_1$ peak center is limited by the angular increment involved.

(iii) The separation $\underline{S}$ between $K\alpha_1$ and $K\alpha_2$ peak centers is assumed.

This assumption is particularly serious because parameter $\underline{S}$ is a sensitive function of the $K\alpha_1$ peak position.[4] Furthermore, in assuming $\underline{S}$, the $K\alpha_1$ peak center is assumed indirectly because

$$S = 2 \sin^{-1}[R \sin(2\theta_{K\alpha_1}/2)] - 2\theta_{K\alpha_1} \tag{1}$$

where $R = \lambda_{K\alpha_2}/\lambda_{K\alpha_1}$.

It is clear that the disadvantages of Rachinger correction greatly outweigh the advantages when precise determination of peak position is required.

NON-LINEAR REGRESSION METHOD

Alternatively, if a suitable analytical function approximating the $K\alpha_1$ and $K\alpha_2$ profiles is available, then with the aid of a computer, it becomes relatively easy to separate the doublet into its $K\alpha_1$ and $K\alpha_2$ components. Pearson Type VII functions have been found to describe the profiles to a good approximation,[5] the form of which is

$$I_{K\alpha_1}(2\theta) = A[1 + B^2(2\theta - 2\theta_{K\alpha_1})^2/m]^{-m} \qquad (2)$$

where

$I_{K\alpha_1}$ = intensity of $K\alpha_1$ component at any particular 2θ value

A = peak intensity of $K\alpha_1$ component

$2\theta_{K\alpha_1}$ = position of $K\alpha_1$ peak maximum

B = $[2\sqrt{m(2^{1/m}-1)}/\text{Half-width}]$

When the exponent $m = 1$, the profile represented by expression (2) is Lorentzian; $m = 2$, Cauchy; and $m = \infty$, Gaussian.

Optimum values of the parameters A, B, m and $2\theta_{K\alpha_1}$ can be estimated with the least squares method by employing either gradient search or grid search techniques. An algorithm utilizing the gradient search technique has been written combining the Fletcher-Reeves method with the Golden Section Line Search method.[6]

This algorithm is ideally suited for approaching the solution from far away, but it does not converge rapidly in the immediate neighborhood. The Marquardt algorithm[7] on the other hand, combines the best features of the gradient search with the method of linearizing the fitting function and thus is more suited to this particular problem.

To make the computer search easier, the following additional assumptions can be made:

(i) Peak intensity of $K\alpha_1$ is twice that of $K\alpha_2$.

(ii) $K\alpha_1$ and $K\alpha_2$ have identical shapes and half-widths.

It should be noted that these assumptions are not necessary but are made to facilitate the computer search technique by reducing the number of parameters. Then with the Pearson Type VII function, an expression for the combined $K\alpha$ profile becomes

$$I_{K\alpha}(2\theta) = A[1 + B^2(2\theta - 2\theta_{K\alpha_1})^2/m]^{-m}$$

$$+ 0.5A[1 + B^2(2\theta - 2\theta_{K\alpha_1} - S)^2/m]^{-m} \quad (3)$$

where $\underline{S}$ is given by equation (1). If N sets of $(I_{K\alpha}, 2\theta)$ values for the *Kα* doublet are measured, then the algorithm provides the solution for equation (3) by which the parameters A, B, m and $2\theta_{K\alpha_1}$ are computed and the *Kα* doublet is resolved into its $K\alpha_1$ and $K\alpha_2$ components.

The algorithm also yields the r.m.s. error involved in this curve fitting procedure;

$$\text{r.m.s. error (pct)} = 100 \cdot [\{\sum_{i=1}^{n} (I_i^{obs} - I_i^{calc})^2\}/\{\sum_{i=1}^{n} (I_i^{obs})^2\}]$$

$$(4)$$

The r.m.s. error in each case involving the Fletcher-Reeves algorithm has been found to be less than 10 percent, whereas that from the Marquardt algorithm has been found to be less than 1 percent.

A brief comparison of results was made in which the Rachinger and Nonlinear Regression methods were used. Least squares analysis of results indicated that the value of $\underline{m}$ obtained by the Nonlinear Regression method was between 1 and 2 for all line profiles, implying that the profiles can be considered as a combination of Lorentz and Cauchy types. As indicated in Table 1, the slope of the line of unit cell parameter vs. $\cos^2\theta/\sin\theta$ obtained from $K\alpha_1$ points of the Nonlear Regression technique was closer to that for the *Kβ* points than was the slope for $K\alpha_1$ points obtained by the Rachinger technique; the comparison thus corroborates Nonlinear Regression as the best technique for use for stress analysis, although the few data from the experiment cannot be construed as sufficient for a rigorous test of our conclusions.

CONCLUSIONS

(i)   For residual stress and lattice parameter determinations, a complete resolution of *Kα* doublet is essential.

(ii) The weighted average $\lambda_{K\alpha}$ used in calculation of $\underline{d}$ spacing is improper when a single peak is measured (i.e., unresolved doublet).

(iii) The three-point and nine-point parabola methods are not sufficiently accurate in calculating peak centers of an incompletely resolved *Kα* profile.

Table 1.   Results of Least Squares Analysis of Data in Fig. 1

Method	Slope (Å)	δ(cm)	a_o(Å)	γ^2
Kβ(3pt parabola)	2.52×10^{-3}	6.75×10^{-3}	5.42760	0.8385
Kβ(9pt parabola)	1.99	5.33	5.42821	0.8895
Kα(3pt parabola)	1.87	5.00	5.43014	0.7365
Kα(9pt parabola)	1.93	5.17	5.43035	0.7985
Kα_1 (Rachinger)	3.14×10^{-3}	8.40×10^{-3}	5.42884	0.9087
Kα_2(Rachinger	0.65	1.75	5.42891	0.1290
Kα_1(Nonlinear Regression)	2.35	6.29	5.42934	0.9710

(iv)   The Rachinger method for resolving a Kα doublet is insufficiently accurate for determining the Kα_1 peak center.

(v)   The Nonlinear Regression method is better than the Rachinger or the weighted average methods for Kα_1 peak center determination, and the Marquardt algorithm is recommended for it. A peak shape description is also produced by the algorithm.

(vi)   The r.m.s. error calculated during the nonlinear regression provides a check for accuracy of curve fitting.

ACKNOWLEDGMENTS

    The assistance of Professor C. W. Allen in the preparation of this paper and of Professor C. S. Barrett for presenting it at the 28th Denver X-Ray Conference and for providing useful criticism is gratefully acknowledged.

REFERENCES

1. "Residual Stress Measurement by X-Ray Diffraction," 1971, SAE
   Information Report, J784A, pp. 51-52.

2. "Residual Stress Measurement by X-Ray Diffraction," 1971, SAE
   Information Report, J784a, pp. 34-35.

3. H. P. Klug and L. E. Alexander, 1974, "X-Ray Diffraction Procedures," John Wiley and Sons, New York, pp. 625-628.

4. R. Delhez and E. J. Mittemeijer, An Improved α_2 Elimination,
   J. Appl. Cryst., 8:609 (1975).

5. M. M. Hall, V. G. Veeraraghavan, Herman Rubin and P. G. Winchell,
   The Approximation of Symmetric X-Ray Peaks by Pearson VII
   Distributions. J. Appl. Cryst., 10:66 (1977).

6. David M. Himmelblau, 1972, "Applied Non-Linear Programming,"
   McGraw Hill Company, pp. 433-468.

7. D. W. Marquardt, An Algorithm for Least-Squares Estimation
   of Non-Linear Parameters, J. Soc. Indust. Appl. Math., 11:431
   (1963).

INCLINATION OF PRINCIPAL RESIDUAL STRESS AND THE DIRECTION

OF CRACKING IN CONTACT-FATIGUED BALL BEARING STEEL

Kikuo Maeda, Noriyuki Tsushima,
Masatoshi Tokuda, and Hiroshi Muro

NTN Toyo Bearing Co. Ltd.

511 Kuwana, Japan

INTRODUCTION

Peeling is a surface fatigue failure of a roller bearing that consists of many shallow pits less than 10 μm in depth and cracks that link the pits. Peeling occurs rather easily on a smooth surface when in contact with a rough surface under insufficient thickness of the lubricating oil film.

X-ray residual stress measurements on and under the contact surface after a peeling test revealed that the 2θ versus $\sin^2\psi$ curve is not linear and that it curves depending upon the rolling contact condition and especially upon the existence of slip. Non-linearity of the 2θ-$\sin^2\psi$ curve has been reported by Wakabayashi[1] in a study of residual stress accompanying the grinding of soft steel and by Faninger[2] in a study of residual stress due to rolling contact with annealed steel, but not in the case of high hardness steel such as ball bearing steel. No complete explanation of this non-linearity has been made as yet.

Recently Peiter and Lode[3,4] reported that the 2θ-$\sin^2\psi$ curve is not linear but curves differently according to the residual stress state, such as the inclination of the principal residual stress and the gradient of residual stress with depth. They proposed an integral method to evaluate the residual principal stress states for such cases.

Peeling tests were carried out with various slip ratios and the relation between the appearance of the cracking and the residual stress state, obtained by Peiter's method, was investigated.

341

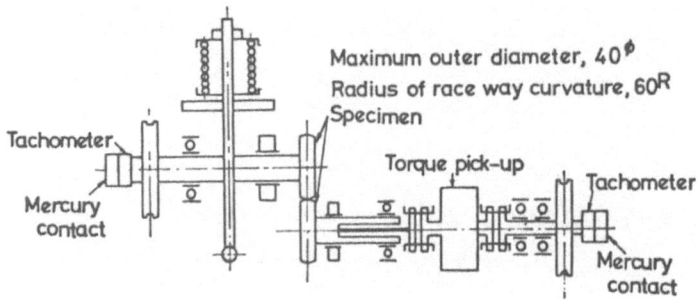

Fig. 1.   Schematic construction of test rig

EXPERIMENTAL PROCEDURE

    The outline of the test rig is shown in Fig. 1.   Two ring
specimens, super-finished and ground, 40 mm in diameter and 60 mm
in radius of curvature in the axial direction, were rolling-
contacted and driven separately so that an arbitrary slip could be
given to the specimens.   Specimens were made of SUJ 2.   Their
hardness was HRC 62.   The slip ratios were changed from 0 to 20%
by increasing the rotational speed of the ground specimen.

    Residual stresses on and under the contact surface of the
super-finished specimens (peeling specimens) were measured by X-
rays.   To investigate the relation between the rolling conditions
and the residual stress state, or the direction of cracking and
the residual stress state, an X-ray beam was projected in the
rolling direction ($\psi > 0$) and its opposite direction ($\psi < 0$) in
the circumferential direction and in both $\psi > 0$ and $\psi < 0$ directions in the axial direction, as shown in Fig. 2.

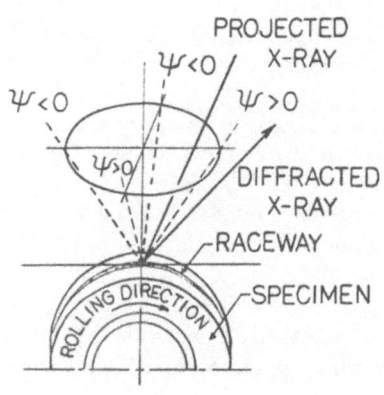

Fig. 2.   X-ray beam projection

RESULTS

Typical views of peeling on tested specimens are shown in Fig. 3 and 4. External and sectional appearance of peeling are changed by the existence of slip.

$2\theta-\sin^2\psi$ Curves of Specimens

Typical $2\theta-\sin^2\psi$ curves measured with the above-mentioned specimens are shown in Figs. 5, 6 and 7.

Fig. 5 shows $2\theta-\sin^2\psi$ curves of the non-used super-finished specimen. Only slight non-linearity is recognized at the surface in both directions. But this tendency vanishes in the surface layers.

Fig. 6 shows $2\theta-\sin^2\psi$ curves of the peeled specimen which was rolled under the no-slip condition. In both circumferential and axial directions, the differences in the 2θ values between the two measuring directions ($\psi > 0$ and $\psi < 0$) and the non-linearity in $2\theta-\sin^2\psi$ relations are recognized at the surface. This tendency remains even at 5 μm depth.

Fig. 7 shows $2\theta-\sin^2\psi$ curves of the peeled specimen which was rolled under 20% slip conditions. In the circumferential direction, the differences in 2θ values between the two measuring directions and the non-linearity are recognized at the surface and 0.005 mm depth. In the axial direction, however, only a slight non-linearity is recognized.

Since peeling test specimens have non-linearity and a difference in 2θ values in their $2\theta-\sin^2\psi$ curves between the two measuring directions, it is impossible to calculate the true stress using the conventional differential method. For this problem, an integral method developed by Peiter and Lode looks effective.

DISCUSSION

Relation Between the State of the Principal Residual Stress and the Sectional Form of Cracks

The principle and a sample calculation have been reported in detail by Peiter and Lode.[3,4] Calculations of the states of principal residual stresses for the present specimens were carried out in accordance with their method.

The principal residual stress values, σ_M, (see Fig. 8), the

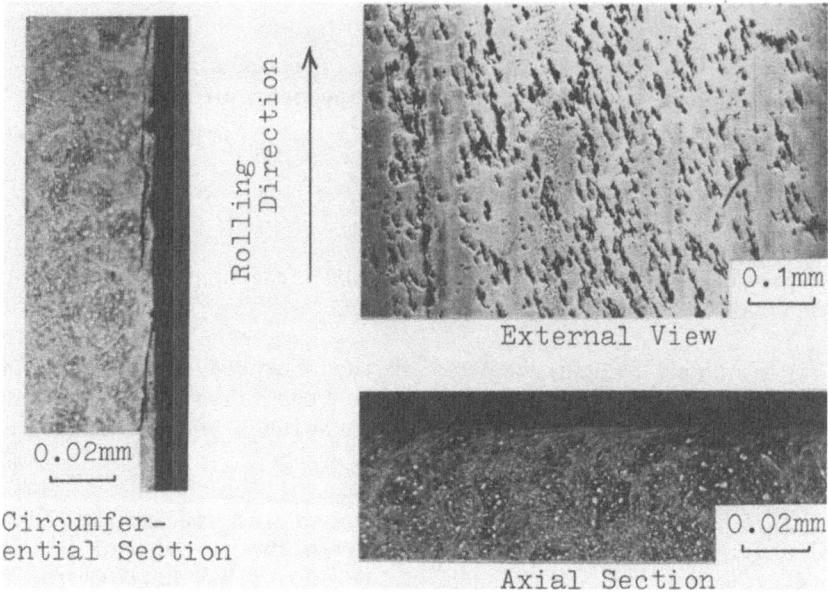

Fig. 3. External and sectional views of peeling on
        the specimen rolling-contacted with 0% slip.

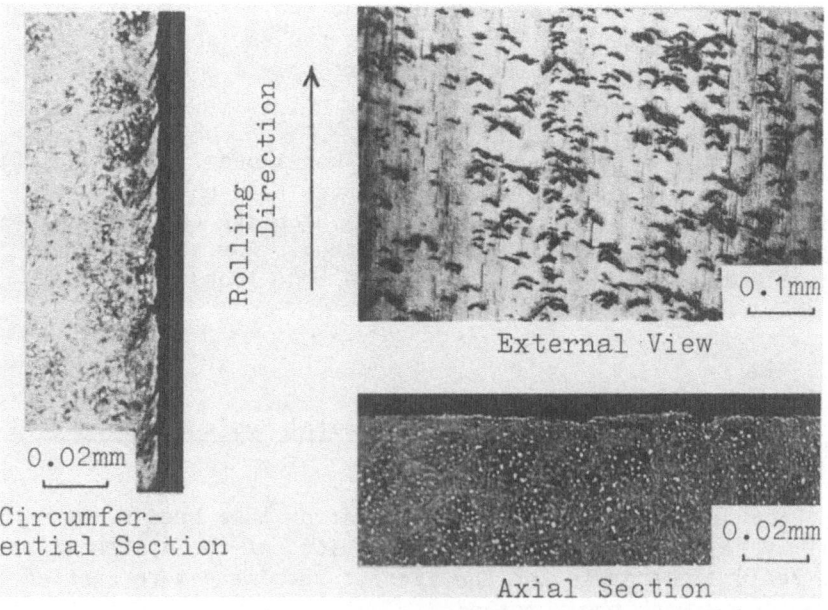

Fig. 4. External and sectional views of peeling on
        the specimen rolling-contacted with 20% slip.

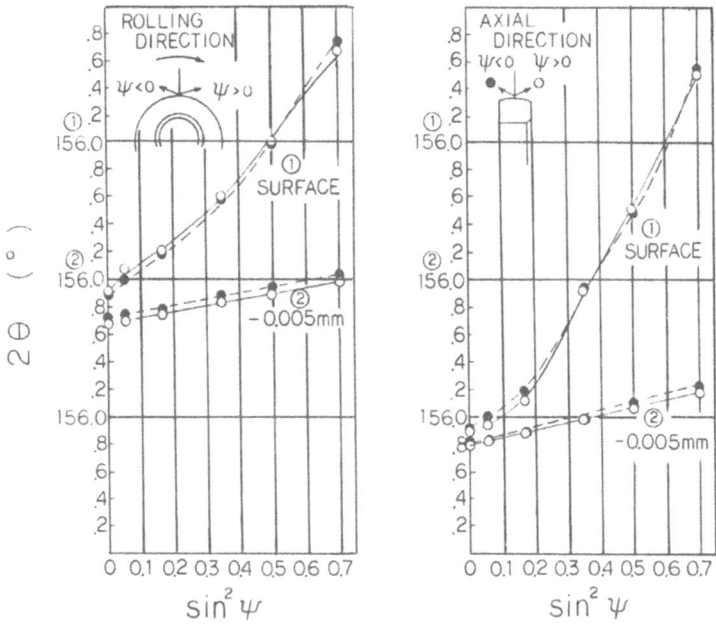

Fig. 5.   2θ-sin²ψ curves of non-used super-finished
specimen.

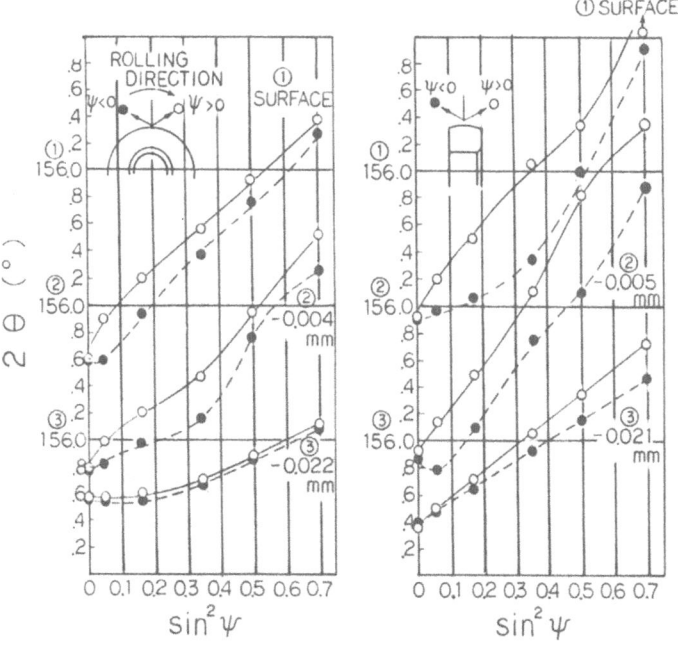

Fig. 6.   2θ-sin²ψ curves of the specimen rolling-
contacted under 0% slip condition.

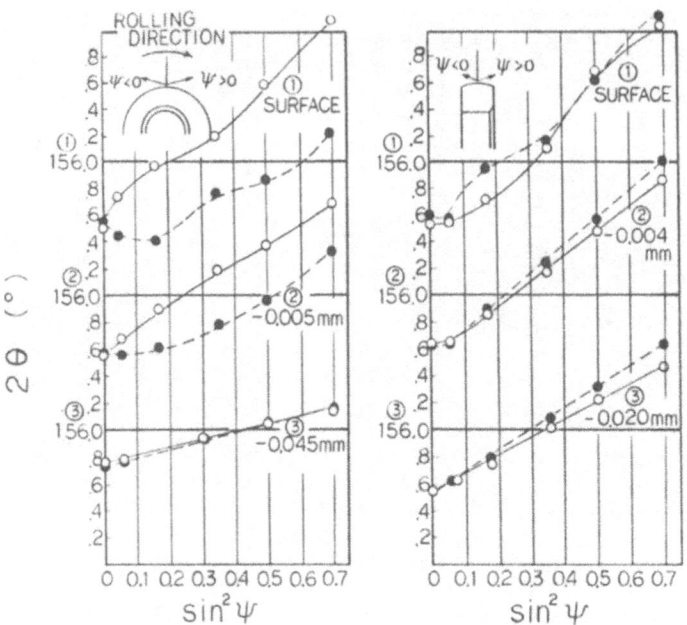

Fig. 7. $2\theta - \sin^2\psi$ curves of the specimen rolling-
contacted under 20% slip condition

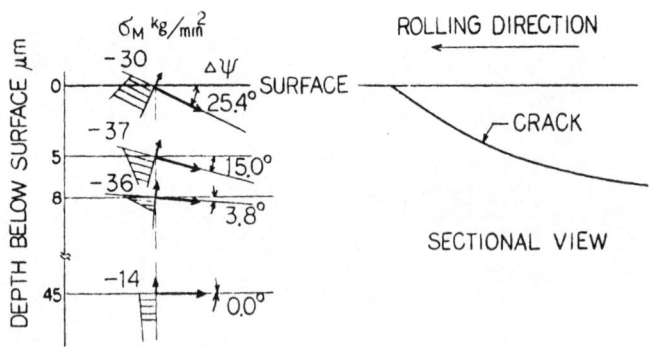

Fig. 3. Relation between changes of σ_M, $\Delta\psi$, and inclination
angle of a crack in a typical specimen.

inclination angles of the principal stress to the surface, $\Delta\psi$, and gradients of principal residual stress within the X-ray penetration depth, $\alpha\sigma$, which were calculated from the $2\theta\text{-}\sin^2\psi$ curve at the surfaces of specimens by the integral method, are shown in Table 1. The calculated inclination angles of the principal residual stress ($\Delta\psi$) correspond well to the observed inclination angles of cracks (θ).

Fig. 8 shows the relation between the changes of σ_M, $\Delta\psi$, $\alpha\sigma$ and the inclination angles of plastic flow or cracks toward the depth direction of a typical specimen. $\Delta\psi$ at each depth corresponds well to the inclination angle of plastic flow or cracks at that depth. Therefore it is considered that Peiter's integral method is valid.

Initiation Mechanism of Peeling Cracks

Since the residual stress is compressive in both circumferential and axial directions, as mentioned above, the principal residual stress normal to the plastic flow or cracks is tensile.

Table 1. Surface residual stress σ_M, inclination angle of residual stress $\Delta\psi$, and gradient of residual stress $\alpha\sigma$, calculated from $2\theta\text{-}\sin^2\psi$ curves at the surfaces of specimens calculated by the integral method.

Specimen	Stress Measuring Direction	Residual Stress State σ_M	$\Delta\psi$	$\alpha\sigma$	Crack Inclination Angle (θ)
% Slip	Circumferential	−45	4.4°	−11.4	$\approx$ 5°
	Axial	−75	12.0°	+ 4.8	10° $\sim$ 20°
10% Slip	Circumferential	−30	25.4°	− 5.7	25° $\sim$ 30°
	Axial	−71	0.05°	+12.7	$\approx$ 5°
20% Slip	Circumferential	−66	13.7°	+15.0	10° $\sim$ 15°
	Axial	−76	0°	+15.2	< 5°

σ_M: in kg/mm^2

$\Delta\psi$: in degrees

$\alpha\sigma$: in $kg/mm^2/\mu m$

The tensile principal residual stresses calculated from σ_M and $\Delta\psi$ using Mohr's stress circle are shown in Fig. 8. The calculated tensile principal stresses of the slightly peeled specimen shows its highest value at 18 kg/mm^2. This stress is almost enough for crack initiation based on stress cycles of small tensile stress, due to this principal residual tensile stress and large compressive stress, due to rolling contact stress.[5]

REFERENCES

1. M. Wakabayashi, M. Nakayama, and A. Nagata, J. of the Japan Society of Precision Engineering, 43:661 (1977).
2. G. Faninger and H. Walburger, Härterei-Technische Mitteilungen, 31, S 79 (1976).
3. W. Lode and A. Peiter, *ibid.*, 32, S 235 (1977).
4. W. Lode and A. Peiter, *ibid.*, 32, S 308 (1977).
5. H. Muro, T. Tsushima, and M. Nagafuchi, Wear, 35:261 (1975).

THE GENERALISATION AND REFINEMENT OF THE VECTOR METHOD FOR THE

TEXTURE ANALYSIS OF POLYCRYSTALLINE MATERIALS

Daniel Ruer, Albert Vadon, and Raymond Baro

Laboratoire de Métallurgie Structurale

Faculté des Sciences – Université de Metz

Ile du Saulcy – 57000 METZ (France)

ABSTRACT

A so-called "Vector Method" for the texture analysis of cubic materials was presented for the first time at this conference in 1976. Since then this method has been refined and applied success-fully to non cubic-materials. It is shown in this paper that the Vector Method provides several advantages over series methods of texture analysis, the most important of which being the relatively small amount of experimental data which are needed for the determi-nation of the entire crystallite orientation distribution.

INTRODUCTION

In this paper it is assumed that the basic elements of the Vector Method of texture analysis are known. Otherwise the reader is invited to consult previously published papers (1 to 3). In these publications it has been shown how the angle space can be partitioned into 2592 classes of orientations and how the same number of corresponding elementary pole figures can be calculated. The present article describes recent improvements of the method, namely generalization to crystal systems of low symmetry in part II. Part I considers improvements concerning the general computa-tional process with application to cubic crystals.

PART I

GENERAL IMPROVEMENTS

The crystallite orientation distribution is represented in the Vector Method by vector Y of N components y_n and a complete pole figure by a vector X of P components with $P \geqslant N$.

The fundamental equation :

$$X = [\sigma]\ Y \qquad\qquad (1)$$

is solved by an iterative process that can now be programmed for computers with small central unit memory of 32 K-words of 16 bits. The time necessary for an analysis is greatly reduced if a large computer memory is available (from a few hours to a few minutes). Furthermore only one program is necessary to calculate any arbitrary matrix $[\sigma(hkl)]$ in the case of cubic materials.

Each column vector of a matrix $[\sigma]$ represents an elementary pole figure which can be regarded as characteristic of the resolving power of the method. This is why we doubled the number of classes G_n (5184 instead of 2592).

Because of the time necessary to perform an analysis it is desirable to accelerate the convergence of the computing process as much as possible. Bearing this in mind, the preliminary determination of the empty classes of orientations before calculation results in increased accuracy. Programming is very simple and is realized by a subroutine called YNUL. This subroutine determines orientations which are not present from regions of zero pole density in a pole figure. The zero components y_n are solutions of the equations :

$$x_p = \sum_{n=1}^{N} \sigma_{pn}\ y_n \qquad\qquad (2)$$

of the linear system (1).

If a pole density $x_p = 0$ and if the matrix element σ_{pn} is positive, this implies that y_n must be zero.

This computing process has already been used in the method developed recently by Imhof (4). But it must be pointed out that in this method equation (2) appears as fundamental whereas in the Vector Method it represents merely an improvement whose efficiency is particularly important when one is dealing with very sharp textures.

EFFECT ON THE RESULTS

In order to show the effect of the improvements we present an example of texture analysis concerning an aluminum sheet of great purity, cold rolled with 95 % reduction rate. The experimental data were kindly communicated to us by K. Lücke. These data consist of three incomplete pole figures measured from 5° to 85° with reference to pole figure center. These pole figures, completed and normalized after the analysis, are presented in the left part of figure 1. The right part shows the recalculated pole figures after improvement of the method. It can be seen that the agreement with measured pole figures is very good. The amelioration of the results appears in Table 1, where the accuracy criterion R characterizes the difference between one measured and the corresponding recalculated pole figure (see ref. 1).

Table 1. Accuracy Criteria

	R(100)	R(111)	R(110)	$\bar{R}$
Before improvement	43.73 %	38.50 %	44.31 %	42.18 %
After improvement	16.78 %	11.75 %	16.04 %	14.86 %

The bad results obtained before improvements can be partly explained by the fact that the (111) pole figure was accidentally measured with the rolling direction in the wrong way. This error was discovered when we tried to perform the analysis from one incomplete pole figure. Thus the Vector Method has allowed us to verify the compatibility of the experimental pole figures. Figure 2 shows three times the (111) recalculated pole figure (with rolling direction in the right way and after the improvements mentioned before) when the analysis is performed from the incomplete (100), (111) or (110) measured pole figure alternatively. After this first correction the mean R value decreases from 42.18 % to 35.98 %.

Consulting Table 1 we can see that the better resolving power and the use of the subroutine YNUL do affect in a very significant manner the accuracy of the results.

CONCLUSION

The refinement of the Vector Method presented here was tested successfully by analysing cubic polycrystals. Of course, the same refinement must lead to similar success if the method is applied to materials of lower crystal symmetry. This will be verified in part II.

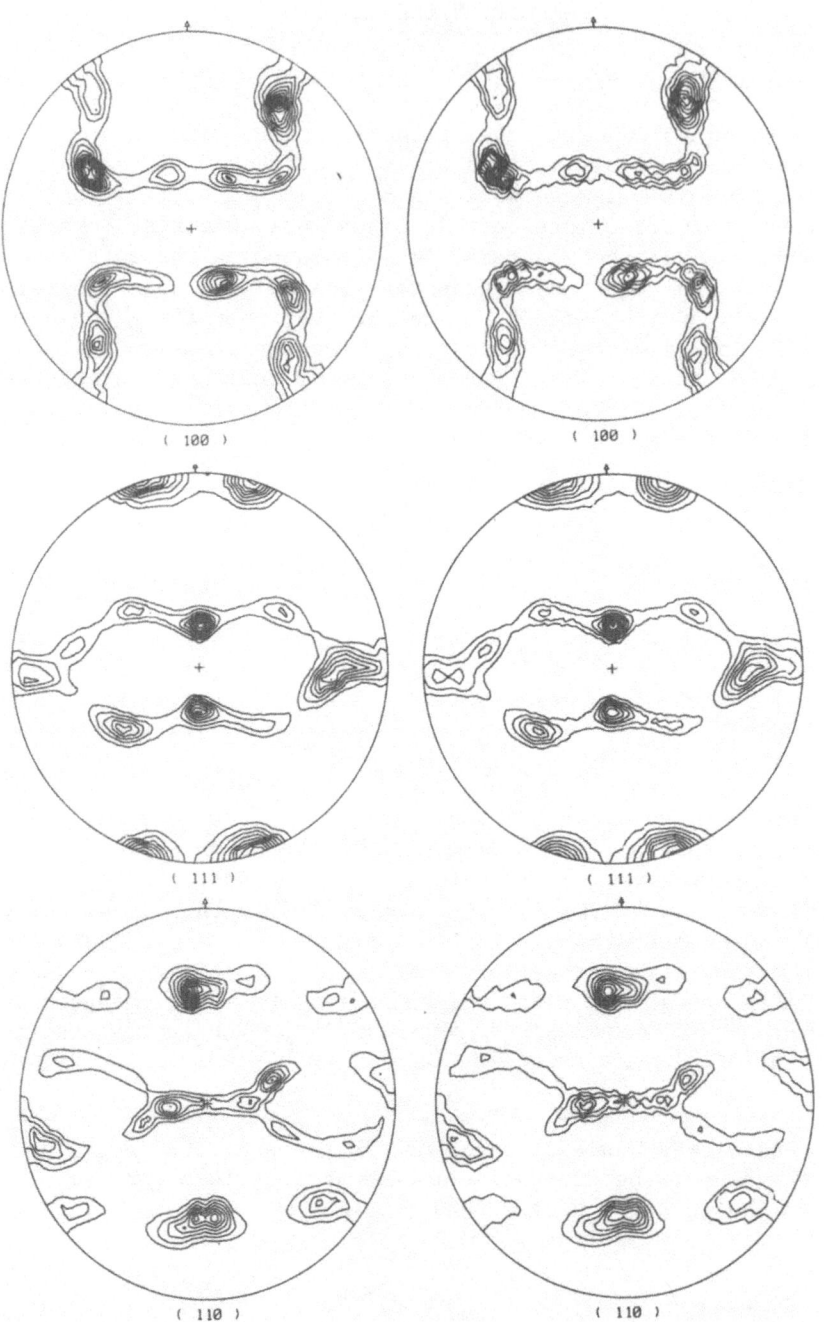

Figure 1
Left part : experimental pole figures completed and normalized after
the analysis.
Right part : recalculated pole figures. Level 1 =1.50 for all figures

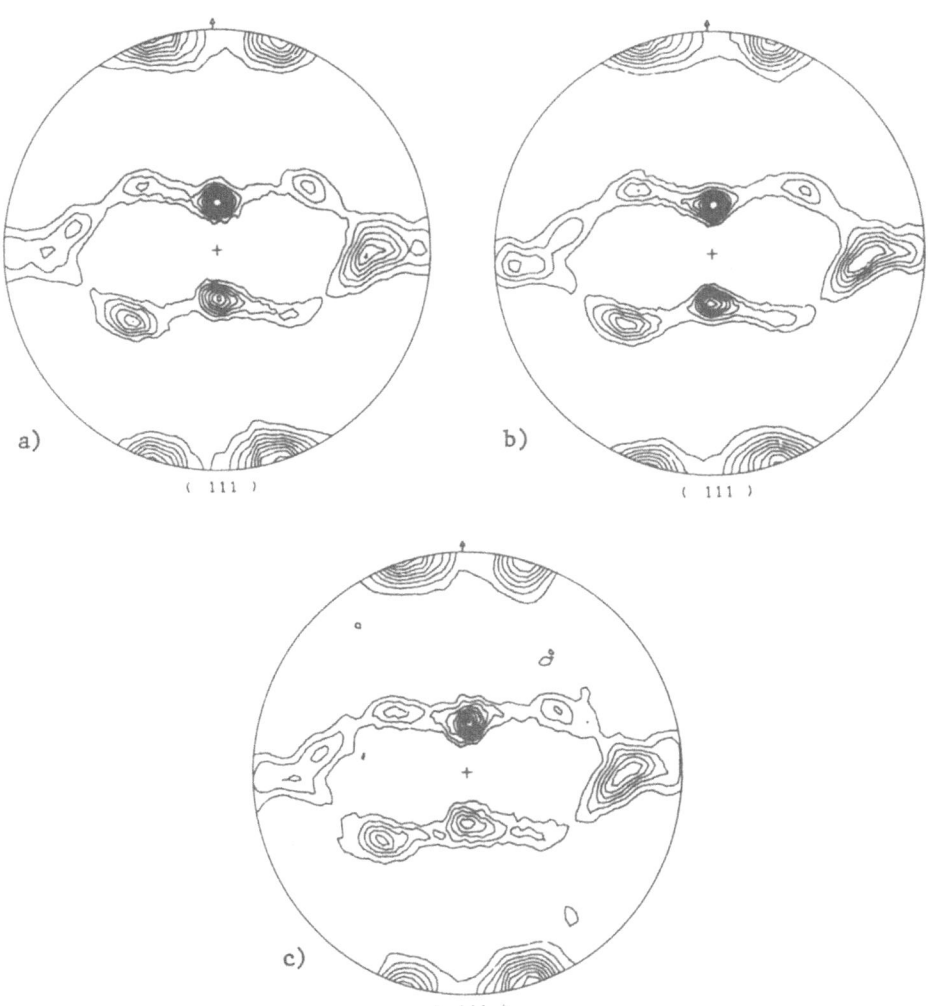

Figure 2
The same recalculated (111) pole figure when the analysis has been
performed from one incomplete measured pole figure :
a) from (100)
b) from (111)
c) from (110)

PART II

DIFFICULTIES IN THE TEXTURE ANALYSIS
OF TRICLINIC POLYCRYSTALLINE MATERIALS

The quantitative analytical procedures for determining the
O.D.F. or the axis distribution function for samples having an
axial symmetry, given by Bunge (5), Roe and Krigbaum (6) and others,
have the following unfortunate common feature : the resolution
obtained depends on the number of experimental pole figures necessary for the mathematical treatment. The number of terms in the
series depends primarily upon the symmetry class of the crystal.
Thus, the number of pole figures required to evaluate the coefficients up to $\ell = 22$ is 45 for triclinic crystals. Moreover,
according to Krigbaum (7), (8), in practice, the polynomial series
must be truncated prior to this theoretical limit because the
simultaneous equations become ill-conditioned for the higher coefficients, presumably because of experimental errors in the pole
figures. If the distribution of the plane normals is sharply peaked,
the representation becomes very bad.

To improve the resolution or reduce the number of necessary
experimental figures, Krigbaum (7) proposes a refinement which,
from 17 experimental figures, allows the computation of 28 additional figures.

All these difficulties, both experimental and theoretical, led
us to generalize the new vector method - applied so far only to the
cubic system - to any crystal system, and to begin with, to the
triclinic system.

GENERALISATION OF THE CONCEPT OF POLE FIGURES

A pole figure represents the distribution of the normals to a
given crystallographic plane (hkl) on the positive hemisphere of
the pole sphere. In the triclinic system, there is only one pole
in the pole figure for each orientation of the crystallite in the
sample.

Since to define an orientation the minimal number of poles
cannot be lower than three, and as in the triclinic system there
is only one pole for each figure, three figures are needed.

Any set of three distinct and complete pole figures will be
called "minimal pole density set " (M.P.D.S.). Any M.P.D.S. allows
the computation of the texture.

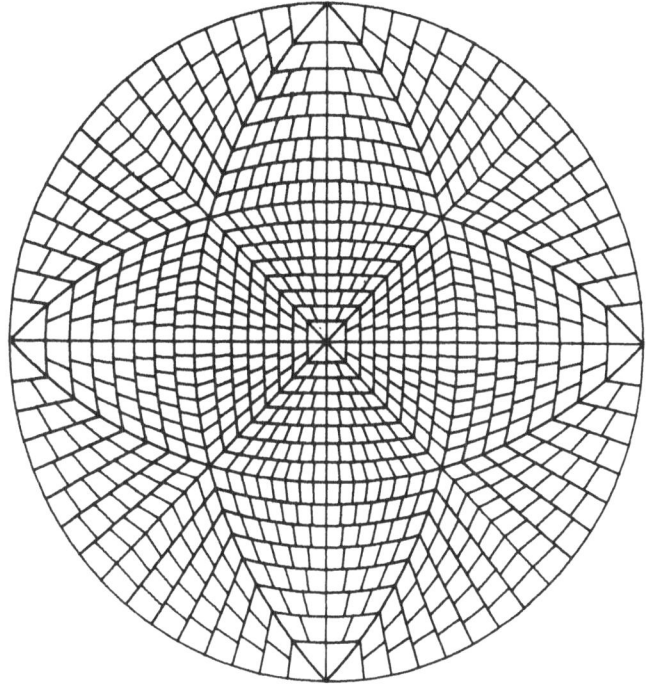

Figure 3
Partition of the space of the fiber axis for triclinic polycrystals

PARTITION OF THE ORIENTATION SPACE

To ensure an analogy in the treatments and an easy correspon-
dence between the orientations of different crystal systems we have
chosen the same partition of the unit triangle as for the cubic
system. It has then been extended to all crystal systems including
the triclinic system which has the largest orientation space. In
this case, the inverse pole figure covers the whole hemisphere,
that is 24 unit triangles (figure 3).

COMPUTATION

The hemisphere of poles is partitioned into 864 concentric
domains corresponding to the macroscopic axial symmetry. Thus, 864
experimental values taken from a M.P.D.S. are required. Each experimental pole figure must then be partitioned into 288 spherical
zones, each having an angular width of 0.3125 degree.

To calculate the matrices of probability density σ(hkl) of
each (hkl) pole figure, we have used a fast and accurate new method

of scanning the pole sphere. The mean errors are respectively
0.0001 % on the rows and 0.01 % on the columns. The computing time
is 30 min on a UNIVAC 1100. These matrices are computed once and
stored.

The matrix of one M.P.D.S. is built by appending three
$\sigma(h_i k_i l_i)$ matrices. The system obtained reads :

$$
\begin{bmatrix} X\ (h_1\ k_1\ l_1) \\ X\ (h_2\ k_2\ l_2) \\ X\ (h_3\ k_3\ l_3) \end{bmatrix} = \begin{bmatrix} \sigma\ (h_1\ k_1\ l_1) \\ \sigma\ (h_2\ k_2\ l_2) \\ \sigma\ (h_3\ k_3\ l_3) \end{bmatrix} \begin{bmatrix} Y^* \end{bmatrix} \tag{3}
$$

APPLICATIONS

We analysed the texture of a polyethylene teraphthalate sample
thanks to experimental data kindly given to us by W. R. Krigbaum.
Since the sample had an axial symmetry, only that case was considered.

The program allows the calculation of the texture vector Y^*
and the comparison between the experimental figures and the figures
recalculated with that vector Y^*.

Computing time : 6 min 30 s

Average number of iterations : 60 to 70 (according to the
number of nul components).

Then, knowing Y^*, to check the validity of the method and the
accuracy of the results, we calculated and re-drew several (hkl)
figures and compared them to the experimental figures.

RESULTS

The first M.P.D.S. chosen was $\{(1\bar{1}1), (100), (010)\}$ for sample
S1. It was chosen at random, among C_3^{17} combinations at our disposal.
We had no criterion of choice, a priori.

The results of the analysis can be seen in figure 4, left
column. The three experimental curves (thin line) and computed
curves (thick line) number 1, 2, 3 show the good agreement between
the experimental data and the computed values. On the contrary,
the curve number 4 corresponding to the experimental and computed
$(0\bar{2}4)$ pole figures show a not very good agreement.

We insist on the fact that this $(0\bar{2}4)$ pole figure is not

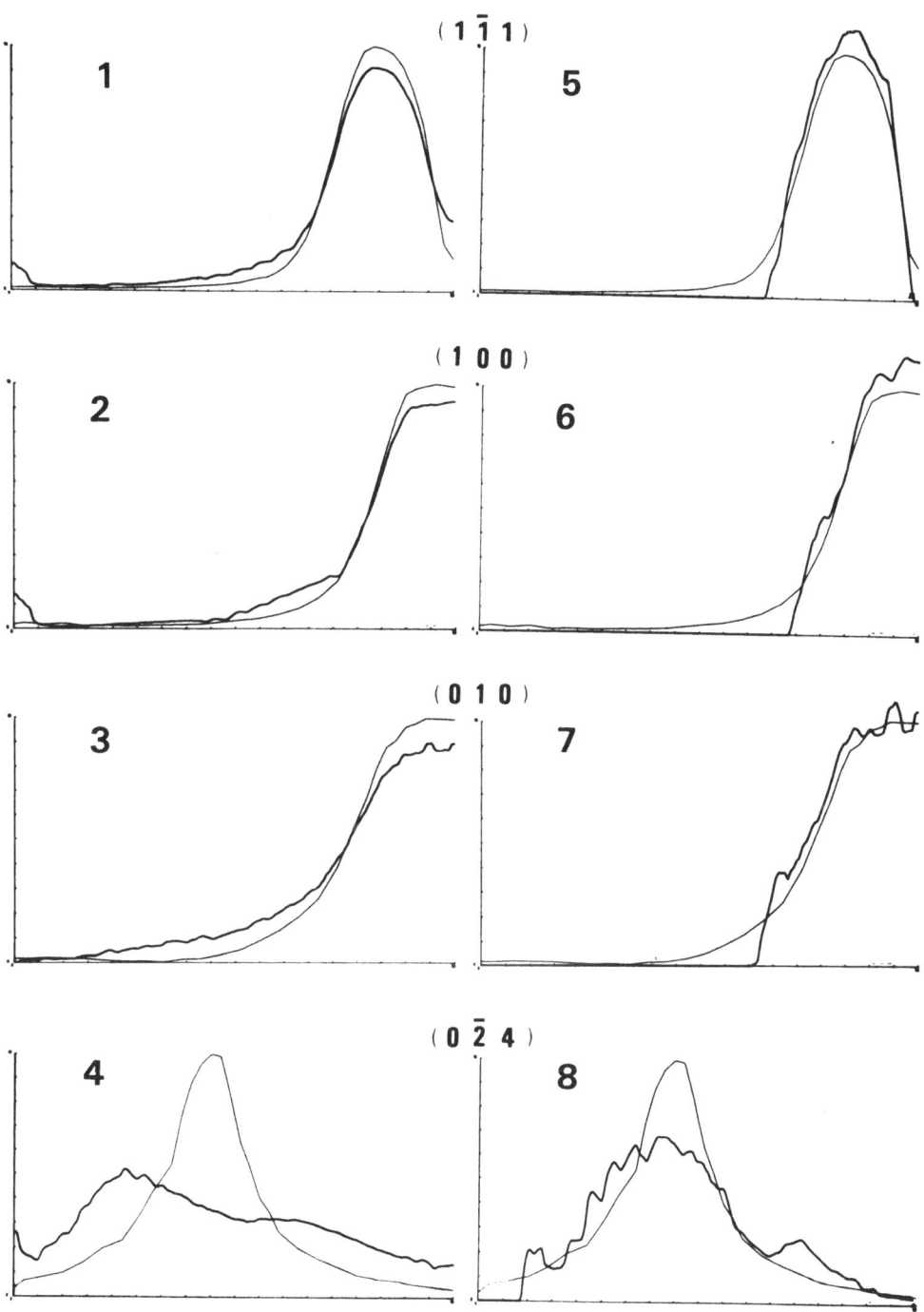

Figure 4 : Sample S1 - M.P.D.S. = {(1$\bar{1}$1), (100), (010) }
    thin line   experimental curve - thick line : computed curve
        Left : deletion ε = 0 %   Right : deletion ε = 15 %

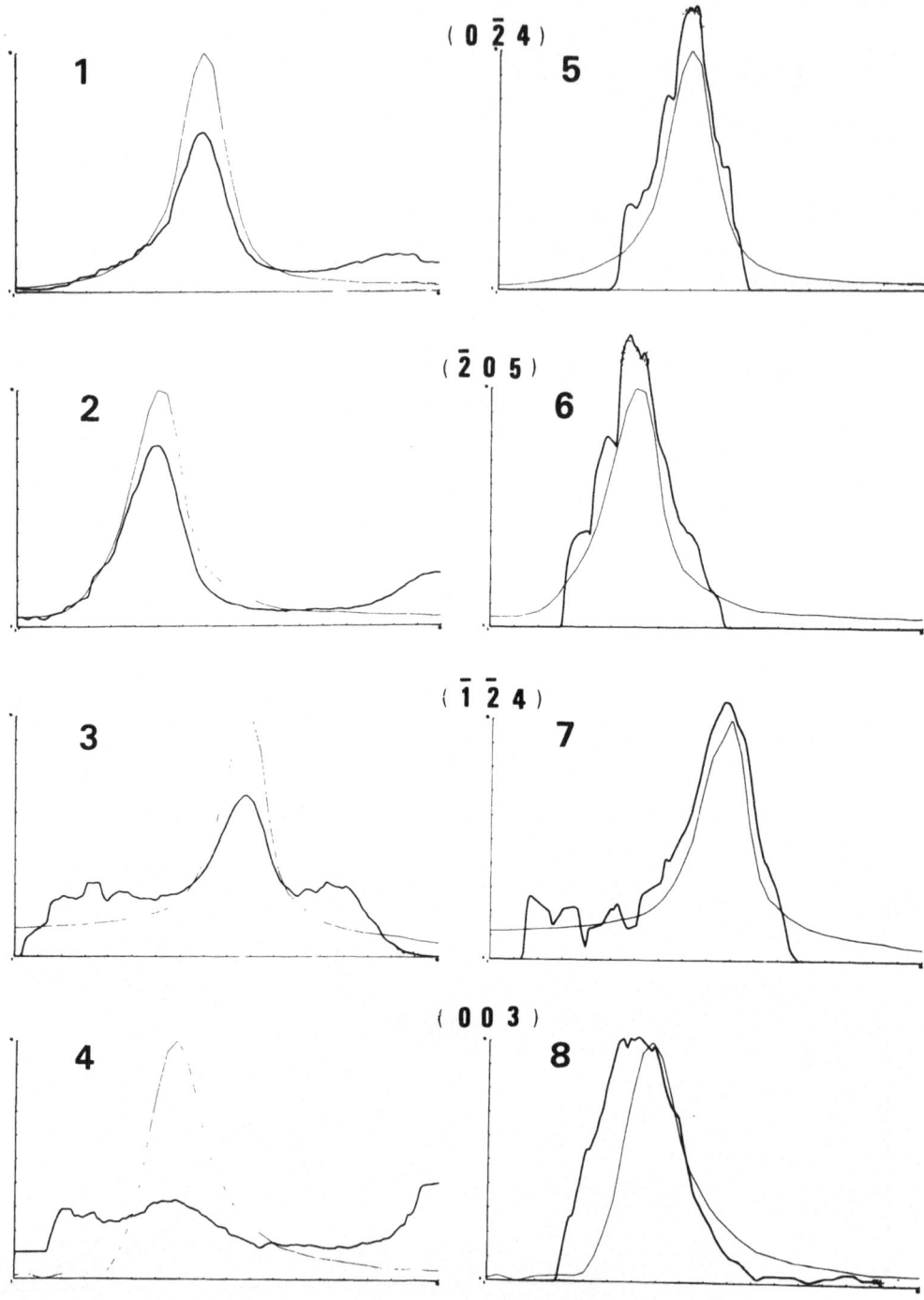

Figure 5 : Sample S2 - M.P.D.S. = {(0$\bar{2}$4), ($\bar{2}$05), ($\bar{1}\bar{2}$4)}
thin line : experimental curve - thick line : computed curve
Left : deletion ε = 0 %   Right : deletion ε = 12 %

included in the M.P.D.S. chosen for the computation.

To improve the results, we replace the condition $x_p = 0$ of relation (2) by condition $x_p \leqslant \varepsilon_{hkl}$, where ε_{hkl} is a given ratio of the experimental peak value of the {hkl} pole density. Thus, the nearly nul components are deleted.

The best result can be seen in the right part of figure 4. It has been obtained with a deletion $\varepsilon = 15\%$ of the experimental peak value. The three pictures number 5, 6, 7 show as good an agreement as pictures 1, 2, 3, between the data and the computed values. For the $(0\bar{2}4)$ computed pole figure, the peak is correctly situated and the computed maximum is 68 % of the experimental peak value.

It then occured to us that we would get better results by choosing the M.P.D.S. not at random but according to a certain criterion. Two criteria could be considered : the reflecting coefficients of the (hkl) planes and the geometrical separation of the BRAGG reflections. The second M.P.D.S. chosen was $\{(0\bar{2}4), (\bar{2}05), (\bar{1}24)\}$ with BRAGG angles of 21.36°, 24.27°, 23.69° respectively, which are larger than for the first M.P.D.S. (13,92°, 12.84°, 8.76° respectively). This second M.P.D.S. concerns sample S2.

Figure 5 shows the results obtained.

On the left, although the double curves number 1, 2, 3 show a good agreement, as concerns the additional {003} pole figure, the agreement is very bad. On the contrary, it becomes very good with a 12 % deletion (right part of the figure).

The last picture shows that :
- the peak is correctly situated, only slightly shifted.
- the computed peak value now reaches 100 % of the experimental peak value.

CONCLUSION

The results obtained show that :

- neither 45, nor even 17 experimental pole figures are required to calculate the texture of a triclinic material, three figures are sufficient.

- the accuracy of the result depends mainly on the quality of the three figures chosen for the minimal pole density set. Many of the C_3^{17} combinations lead to unsatisfactory results. This proves that in that case the three initial figures are inconsistent. Thus, the vector method is also a consistency test for the experimental pole figures.

- rather than have a large number of experimental figures, it is better to make only a few, well made, and to choose carefully the (hkl) planes.

It appears that the main advantages of the vector method lie in the small number of experimental pole figures required and in the accuracy of the results obtained.

ACKNOWLEDGMENTS

The authors gratefully express their thanks to Professor K. Lücke of Aachen (West Germany) and Professor W.R. Kriegbaum of Durham (USA) for the communication of experimental data as well as to the Director of the Station d'Essais in Maizières-les-Metz (France) and the Head of the Ecole Nationale d'Ingénieurs of Metz (France) for the computation facilities.

BIBLIOGRAPHY

1 - D. Ruer and R. BARO, "A New Method for the Determination of the Texture of Materials of Cubic Structure from Incomplete Reflection Pole Figures", Adv. in X-Ray Analysis, 20 : 187,1977

2 - D. Ruer et R. BARO, "Méthode Vectorielle d'Analyse de la Texture des Matériaux Polycristallins de Réseau cubique", J.Appl. Cryst., 10 : 458, 1977

3 - D. Ruer and R. Baro, "Practical Advantages of the Vector Method of Texture Analysis", Textures of Materials, 1 : 169, Springer Verlag, 1978.

4 - J. Imhof, "An Approximative Determination of the Orientation Distribution Function", Textures of Materials, 1 : 149, Springer Verlag, 1978

5 - H.J. Bunge, "Mathematische Methoden der Textur Analyse", Akademie Verlag, Berlin, 1969

6 - R.J. Roe and W.R. Krigbaum, "Crystallite Orientation in Materials Having Fiber Texture", J. Chem. Phys., 40 : 2608, 1964

7 - W.R. Krigbaum and A.M. Harkins Vasek, "A Test of the Refinement Procedure for Determining the Crystallite Orientation Distribution : Polyethylene Terephthalate, J. Texture, 1 : 9, 1972

8 - W.R. Krigbaum, "A Refinement Procedure for Determinig the Crystallite Orientation Distribution Function", J. Phys. Chem., 74 : 1108, 1970

THE FITTING OF POWDER DIFFRACTION PROFILES TO AN ANALYTICAL EXPRESSION AND THE INFLUENCE OF LINE BROADENING FACTORS

Allan Brown

Studsvik Energiteknik

S-611 82 Nyköping, Sweden

J. W. Edmonds

Dow Chemical Company

Midland, MI 48640, USA

ABSTRACT

Powder diffraction profiles of well crystallized compounds can be fitted to distributions of the type $I_\theta = I_o (1 + k^2 x^2)^{-n}$ where k is a scale factor related to the half width of the profile. The value of n varies with the diffraction angle, 2θ, and is generally different for the low-angle and high-angle sides of the same profile. Limiting values of n for a specific Guinier camera-microdensitometer combination are $1.2 \leq n \leq 2.3$. Similar values are obtained for diffractometer profiles after $K\alpha_2$ stripping. Line broadening due to departure from perfect crystallinity in the specimen affects the value of n as well as that of k.

The above observations are interpreted in terms of the convolution of a Gaussian with a Lorentzian distribution, the exponent n of the convolute being dependent upon the relative half widths of these two functions, expressed as the ratio b_L/b_G.

1. INTRODUCTION

Analytical expressions for the observed diffraction profiles of powder specimens date from Jones who gave

$$I_\theta = I_o (1 + k^2 x^2)^{-2}$$

as an approximation to a microphotometered line profile at $160°$, (2θ) where the $K\alpha_1$ and $K\alpha_2$ components were well resolved (1). In this expression

I_θ = intensity at angle θ

I_o = peak maximum intensity at θ_o

x = angular distance $\theta_o - \theta$ from peak position

k = a scale factor which can be related to the half
        width b_2 of the profile.

Many workers have since agreed with Jones by adopting this
expression for the experimental profile, at least for well crystallized materials. Its popularity is such that it has been designated
the modified Lorentzian or ML distribution as distinct from the
simple Lorentzian distribution described by

$$I_\theta = I_o(1 + k^2x^2)^{-1}$$

Thus, a good fit to the ML distribution was obtained for
diffractometer measurements on the low angle side of the $K\alpha_1\alpha_2$
doublet where it was assumed that the $K\alpha_2$ contribution was negligible (2). More recently, microdensitometry of Guinier powder
patterns, recorded with subtractive geometry and $K\alpha_1\alpha_2$ radiation,
gave good agreement with the ML function in the angular region
about 30° (2θ), (3). With additive Guinier geometry and strictly
monochromatic radiation the ML function was found to give a more
satisfactory fit to powder profiles than either the pure Lorentzian
or the pure Gaussian (normal) distribution (4).

In the present context the Gaussian distribution is written

$$I_\theta = I_o \exp (-x^2/a^2)$$

which can be approximated by an expression

$$I_\theta = I_o(1 + k^2x^2)^{-n}$$

where the scale factor $k = 1/a\sqrt{n}$ and n is a large number.

In practice the difference is marginal between distributions
for which $n > 10$ and is detectable only in the tail region, as can
be verified by evaluating the half width at $1/p$ of the peak maximum
from

$$b_p = a[n(p^{1/n} - 1)]^{\frac{1}{2}}$$

Thus, for the half width at half maximum height, $p = 2$ and
for $n = 1, 2, 10$ and 50 the corresponding values of b_2/a are 1.0,
0.9101, 0.8472 and 0.8358 respectively while for the Gaussian
distribution $b_2/a = 0.8325$. In terms of the half width b_2 it seems

fair therefore to regard the ML function, with n = 2, as lying
midway between the Lorentzian and Gaussian extremes, as illustrated
in Figure 1.

Alternative expressions related to the ML function have been
put forward as more suited for profile fitting. Thus, the distribution with n = 1.5 was recently described as giving a more satisfactory fit for an accurate diffractometer step scan in which
monochromator selected CuKβ radiation was used (5). Then again the
function with n = 3 has been fitted to both sharp and broadened
002 profiles of martensite (6).

2.   CHANGES OF PROFILE SHAPE WITH
θ IN GUINIER POWDER PATTERNS

The present work began with unsuccessful attempts to fit
powder profiles to the ML function using the method described in
(4). The patterns had been recorded photographically in a Guinier-
type camera with subtractive geometry and aligned for strictly
monochromatic $CuK\alpha_1$ radiation. The pattern was retained on only
one side of the film. The specimens were well crystallized
$Pb(NO_3)_2$, α-quartz obtained by crushing a defunct monochromator
crystal, and a mixture of silicon and ThO_2. Details of the microdensitometer technique used to record, evaluate and display the
results have been reported earlier (7).

It was quickly observed that while reflections in the middle
region of the pattern were reasonably symmetrical the brush effect
led to pronounced asymmetry for profiles with 2θ < 25°. This is
reflected in the appearance of a marked tail which stretches
towards the low angle end of the profile. A similar but less pronounced tail was observed for reflections with 2θ > 70°. In this
instance the tail stretches towards the high angle cut off. It was
accordingly decided to fit the two halves of each profile separately,
using independent values of the half width. This measure did not
prove altogether effective, however, in reducing the error to
acceptable levels for all the reflections. The exponent n was
therefore allowed to vary freely for the separate halves of each
profile.

Good agreement was now obtained with R factors lying in the
range 1.5 to 5 % with R defined as

$$R = 100 \sqrt{\frac{\sum_i w_i^2 \, (I_{Obs} - I_{Calc})^2}{\sum_i w_i^2 \, I_{Obs}^2}}$$

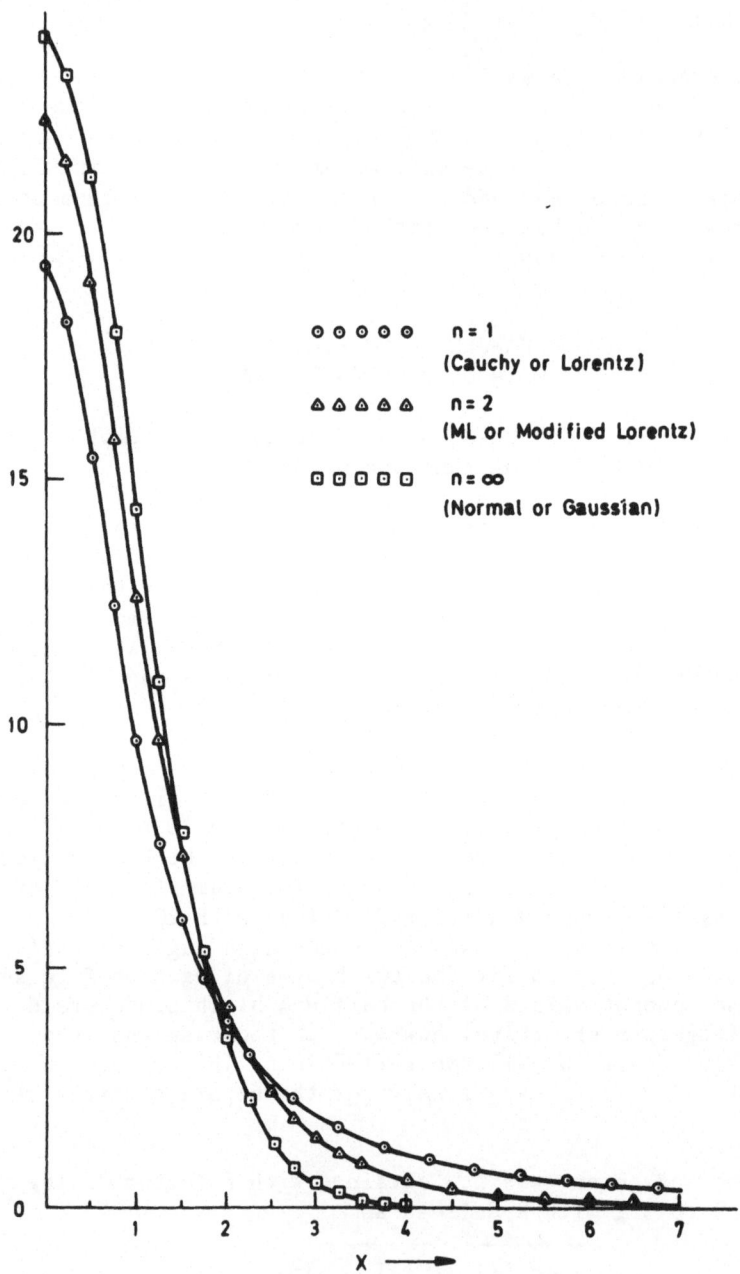

Figure 1

Distribution curves of the form $I_\theta = I_0(1 + k^2x^2)^{-n}$. Curves are plotted for constant area.

where i, the number of points measured in the half profile, is
15 - 30 depending upon the half width of the line, and ω was set
equal to $1/I_{Obs}$ up to a limit of $I_o/100$.

Figure 2 shows the plot of the exponent n and the half width
b_2, as functions of 2θ for the pattern of $Pb(NO_3)_2$. The reflections
used to give this plot were generally of medium to strong intensity.
It was noticeable that five weak reflections in the pattern were
associated with much higher values of n (> 3) than those shown
here. This can be attributed to poor development in the tail
region of the profile which, in combination with the response
characteristics of the film, leads to a sharp cut off in the
measurable intensity at distances of about 2 half widths from the
peak maximum.

The profile of the 111 reflection of $Pb(NO_3)_2$ at 19.5^o is
shown in Figure 3. This provides the extreme limits of n with
values of 1.2 and 2.3 for the low and high angle sides respectively.
The results plotted in Figure 2 demonstrate that for the low angle
side n increases as a continuous function of 2θ towards a limit of
1.9. For the high angle side n falls with increasing rapidity to
1.6 at the high angle cut off provided by the 731 reflection at 99.8^o.

The low angle half width is 85 % of the high angle value in
the 111 profile. The two half widths fall to the same value,
within the limits of error, in the angular range 30° - 70° (2θ)
which is the region of maximum line definition. Beyond 70° the
high angle side becomes detectably broader than the low angle side
of the profile so that for the 731 reflection b_ℓ is roughly 90 %
of b_h.

3.   CHANGES OF PROFILE SHAPE
WITH SPECIMEN CONDITION

The results presented in Figure 2 are not generally applicable to all diffraction patterns. In addition to the discrepancies
noted for very weak reflections it was observed that values of n
different from those shown here were given by profiles in powder
patterns exhibiting line broadening. In order to avoid effects
arising from the poor development of profile tails in weakly
exposed photographs,use was made of a step scanning powder diffractometer to record broadened profiles for analysis. The strategy
that proved most suitable as regards the consistency of the results,
was to use uniformly short steps of 0.01^o (2θ) and 1 - 10 s counting
times depending upon the counting rate for peak versus background.
After correcting the background level, the $K\alpha_2$ component was
stripped from the combined profile (8) and the residual profile
divided for analysis into low and high angle halves about the peak
position.

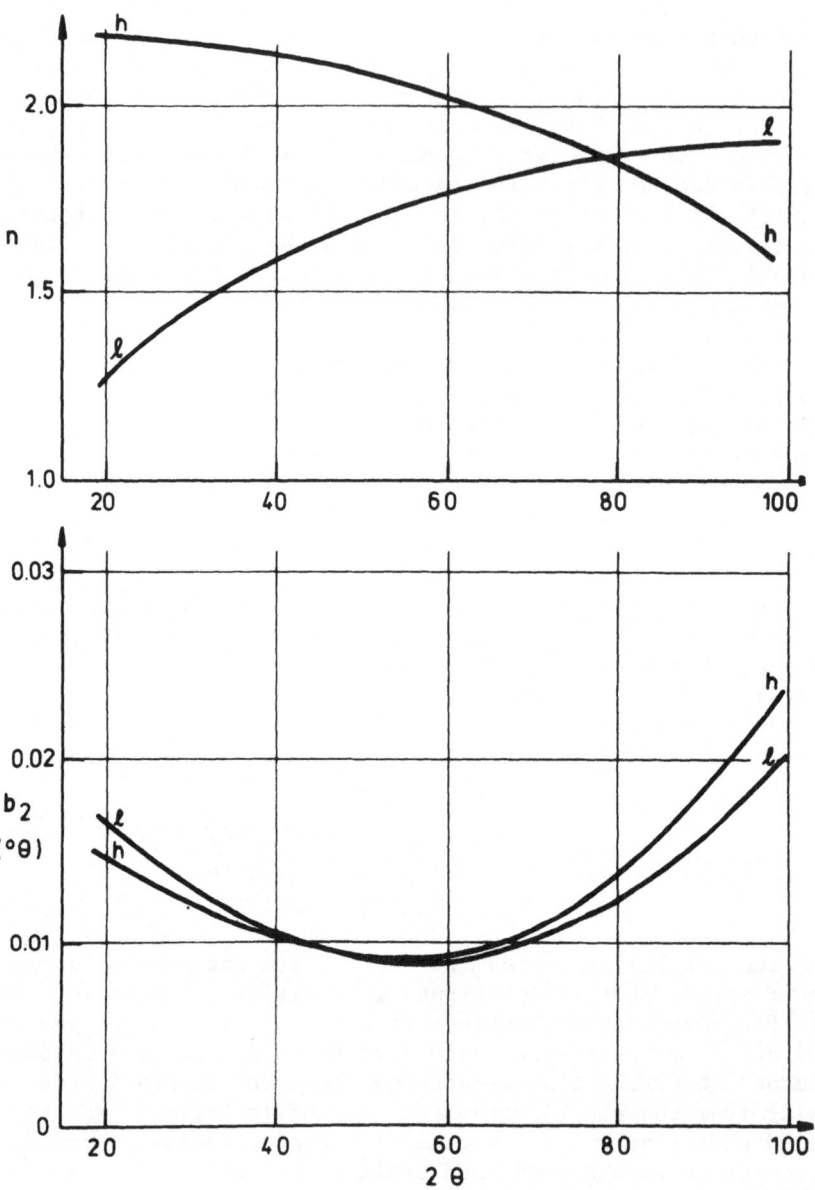

Figure 2

Curves of n and b$_2$ versus 2θ for the low and high angle sides (ℓ and h curves) of Guinier profiles.

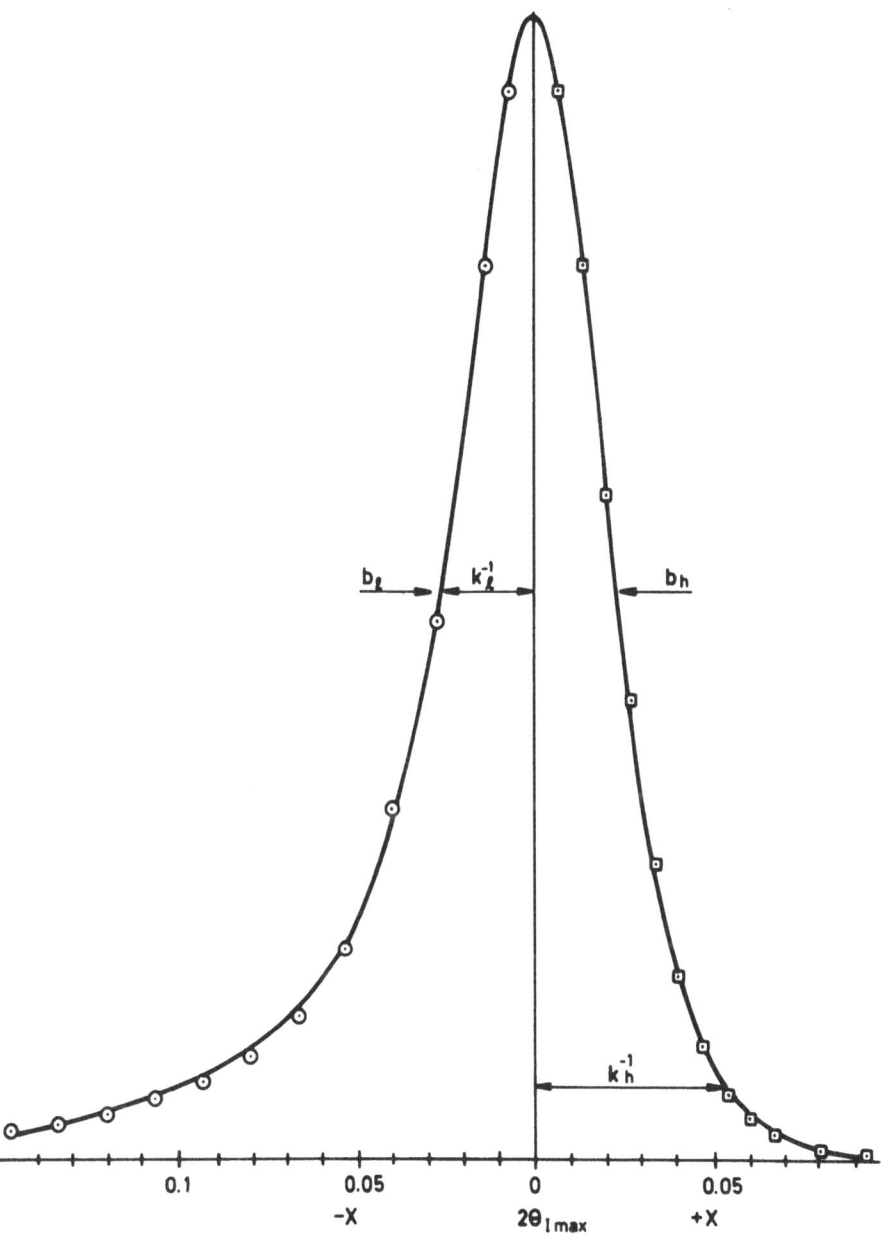

Figure 3

Intensity distribution for the 111 profile of Pb(NO$_3$)$_2$ at 19.5° (2θ). Rings and squares represent values for low and high angle data calculated from b$_L$ and b$_G$ in Table 3.

The material selected for study was the $Pb(NO_3)_2$ used to obtain the data of Figure 2. The specimens were made up by side-filling a special shallow cup which was spun in the Bragg-Brentano focusing plane (9). The specimens were (a) as precipitated from solution, (b) lightly crushed in an agate mortar, (c) heavily ground in the mortar, (d) left in the unfiltered X-ray beam for 48 h at a dose rate of ~ 0.5 Mrad/h and (e) left in the X-ray beam for 96 h at this dose rate.

The comparison data in Table 1 illustrate the similarities and differences of Guinier and diffractometer results at the upper and lower angular limits shown in Figure 2 and at a point roughly midway in this range. The chief difference is that in the range $30 - 70^{\circ}$ (2θ) the Guinier profiles are about half the width of the diffractometer profiles. They also feature a slightly less extensive tail (greater value of n) except in the angular range beyond 70°.

The results listed in Table 2 for the 731 reflection show that line broadening produced by lightly crushing the sample is accompanied by a decrease in n. With continued and more vigorous grinding the exponent increases again to a value which exceeds even the initial value while broadening also increases. A similar but less marked reversal in the value of n is observed as between the 48 h and 96 h irradiations in the unfiltered X-ray beam. In this instance line broadening is initially smaller than that after light grinding whereas the development of the tail region is considerable (n ~ 1). The marked asymmetry which is reflected in the half width ratio after the 96 h irradiation is accompanied by a peak shift of -0.1° (2θ). This corresponds to an increase of 0.0078 Å in the d spacing. In the other cases peak shifts are either very small or within the limits of experimental error.

4.   DIFFRACTION PROFILES AS CONVOLUTES OF A GAUSSIAN AND A LORENTZIAN FUNCTION

Jones proposed that broadened profiles are convolutes of an instrument and an intrinsic function (1). The observed profile is then written

$$h(x) = \int_{-\infty}^{+\infty} f(u) \cdot g(x - u) du$$

where

$f(u)$ = the pure or intrinsic profile given by the crystallites that make up the X-ray specimen

$g(u)$ = the instrument profile which is a function of all the aberrations and physical processes associated with the diffracting system, from source to measuring device.

Table 1

Comparison of profiles recorded in a Guinier-type focusing camera and in a powder diffractometer.
Material: $Pb(NO_3)_2$ precipitated from solution with EtOH.
Radiation: $CuK\alpha_1$.

Reflection and Angle (2θ)	Instrument	Half width in $^o\theta$		Exponent	
		b_ℓ	b_h	n_ℓ	n_h
111 at 19.5^o	Guinier	0.017	0.015	1.2	2.3
	Diffractometer	0.021	0.015	1	2.3
511 at 61.3^o	Guinier	0.008	0.009	1.7	2.0
	Diffractometer	0.018	0.018	1.8	1.7
731 at 99.8^o	Guinier	0.019	0.022	1.9	1.6
	Diffractometer	0.022	0.018	1.6	1.2

Table 2

Data for the 731 profile of $Pb(NO_3)_2$ at 97° (2θ).
Instrument: Step-scanning powder diffractometer with graphite monochromator.
Radiation: $CuK\alpha_1$ after $K\alpha_2$ stripping.

Condition of specimen	Half width in $^o\theta$		Exponent	
	b_ℓ	b_h	n_ℓ	n_h
Freshly precipitated	0.022	0.018	1.6	1.2
Lightly crushed	0.078	0.077	1.3	1.3
Heavily ground	0.122	0.089	2.3	1.5
48 h irradiation, 5 Mrad	0.035	0.038	1	1.1
96 h irradiation, 10 Mrad	0.058	0.127	1.1	1.4

The function f(u) is described as having a Lorentzian distribution when the specimen is made up of crystallites less than 0.3 μm in size and is expected to be Gaussian when the specimen contains microstrains (1, 10). In both cases the contributions are recognized as having half widths that are dependent upon θ.

The above treatments assume that g(u) is an ML function with a half width dependence upon 2θ. Convolution synthesis to simulate diffractometer profiles at various angles indicates, however, that even its shape varies with 2θ (11). Contributions to g(u) are made by seven functions g_I - g_{VII} which represent different physical processes associated with diffraction. The source is regarded as making a constant, near Gaussian contribution, whereas chromatic dispersion has a Lorentzian distribution with a half width contribution that increases as 0.0424 tan θ for CuKα radiation. Axial divergence is also angle dependent and moreover asymmetric. This function describes the observed change in profile symmetry with 2θ and is associated with the brush effect, seen at low angles in powder photographs recorded with the line focus (12).

Profiles generated by convolution synthesis do not appear to provide such a good fit to observed profiles as those obtained in the present study. This might be explained if not all the instrumental factors had been identified correctly, a possibility that underlines the difficulties of the synthesis approach. This is unimportant, however, in view of the way in which synthesis stresses that a change in the balance of angle dependent aberrations will affect the shape of the convolute.

To test the possibility that fractional values of n, as obtained for the Guinier profiles plotted in Figure 2, might be the result of a convolution process, pairs of functions of the form

$$I_\theta = I_o (1 + k^2 x^2)^{-n}$$

with different values of n and of k were generated in the computer and then convoluted. The convolute was then analysed, using the programme already described, to determine the value of the half width, b_H and of the exponent, n_H. It was soon realized that all the results obtained could be explained by convolution of functions g_I g_{II} with n_I = 1 and n_{II} = 10 as long as the scale factors k_I and k_{II} assumed a sufficiently wide range of values.

Restated, this means that the distributions, for which 10 > n > 1, are hybrids formed by the convolution of a Lorentzian with a Gaussian function. The value of n which results depends upon the half width ratio b_L/b_G for these two input functions and varies continuously as this ratio is varied between chosen numerical limits. The curve of n as a function of b_L/b_G is shown in Figure 4 together with b_H/b_G the half width ratio of the convolute referred to that of the input Gaussian.

A significant feature of Figure 4 is the rapid decrease of n as b_L/b_G increases towards unity. When $b_L/b_G = 1$, $n = 1.4$ which emphasizes the dominant influence of the Lorentzian function on the form of the overall convolute. The ML function, with $n = 2$, is seen to correspond to a half width ratio $b_L/b_G = 1/1.8$ while the Lorentzian limit close to unity is reached for $b_L/b_G > 6$.

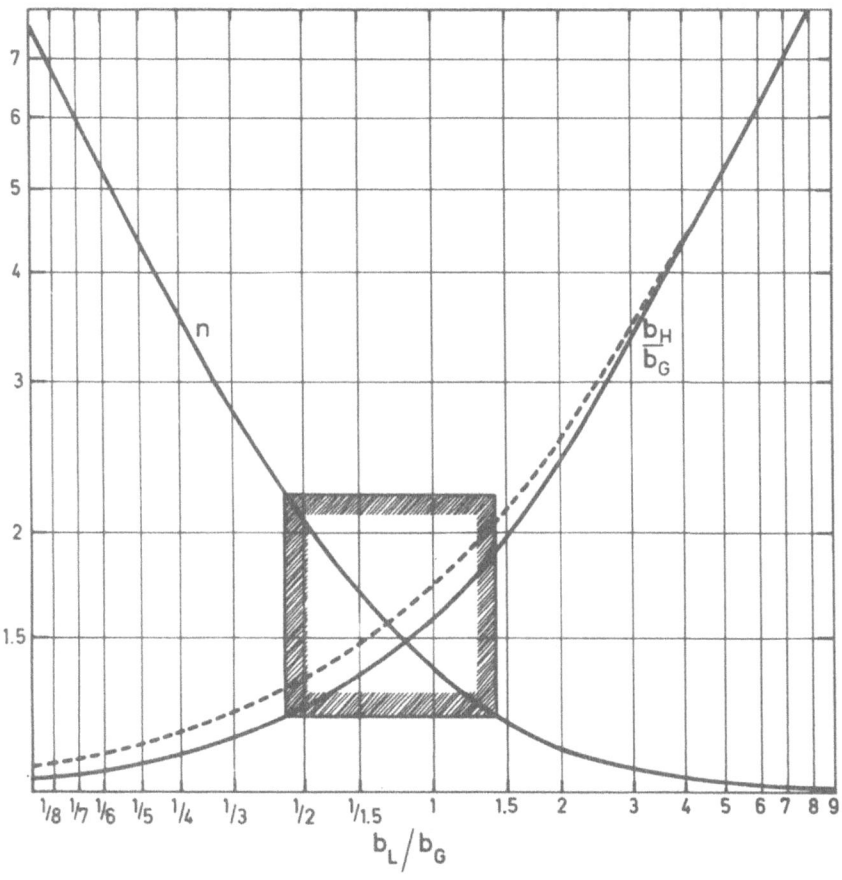

Figure 4

Curves of n and half width ratio bH/bG versus bL/bG for convoluted Lorentzian and Gaussian functions.

The rectangle whose upper and lower edges are defined by
n = 2.3 and 1.2 covers the range of Guinier profiles included in
Figure 2. The low angle side of the 111 profile of $Pb(NO_3)_2$ with
n = 1.2 is seen to correspond to a value of 1.4 for the ratio
b_L/b_G. On the high angle side where n = 2.3 the ratio has the much
lower value of 1/2.1 which accordingly represents a considerable
decrease in the Lorentzian contribution. Values of b_L and b_G
required to synthesize the $Pb(NO_3)_2$ profiles can now be derived
from the curves for n and b shown in Figure 2 and the convolution
data contained in Figure 4. Examples are given in Table 3 for the
111, 511 and 731 reflections for which relevant figures for n and
b are listed in Table 1.

Table 3

Lorentzian and Gaussian half widths, b_L and b_G required for
convolution synthesis of Guinier profiles of $Pb(NO_3)_2$. Data
from Tables 1 and 2.

Profile		n	b_H	b_L/b_G	b_H/b_G	b_L	b_G
111	Low	1.2	0.015	1.4	1.9	0.011	0.008
	High	2.3	0.017	1/2.1	1.2	0.007	0.014
511	Low	1.7	0.008	1/1.6	1.25	0.004	0.006
	High	2.0	0.009	1/1.9	1.20	0.004	0.008
731	Low	1.9	0.019	1/1.8	1.21	0.009	0.016
	High	1.6	0.022	1/1.4	1.4	0.011	0.016

5.   DISCUSSION

The results obtained for Guinier and diffractometer profiles
can now be interpreted in terms of the convolution of instrument
and intrinsic factors, even if not all of these have been identified or correctly described. The goodness of fit to the analytical
expression indicates that such profiles can be described in terms
of the convolute of a basic Lorentzian and a basic Gaussian function.
Even when a factor belongs to neither of these basic types its
contribution to the overall convolute can still be apportioned

between the two limiting cases. Thus, axial divergence appears to
make an essentially Lorentzian contribution, judging by the fit to
the 111 profile of $Pb(NO_3)_2$. The observed increase in n for the
low angle profile halves in Figure 2 reflects the known decrease
in the half width of this function with increasing 2θ (13). Other
factors that contribute to the Guinier profiles are the decrease
to zero (at 34° for this camera) in the half width contribution
from chromatic dispersion and the increase in the half width
beyond 70° due to oblique incidence (14). The first of these
factors is symmetrical Lorentzian while the latter appears to have
unsymmetrical Lorentzian character judging by the sharp turn down
in the h curve beyond 70° in the diagram for n versus 2θ, in Figure 2.

The diffractometer profiles, for which data are listed in
Table 2 can be interpreted in terms of an initial decrease in
crystallite size. When the powder is lightly crushed, however,
there is evidently a contribution from strain broadening. This is
seen in terms of a higher value of n for the magnitude of broadening observed, as compared with the values given by irradiation.
The increase in the value of n with prolonged grinding and irradiation indicates that the continued line broadening is essentially
produced by an increase in the level of strain within the crystallites.

REFERENCES

1.   F. W. Jones, "The measurement of particle size by the X-ray
      method", Proc. Roy. Soc. A 166 16 (1938).

2.   F. R. L. Schoening, J. N. van Niekerk and R. A. W. Haul,
      "Influence of the apparatus function on cyrstallite size
      determinations", Proc. Phys. Soc., London B 65 528 (1952).

3.   E. J. Sonneveld and J. W. Visser, "Automatic collection of
      powder data from photographs", J. Appl. Cryst. 8 1 (1975).

4.   G. Malmros and J. O. Thomas, "Least-squares structure refine-
      ment based on profile analysis", J. Appl. Crys. 10 7 (1977).

5.   C. P. Khattak and D. E. Cox, " Profile analysis of diffracto-
      meter data", J. Appl. Cryst. 10 405 (1977).

6.   M. M. Hall Jr and V. G. Veerarghavan, "The approximation of
      symmetric X-ray peaks by Pearson type VII distributions", J.
      Appl. Cryst. 10 66 (1977).

7.   J. W. Edmonds and W. W. Henslee, "Application of Guinier
      camera, microcomputer controlled film densitometry", Advances
      in X-ray Anal., Vol 22, 143, Plenum Press (1979).

8.   D. T. Keating, "Elimination of the $\alpha_1\alpha_2$ doublet in X-ray patterns", Rev. Sci. Instr. $\underline{30}$ 725 (1959).

9.   A. M. Byström-Asklund, "Sample cups and a technique for sideward packing of X-ray diffractometer specimens. Am. Mineral. $\underline{51}$ 1233 (1966).

10.   R. I. Garrod, J. F. Brett and J. A. MacDonald, "X-ray line broadening and pure diffraction contours", Australian J. Phys. $\underline{7}$ 77 (1954).

11.   L. Alexander, "The synthesis of X-ray spectrometer line profiles with application to crystallite size measurements", J. Appl. Phys. $\underline{25}$ 155 (1954).

12.   L. Alexander, "Geometrical factors affecting the contours of X-ray spectrometer maxima", J. Appl. Phys. $\underline{19}$ 1068 (1948).

13.   P. M. de Wolff, "Multiple Guinier cameras", Acta Cryst. $\underline{1}$ 207 (1948).

14.   E-G. Hofmann and H. Jagodzinski, "Ein neue, hochauflösende Röntgenfeinstruktur-Anlage mit verbessertem, fokussierendem Monochromator und Feinfokusröhre", Z. für Metallkunde $\underline{46}$ 601 (1948).

QUANTITATIVE PHASE ANALYSIS OF SYNTHETIC SILICON NITRIDE BY

X-RAY DIFFRACTION

Z. Mencik, M. A. Short and C. R. Peters

Engineering & Research Staff, Ford Motor Company

Dearborn, Michigan 48121

INTRODUCTION

Synthetically prepared silicon nitride is one of the more promising ceramic materials for structural components of gas turbines. Typical material may contain α-silicon nitride, Si_3N_4 (which is believed to always contain oxygen and therefore, according to Grievson, Jack and Wild[1], is more properly written as $Si_{11.5}N_{15}O_{0.5}$), β-silicon nitride, Si_3N_4, silicon oxynitride, Si_2ON_2, silicon metal, Si, and α-cristobalite, SiO_2. Because the physical properties of the ceramic parts are dependent on their phase composition, it is essential that a technique be available for performing a phase analysis. An X-ray diffraction procedure has been developed for the quantitative phase analysis of synthetically prepared silicon nitride. This procedure converts experimentally measured intensities of selected X-ray diffraction peaks to weight fractions of components using empirically determined intensity coefficients.

It can be shown that the weight fraction of each crystalline component i, W_i, is given by the expression:

$$W_i = K_i I_i \ / \ \Sigma K_j I_j \tag{1}$$

where I_j's are the selected diffraction peak intensities for each of the j components (j includes i) and the K_j's are constants for each component and contain contributions due to structure factors, multiplicities, Lorentz-polarization factors, equipment geometry, phase densities, etc. This equation is dependent upon the provision that the prepared "infinitely thick" powder samples are homogeneous, with small particle size, and that all crystalline phases are measured. Consequently, the analytical procedure is based on a careful determination of the intensity coefficients, K_j, and an accurate measurement of the diffraction intensities of the selected reflections. The procedure was initially established using peak heights. Recently, however, it has been modified and automated to use peak areas.

ANALYSIS USING DIFFRACTION PEAK HEIGHTS

Determination of Intensity Coefficients

From a preliminary examination of various synthetic silicon nitrides, the reflections shown in Table 1 were selected for measurement in the phase analysis.

Table 1.   Selected Reflections for Silicon Nitride Analysis

Phase	Reflection	d (Å)	°2θ, CuKα
α-Si$_3$N$_4$	(201)	2.89	30.95
β-Si$_3$N$_4$	(200)	3.29	27.10
Si$_2$ON$_2$	(200)	4.44	20.00
Si	(111)	3.14	28.42
SiO$_2$(α-crist.)	(101)	4.04	22.00

Inspection of equation (1) shows that it is necessary to obtain only relative values of the intensity coefficients for this analysis. The coefficient for β-Si$_3$N$_4$ has arbitrarily been set at unity and the other coefficients have been obtained relative to this. To determine these, 15 binary mixtures of β-Si$_3$N$_4$ and known amounts of α-Si$_3$N$_4$, Si$_2$ON$_2$, Si and SiO$_2$ were prepared. The phase compositions of the starting materials used were as follows:

α-Si$_3$N$_4$	:	99.4% α-Si$_3$N$_4$ + 0.6% β-Si$_3$N$_4$
β-Si$_3$N$_4$	:	98.5% β-Si$_3$N$_4$ + 1.5% α-Si$_3$N$_4$
Si$_2$ON$_2$	:	80.0% Si$_2$ON$_2$ + 13.8% β-Si$_3$N$_4$ + 6.2%α-Si$_3$N$_4$
Si	:	100% Si
SiO$_2$	:	100% α-cristobalite

The 15 binary mixtures were each prepared for diffraction by grinding in a tungsten carbide ball mill for ten minutes and sieving through a 140 mesh standard sieve. The resulting powders were packed in standard Philips diffractometer sample holders by gentle packing against a flat glass surface.

All X-ray diffraction measurements were made using a Philips diffractometer equipped with a diffracted beam graphite crystal monochromator. A fine focus copper target X-ray tube, run at

45 kV and 28 mA, was used together with a 1° beam slit and a 0.006"
receiving slit. The peak heights of the binary calibrating mixtures
were determined by subtracting the background from the count rate
established by a scaler-timer at the peak of the diffraction line.
The background value used was the average of the readings on both
sides of the peak unless the presence of an interfering line
prohibited such measurement on one side.

For each of the 15 binary mixtures the net peak intensities
of the selected diffraction lines were obtained. From these data
and the known compositions, $K_{\alpha-Si_3N_4}$, $K_{Si_2ON_2}$, K_{Si}, and K_{SiO_2}
were determined as the ratio of the count rate per centum of each
of these phases to the count rate per centum for $\beta-Si_3N_4$ in each
binary mixture. $K_{\beta-Si_3N_4}$, as stated, is taken as unity.

Table 2 shows the individual intensity coefficients found for each
of the 15 binary mixtures and also the average values used in
subsequent analyses.

Table 2. Individual and Averaged Intensity Coefficients

$K_{\alpha-Si_3N_4}$	$K_{Si_2ON_2}$	K_{Si}	K_{SiO_2}	$K_{\beta-Si_3N_4}$
1.70	0.59	0.20	0.26	1.00
1.70	0.57	0.28	0.24	
1.74	0.56	0.23	0.23	
1.73		0.20	0.23	
average: 1.72	0.57	0.23	0.24	1.00

Analysis of Synthetic Silicon Nitrides

Samples of synthetic silicon nitride were prepared for analysis
and were run on a Philips diffractometer in the manner described
for the binary mixtures above. In making a quantitative phase
analysis there is one diffraction line interference which must be
taken into account. It involves the interference between the (111)
line of SiO_2 and the (111) line of Si. Because of this, the
measured intensity of the Si (111) line must be corrected by
subtracting 5% of the intensity of the SiO_2 (101) line before
multiplying by K_{Si}.

The weight fractions of the various component phases may be
then calculated using the measured diffraction intensities, the
intensity coefficients given in Table 2 and Equation (1). For

example, the weight fraction of α-Si_3N_4 would be given by:

$$W_{\alpha\text{-}Si_3N_4} = 1.72\ I_{\alpha\text{-}Si_3N_4} / (1.72\ I_{\alpha\text{-}Si_3N_4} + I_{\beta\text{-}Si_3N_4} + 0.57\ I_{Si_2ON_2}$$
$$+ 0.23\ I_{Si}^* + 0.24\ I_{SiO_2})$$

where $I_{Si}^* = I_{Si} - 0.05\ I_{SiO_2}$.

ANALYSIS USING DIFFRACTION PEAK AREAS

The phase analysis has recently been automated using the same diffractometer coupled to a PDP 11/34 computer, through Canberra Datanim electronics, operating under DEC's RSX-11M system[2]. The program for the analysis is written in FORTRAN IV. Because this program is designed to measure peak areas rather than peak heights, some changes were necessary to the procedures outlined above.

First, an alternative Si_2ON_2 reflection, the (110) located at $2\theta = 18.96°$, was selected to avoid the interference of the original (200) reflection with the neighboring α-Si_3N_4 (101) reflection. Second, the β-Si_3N_4 (200) reference requires a correction due to overlap with reflections (111) of Si_2ON_2 and (200) of α-Si_3N_4. The intensity coefficients necessary to effect these two changes were empirically derived using a number of synthetic silicon nitride samples which had been analyzed by the peak height method. Table 3 shows the intensity coefficients, Miller indices, 2θ integration scan limits, and angles for background measurement.

Table 3. Parameters for Analysis Using Peak Areas

Phase	Reflection	K_j	2θ Scan, CuKα	Background(60 sec)
α-Si_3N_4*	(201)	1.72	30.3 - 31.55°	30.3°
β-Si_3N_4	(200)	1.00	25.9 - 27.7°	25.9°
Si_2ON_2	(110)	0.95	18.5 - 19.5°	18.5°, 19.5°
Si*	(111)	0.23	28.0 - 29.0°	28.0°, 29.0°
SiO_2	(101)	0.24	21.6 - 22.2°	21.6°, 22.2°

*peak areas used to calculate phase composition are the measured areas after correcting for overlap:

$$I_{Si(111)}^* = I_{Si(111)} - 0.05\ I_{SiO_2(101)}$$
$$I_{\beta\text{-}Si_3N_4}^* = I_{\beta\text{-}Si_3N_4} - 0.266\ I_{\alpha\text{-}Si_3N_4(201)} - 1.4\ I_{Si_2ON_2(110)}.$$

The computer program controls operation of the diffractometer and the scaler-timer so that the various reflections and backgrounds shown in Table 3 are properly measured. The program then processes the data obtained to determine the weight fractions of the phases present. Excellent agreement has been obtained between the peak height and peak areas methods.

REFERENCES

1. P. Grievson, K. H. Jack and S. Wild, U. K. Ministry of Defense Contract, N/CP.61/9411/67/4B/MP.387, Progress Report no. 1.

2. B. E. Artz, E. C. Kao and M. A. Short, "Using DEC Operating System RSX-11M for X-Ray Diffraction and X-Ray Fluorescence Analysis", in G. J. McCarthy et al, Editors, Advances in X-Ray Analysis, Vol. 22, pp. 425-431, Plenum Press (1979).

AUTHOR INDEX

A

Abplanalp, H. J., 157
Ahlgren, L., 185
Auermann, R., 65
Ayers, G. L., 313

B

Bahgat, A. A., 203
Baird, A. K., 249
Barker, M., 81
Baro, R., 349
Barrett, C. S., 331
Boettinger, W. J., 209
Bouéres, L. C., 143
Brown, A., 361
Brown, F. V., 57
Burdette, H. E., 209

C

Camp, D. C., 163
Claisse, F., 87
Clark, B. C., 249
Clayton, C. G., 1
Courtney, W. J., 149
Criss, J. W., 93, 99, 111
Cullity, B. D., 333

D

Dabrowski, A. J., 249
Dalheim, P., 71
Das Gupta, K., 203
Doster, J. M., 117
Drane, E. A., 149
Dzubay, T. G., 149

E

Ebel, H., 223
Ebel, M. F., 223
Edmonds, J. W., 361

G

Gardner, R. P., 117
Goehner, R. P., 305
Grönberg, T., 185
Gupta, S. K., 333
Gurker, N., 263

H

Hahm, Y., 279
Hasegawa, K., 325
Huang, T. C., 313
Huth, G. C., 249

I

Iwanczyk, J. S., 249

J

James, G. W., 77
Jenkins, R., 279, 287
Johnson, Q., 273
Jones, S. A., 57

K

Kuriyama, M., 209

SUBJECT INDEX

A

Absorption coefficients in XRF, 89

Absorption in XRF, 87, 94

Al
  in coal, 66
  in ink, 221

Algorithm, second derivative, 287

Alloy segregation, XRF of, 209

Alloys, binary phase, 209

Alpha-2 ($K\alpha_2$) stripping, 280

Alpha coefficients or factors, 87, 93, 96

Aluminum pole figures, 351

Anode absorption, 95

APD 3600, 280

Ash in coal, 45, 59

Atmospheric aerosol analysis, 143, 149

Attenuation correction in XRF, 154

Automated XRD analysis, 378

Automation, 305
  minicomputer, 313

Ba, 133
  in ink, 221

Background
  filter function, 126
  in EDXRF, 151
  in XRF, 231
    from photoelectrons, 253
  XRF, 127
    from incomplete charge
    collection, 254

B

Backscatter
  from coal, 59
  in EDXRF, 40

Back projection of XRF images, 265

Ball bearing steel, 317

Band structure of SnO_2, 203

Basalt, 255

$BaSO_4$, 133

Bearing steel, 341

Boeing's XRF practice, 157

Borehole XRF probe, 19

Brass, XRF analysis, 33